The Sounds of the Diseased Heart

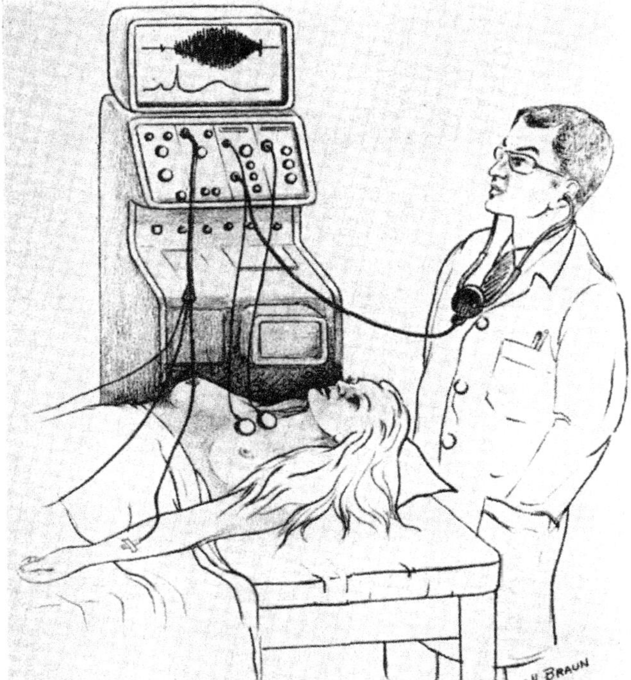

......Beautiful........ really Beautiful !!!

The Sounds of the Diseased Heart

By

ALDO A. LUISADA, M.D.

Distinguished Professor of Physiology and Medicine
The Chicago Medical School
University of Health Sciences
Chairman, Department of Cardiology
Oak Forest Hospital
Oak Forest, Illinois

BRISTOL: JOHN WRIGHT & SONS LTD.

Published by

WARREN H. GREEN, INC.
10 South Brentwood Blvd.
St. Louis, Missouri 63105, U.S.A.

Library of Congress Catalog No. 73-110806
ISBN 0 7236 0359 6

Printed in the United States of America

To little Andréa

Introduction

This book represents the second part of a comprehensive work on auscultation and phonocardiography.

Based on the previous description of the sounds of the normal heart and of routine methods of study, this book first examines the abnormal heart sounds in Section I. Changes of the first and second heart sounds in various clinical conditions, and the increase of the third or fourth heart sound (gallop rhythms) are discussed. Then the systolic and diastolic clicks and the friction rubs are presented. A subsequent chapter discusses the dynamic interpretation of cardiovascular murmurs.

In Section II, one can find a systematic description of the alterations of sounds and of murmurs in the various clinical conditions. Mitral, tricuspid, aortic and pulmonic defects are the subject of two chapters. Left-to-right shunts and complex congenital malformations follow. Then arrhythmias, myocardial and coronary disease, systemic and pulmonary hypertensions are discussed. Syphilitic heart disease, bacterial endocarditis, the continuous murmurs, the vascular murmurs and the effect of heart failure are then presented.

In Section III, the borderline syndromes are discussed. These include the straight back syndrome, the senile heart, the heart in pregnancy, the functional and innocent systolic and diastolic murmurs, certain unusual sound phenomena, and the sounds caused by either prosthetic valves or a pacemaker.

This book is illustrated by numerous graphic tracings obtained in the author's laboratory and is completed by an extensive bibliography.

The excellent work done by Dr. Beatrice Argano in the last two years as Chief of the Phonocradiography Laboratory at Mount Sinai Hospital has greatly helped in the preparation of this volume.

The help of Mrs. Wanda Motton, Phono Technician, is also gratefully acknowledged.

Contents

Page

Part III — Borderline Conditions

The Sounds of the Diseased Heart

Part I

Systematic Phonocardiography

Chapter 1

The Abnormal First Heart Sound

IN A PREVIOUSLY PUBLISHED volume (A. A. Luisada: The Sound of the *Normal Heart*), a new concept of the production of the heart sounds was presented in detail. This concept differs from those currently accepted. Being based on new experimental evidence, it now represents the only possible assumption that can be used for interpreting the abnormalities of the heart sounds.

Experimental studies exclude "valve closure" as a cause of the first heart sound and recognize "valve tension" only as a minor contributing factor. They accept the exclusive importance of *left ventricular* events in causing the first heart sound and attribute the vibrations of this sound to accelerations and decelerations of the left ventricular-aortic system plus the blood it contains.

Three main components of higher frequency can be recognized within the normal first sound. The *first component* occurs long after closure of the A-V valves, during the first part of the rapid rise of pressure within the left ventricle. The *second component* coincides with the opening of the aortic valve and is related to the dynamic changes resulting from this event. The *third component* is chiefly of aortic origin and coincides with the end of rapid distension of the ascending aorta.

The left ventricular pressure rise may be regarded as a manifestation of a contractile force which, except for the resistance offered, would have accelerated the blood into the aorta. Prevented from doing this by the closed valves, the left intraventricular blood mass is instantaneously decelerated. Thus, the contractile energy tending to accelerate the blood is converted into the potential energy of pressure during isovolumic contraction.

The percentage change in the magnitude of the first component of the first sound has been correlated with the level of left ventricular pressure and with the percentage change in the amplitude of the first systolic wave of the first derivative of this pressure. A linear correlation was found between the first component and the first derivative of LV pressure suggesting a strong relationship between the first sound and the rapidity of left ventricular contraction (Sakamoto *et al.*, 1965).

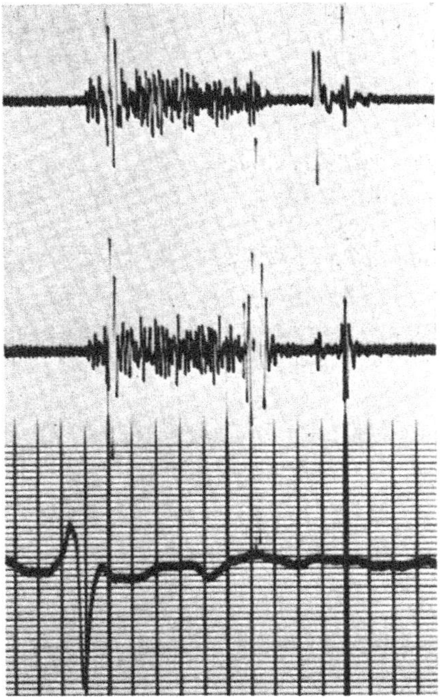

FIGURE 1. Case of pure mitral insufficiency proven by cardiac catheterization (pansystolic murmur, late opening snap, large third sound).
The first sound has a normal amplitude.
above = second left i.c.s.
center = apex.
High pass filter at 200 Hz, slope 24 db/oct.

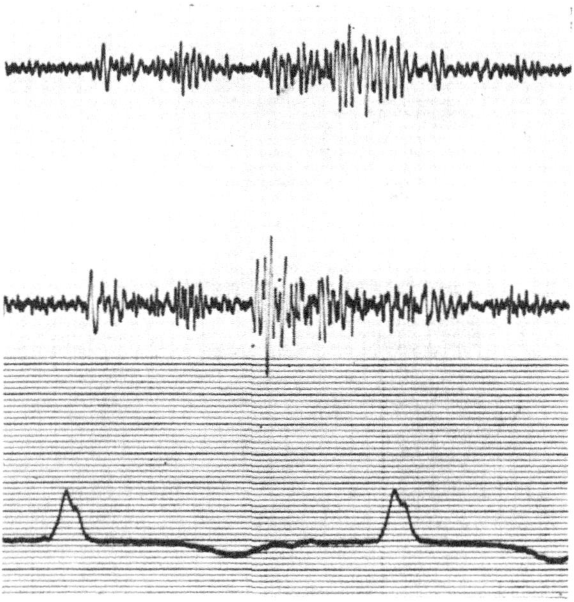

FIGURE 2. Case of syphilitic heart disease with aortic flow murmur and aortic insufficiency. (Diamond-shaped systolic murmur, diastolic murmur) The first sound is markedly decreased, even in the low frequency bands.
above = third right i.c.s.
center = third left i.c.s.
High pass filter at 50 Hz, slope 6 db/oct.

DECREASE IN AMPLITUDE, LOUDNESS, OR BOTH

Hypertrophy or Dilatation of the Left Ventricular Wall. The effect of pure *hypertrophy* of the wall is exemplified by the *athlete's heart.* In these individuals, the first heart sound is often fainter on auscultation on account of the fact that the maximum amplitude is in a lower frequency range (Adolph *et al.*, 1970). On the other hand, compensatory hypertrophy has more than one effect because it tends to increase the rapidity of pressure rise. Thus clinical tracings may show a normal or increased first heart sound. *Dilatation* of the left ventricle seems to decrease the amplitude of the first heart sound in experimental animals (Sakai *et al.*, unpublished observations).

Mitral Insufficiency. Dilatation of the left ventricle would contribute to a decrease of the first heart sound, which is often found in these patients. In addition, incomplete closure of the

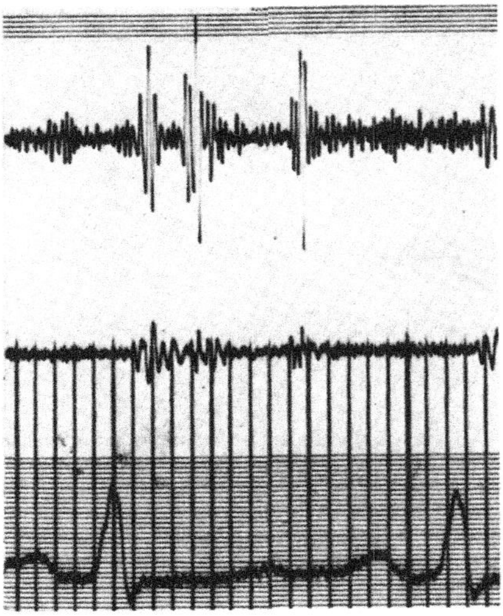

FIGURE 3A. Case of rheumatic heart disease with aortic insufficiency. Presystolic murmur (A. Flint) and diastolic murmur.
The first heart sound has a normal amplitude and there is a large ejection sound.
above = third left i.c.s.
below = apex.
High pass filters at 100 Hz, filter, slope 12 db/oct.

mitral valve will cause a slower rise of pressure (smaller dp/dt), and this will result in a smaller first sound. This was clearly demonstrated in experimental animals by van Bogaert *et al.* (1962). However, compensatory elements may cause a normal or even increased first sound.

Myocarditis. It is classic to admit that the first sound becomes faint and muffled in patients with myocarditis. Phonocardiography has confirmed this fact (Fig. 5). Until recently, however, it was not clear why the first sound should decrease in magnitude when the left ventricle was able to maintain a normal, or near normal, pressure in the aorta. The observation that the magnitude of the first component of the first sound is closely related to the rapidity of contraction (first derivative of left ventricular pressure) (Sakamoto *et al.,* 1965) gives an explanation for the weakening of the sound.

Myocardial Infarction. In myocardial infarction, the first sound is often weak and formed by vibrations of low frequency, even in patients with normal systemic pressure.

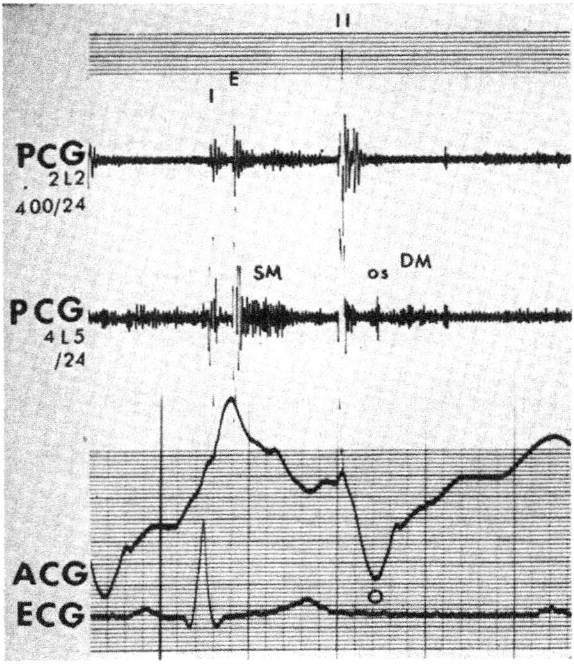

FIGURE 3B. Large ejection sound in a child with coarctation of the aorta. Opening snap.

The experimental and clinical observation that the infarcted area ceases to contract and tends to balloon outwards at each systole explains this finding. Part of the energy developed by the normal sections of the left ventricle is absorbed by the elastic distension of the infarcted area; therefore, the rise of pressure will occur more slowly, as shown by the first derivative of LV pressure.

Since the magnitude of the first component of the first sound is closely related to the rapidity of rise (as shown by Sakamoto *et al.*, 1965 in animal experiments), this will cause a decrease of the first sound. Sakamoto proved the existence of a definite correlation between the slower rise of LV presure in myocardial infarction and weakening of the first sund. In addition, the maximum amplitude of the first sound occurs at a lower frequency (Adolph *et al.*, 1970). For this reason, a softer first sound in myocardial infarct does not necessarily define left ventricular failure. On the other hand, such failure is often present and may be a concurrent cause of decrease of the first heart sound by further decreasing the rapidity of pressure rise.

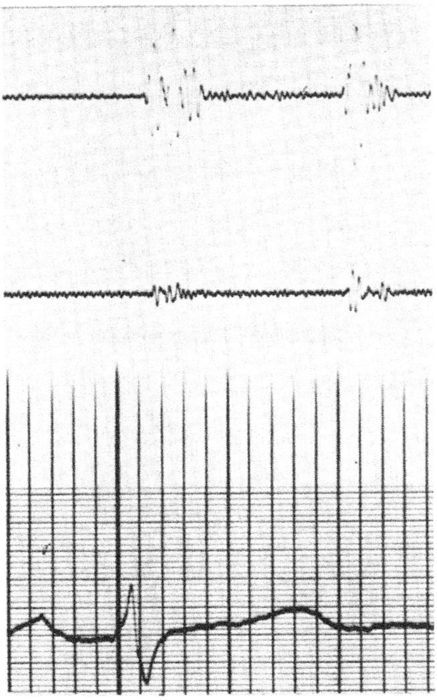

FIGURE 4. Case of syphilitic heart disease with aortic insufficiency. The first heart sound shows three components with a specially developed component *b*.
The murmurs are not visible on account of the type of filter.
above = second left i.c.s.
below = apex.
High pass filters at 100 Hz; filter slope 12 db/oct. Same amplification.

Left Ventricular Failure. Weakness of the left ventricle is found in left heart failure, where the pressure has a slower course and may reach a lower peak. If the ventricle is markedly dilated on acount of failure, there will be three causes for a decrease of the first sound (lower peak, slower rise of pressure, larger ventricle).

Hypothyroidism. The first heart sound is usually decreased. Such a change is probably related to a decrease in rapidity of left ventricular contraction.

INCREASE IN AMPLITUDE, LOUDNESS, OR BOTH

Clinically, increased loudness of the first heart sound is found in *tachycardia,* like that related to exercise, fever, emotional

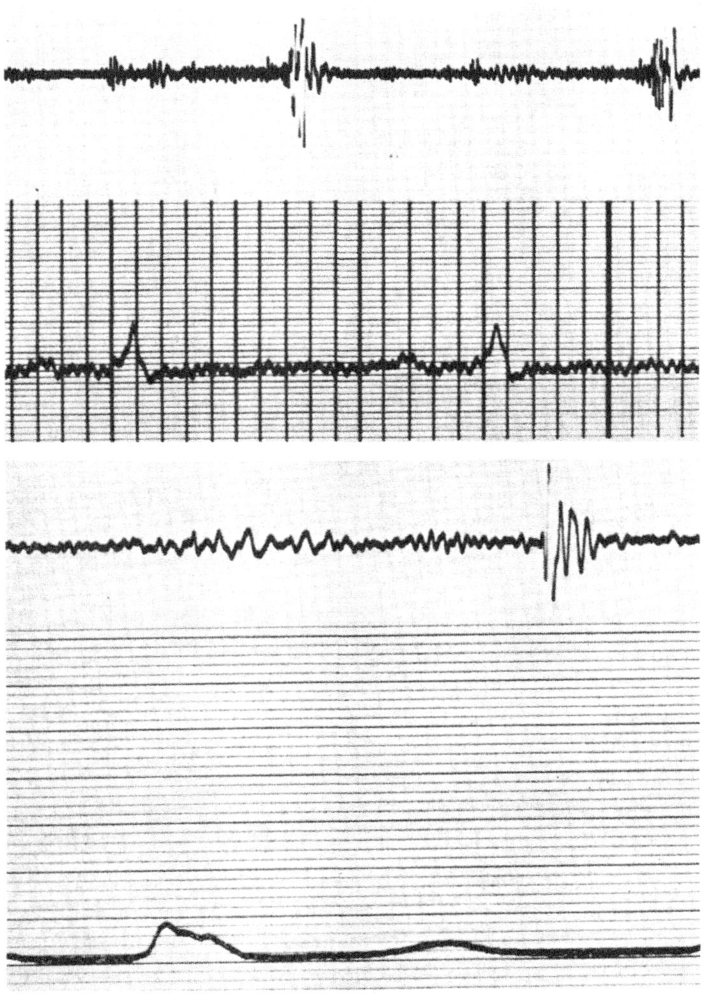

FIGURE 5. Two cases of viral myocarditis. (A,B)
The first sound is barely visible.
Tracings at apex. High pass filter at 100 Hz; filter slope 12 db/oct. (B) is recorded at higher speed.

upset, anemia, or hyperthyroidism). It is found after administration of atropine, epinephrine (W. P. Harvey), isoproterenol, or amyl nitrite. Increase loudness has been reported in the *Wolff-Parkinson-White syndrome*, in relation to a short P-R interval.

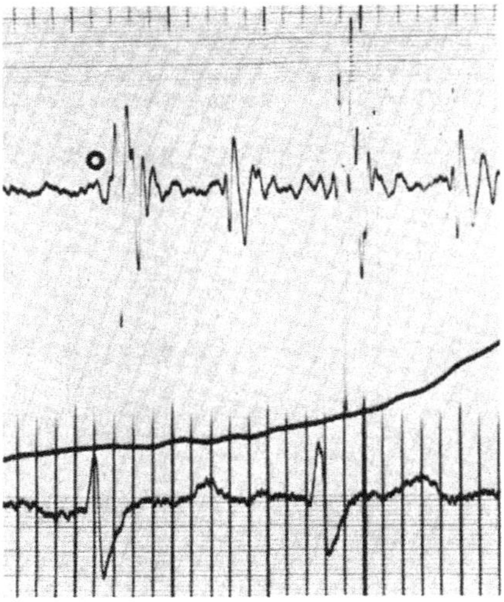

FIGURE 6. Case of thyrotoxicosis with atrial fibrillation. Large first heart sound with varying amplitude. Velocity tracing at apex, unfiltered.

Recent studies have revealed that increase in amplitude of the first heart sound ocurrs whenever *the rapidity of left ventricular contraction* has increased, as typically observed with injection of catecholamines or stimulation of the sympathetic center. This would especially apply to the large first sound of *thyrotoxicosis* (Fig. 6).

Ventricular hypertrophy *per se* tends to decrease the sound on account of greater thickness of the wall. On the other hand, if the increased thickness is due to muscular hypertrophy alone and not to increase of connective tissue, a greater power of contraction will occur resulting in a larger first heart sound. This will occur in systemic hypertension, coarctation of the aorta, aortic stenosis, or aortic insufficiency, even though other factors (failure, fibrosis, valve leaks) may tend to cause an opposite effect.

Left Ventricular or Systemic Hypertension. The Q-I interval is prolonged in *systemic hypertension* (Weissler *et al.*, Puchner *et al.*, Sakamoto *et al.*, 1960). This prolongation seems to be more severe in patients with left ventricular failure or at least with severely increased left ventricular load. In such cases, it has been stated that it is caused by an increase of left atrial pressure. Considering that the level of left atrial pressure is still similar to that of left ventricular pressure at end-diastole, it is inaccurate to compare these cases with cases of mitral stenosis. A wide splittting of the first sound was noted long ago in *aortic stenosis* or *systemic hypertension* (Potain, 1866, McKusick, 1958; Bellettini). This phenomenon is usually caused by increased loudness and higher frequency of the third component of the first sound (so-called *ejection sound* or *early systolic click*). In *aortic stenosis,* it is typical to observe it in patients with medium narrowing of the valve; when the stenosis is severe, usually the rise of pressure in the aorta is too slow for it to cause a loud vibration (Fig. 7). Patients with tetralogy of

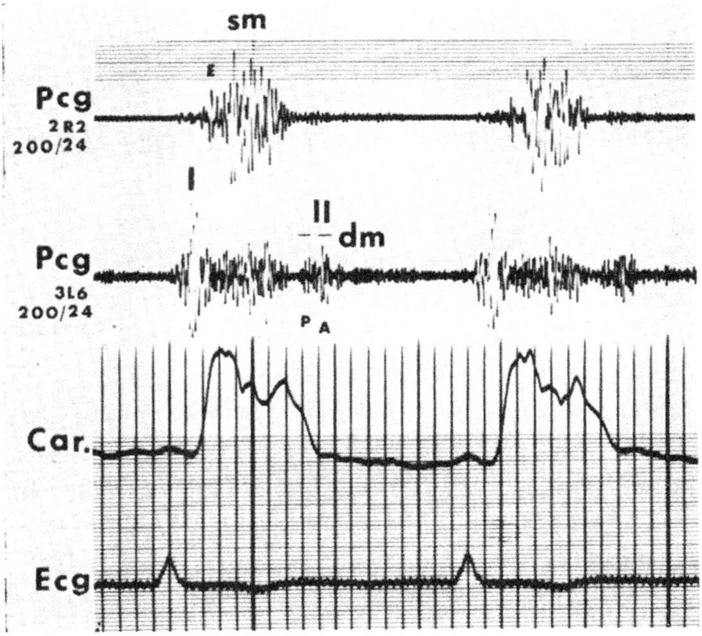

FIGURE 7. Case of aortic stenosis and minor insufficiency. An ejection sound is present at the 2nd right i.c.s. (above), not at the 3rd left i.c.s. Reverse splitting of the second sound. Pulsus bisferiens. Tracings recorded with high pass filter at 200 Hz; filter slope 24 db/oct.

Fallot or transposition often present a large ejection sound that sometimes causes a wide splitting of the first sound resulting in a triple rhythm (Fig. 8). The lound third component of the first sound (so-called ejection sound) of these patients should be attributed to the aorta and *not* to the pulmonary artery.

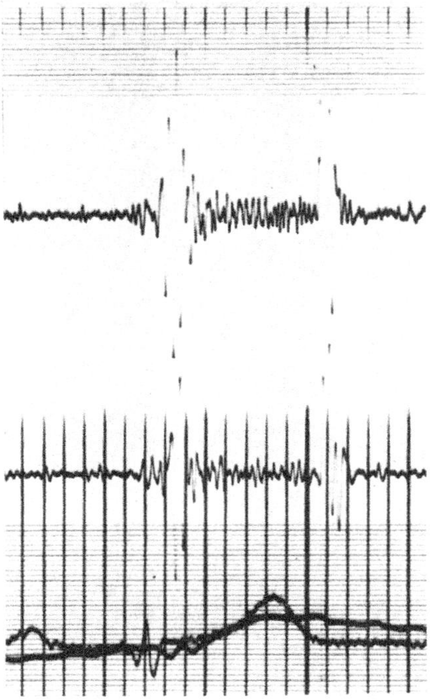

FIGURE 8. Case of transposition with ventricular septal defect (age 22)
above = second left i.c.s.
center = apex.
High pass filters at 100 Hz, filter slope 12 db/oct.
Large ejection sound over both areas.

Right Ventricular or Pulmonary Hypertension. When the pressure rises considerably in the right ventricle, right ventricular components *may* contribute to the character of the first sound. Sound vibrations are proportional in magnitude to the first derivative of ventricular pressure (Sakamoto *et al.*, 1965); therefore, only cases in which right ventricular pressure becomes similar to or higher than systemic pressure *might* produce important right ventricular vibrations. On the other hand, Sakamoto *et al.* have proved that, in acute right ventricular overload, the right ventricle does not contribute much to the first sound. The latter decreased (due to the slower rise of pressure in the left ventricle) even though both the pressure and the rapidity of its rise increased in the right ventricle, and right ventricular pressure was slightly higher than left ventricular pressure. This fact seemed to be the result of decreased cardiac output and decreased left ventricular dynamics. Therefore, the presence of right ventricular components in the first sound still lacks demonstration.

Dilatation of the Pulmonary Artery, Moderate Pulmonary Stenosis, Atrial Septal Defect. Dilatation of the pulmonary artery or moderate pulmonary stenosis is often associated with a wide splitting of the first sound. This phenomenon is the result of audibility of the first component plus a loud third component (so-called *ejection sound*). The latter is most likely the result of an abnormally rapid jet of blood in the pulmonary artery causing sudden distension of the wall. Normally, a similar small vibration would occur between the first and second, or second and third components, of the first sound and would not be

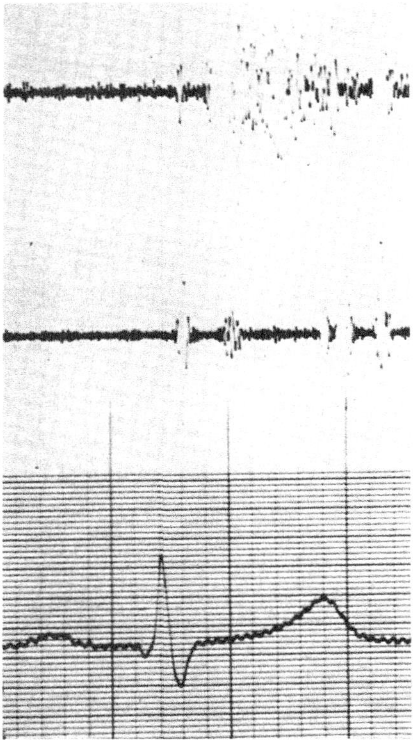

FIGURE 9. Case of atrial septal defect. Large ejection sound at 4th left i.c.s. (center). Large ejection murmur at second left i.c.s. (above). High pass filters at 400 Hz; filter slope 24 db/oct.

identifiable. The abnormal conditions of either the pulmonary valve or the pulmonary artery cause both an increase in magnitude and a delayed occurrence of this component. This loud ejection sound is particularly obvious in *atrial septal defect* fig. 9).

Mitral Stenosis. It is classic to quote the loud, snapping first sound and the prolonged Q-I interval of mitral stenosis Fig. 10). The latter, described by Cossio and Berconsky and by Wells (1954), presents a typical variability in patients with atrial fibrillation and irregularity of the ventricular rhythm (Luisada, 1941). Since mitral valve closure occurs only when the left ventricular pressure has risen above that of the left atrium, closure of the mitral valve in patients with mitral stenosis occurs later than in normal subjects, as shown by Braunwald *et al.*

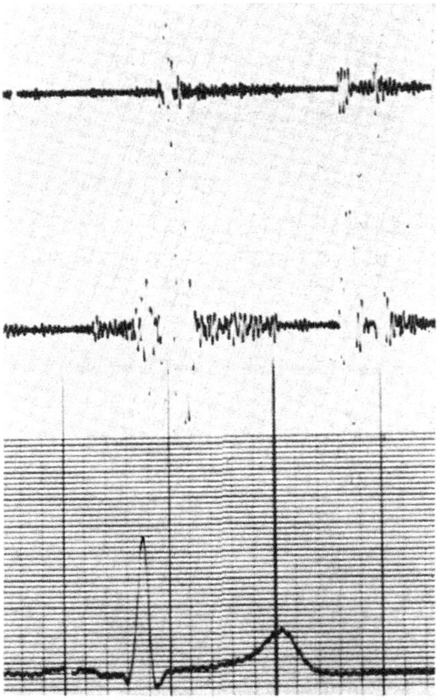

FIGURE 10. Case of mitral stenosis.
Presystolic murmur, large first sound, small systolic murmur, opening snap.
above = second right i.c.s.
center = fourth left i.c.s.
High pass filters at 200 Hz; filter slope 24 db/oct.

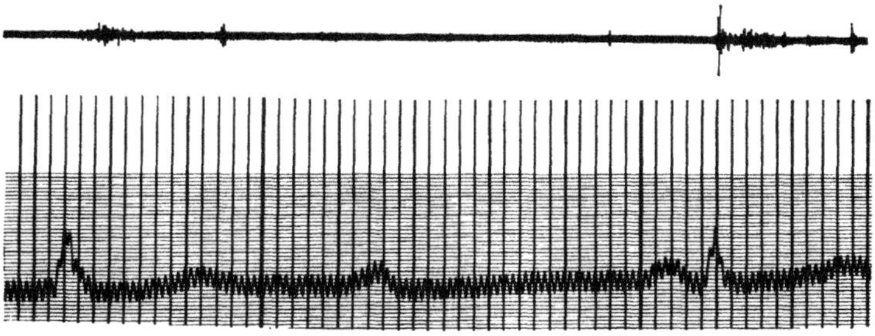

FIGURE 11. Case of complete AV block, tracing at apex; high pass filter at 200 Hz; filter slope 24 db/oct.
Varying amplitude of first heart sound.

1955) and by Moscovitz and Wilder. Its timing will fall, not before the rapid rise of pressure (as in normal hearts), but during the first part of such rise. Thus, mitral valve closure and rapid rise of pressure will occur at the same time. The rise of pressure subsequent to mitral closure is rapid, a fact that will contribute to the loudness of the sound and its snapping quality. The first sound in mitral stenosis was studied clinically by Kurz *et al.* (1967). They demonstrated that patients suffering from narrowing of the mitral valve have a typical alteration of the pressure rise in the left ventricle that is well revealed by the first derivative of this pressure: there is at first a slow rise of pressure (until the level of left atrial pressure is reached), and then an extremely rapid rise. The decreased mobility of the mitral ring and leaflets (fibrosis, calcification) increases the resistance to pressure rise and contributes to the typical increase in loudness of the first sound. However, it should be noted that the first sound of mitral patients may seem louder on account of a shift of the frequency toward a higher band even though its magnitude may not be increased. If the left ventricle is small, there would be another factor causing an increase in magnitude of the first heart sound. If the left ventricle is large (myocarditis, fibrosis, or failure), the opposite trend would be present.

VARYING AMPLITUDE OF THE FIRST HEART SOUND

A varying amplitude of the first heart sound is found in *atrial fibrillation* (Fig. 6), *A-V block* (Fig. 11), and *ectopic beats* (Fig. 12). It can be explained by the combination of vari-

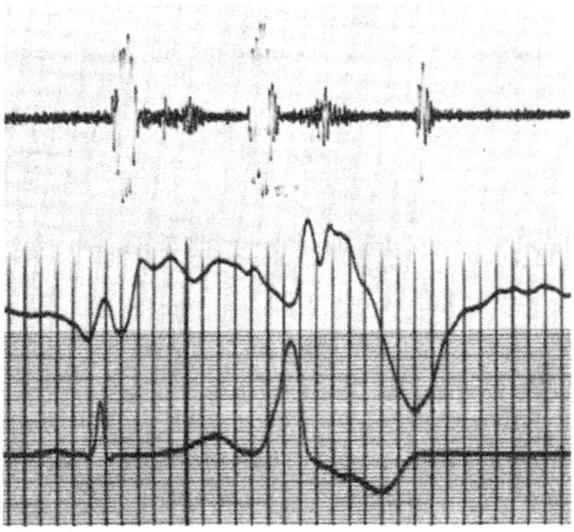

FIGURE 12. Ventricular ectopic beat with small first heart sound. From above = PCG, ACG, ECG.

ous factors: the left ventricle has a different size for each beat on account of different previous filling; it exerts a different force of contraction, largely related to a different degree of filling; and has a longer or shorter isovolumic period according to the level of aortic pressure it has to overcome. In addition, ventricular pressure rise will find the mitral value in different positions and will become the main factor of valve closure. As a result, functional mitral insufficiency is often present and the rapidity of pressure rise will greatly vary. Thus a multiplicity of factors will play a role in causing changes of the first heart sound in arrhythmias and A-V block. For further details see Chapter 11.

A correlation of the amplitude of the first heart sound with the first derivative of left ventricular pressure was made by Sakamoto *et al.* in experimental atrial fibrillation (1965) and in experimental alternans (1966), and more recently by us in ectopic beats and AV block (in preparation).

SPLITTING OF THE FIRST HEART SOUND

As discussed in a previous work (Luisada: *The Sounds of the Normal Heart*), the first sound may be split in normal young subjects because both the first and the second component are audible (Fig. 13). In mature or elderly people, the first sound may be split due to audition of the first and third components. Thus, any statement in regard to splitting of the first sound should exclude "physiological splitting" before assuming that the phenomenon is related to a particular condition.

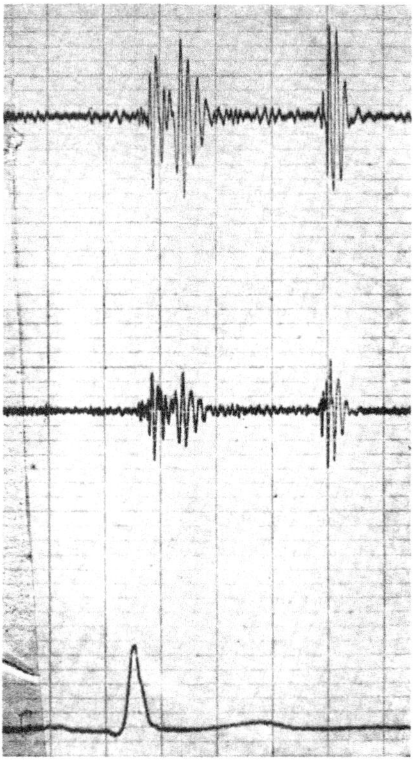

FIGURE 13. Normal 24-year-old subject. Tracings of displacement (above) and velocity (center) recorded over the third left i.c.s. High pass filters at 120 Hz; filters slopes 48 db/oct. Apparent splitting of first heart sound.

Several statements have been made in regard to the first heart sound on the basis of clinically assumed changes of the "mitral" and "tricuspid" components of this sound.

Mitral Stenosis. It was stated that, in mitral stenosis, the "mitral" component is delayed and follows a nornal "tricuspid" component. Simultaneous pressure tracings from the two ventricles show that right ventricular pressure rise still *follows* left ventricular pressure rise. The delay of mitral valve closure caused by high left atrial pressure is much shorter than that necessary for causing reversal in the sequence of the first two components of the first heart sound. Therefore, the latter are still the two normal components (*a* and *b*) with some increase in magnitude of the latter.

Atrial Septal Defect. It has been stated that, in atrial septal defect, the "tricuspid" component becomes louder. More recent studies (Zakrzewski *et al.*, 1969) have demonstrated that the larger component occurs at the time of pressure rise in the pulmonary artery (ejection sound). Therefore, the above view should be modified.

Bundle Branch Block, Ectopic and Paced Beats. Since the study of Wolferth and Margolies (1935), it has been accepted that splitting of the first heart sound is typical of bundle branch block. Leatham found splitting in RBBB but not in LBBB and claimed that the same phenomenon occurred in left ventricular ectopic beats but not in right ventricular ectopic beats. More recent studies in bundle branch block (Oravetz *et al.*, 1967) as well as in ectopic and paced beats (Luisada *et al.*, 1970) have demonstrated the following facts.

In left bundle branch block, right ventricular ectopic beats, and right ventricular pacing, the first heart sound is markedly delayed. Splitting of the first sound may be due to several alternative possibilities.

(1) A simulation of splitting may be the result of a presystolic gallop (IV sound).
(2) A normal close splitting may be due to the audibility of the first and second components (less common).
(3) A wide splitting may result from audibility of the first and third components. This was actually ascertained in some of our cases of left bundle branch block by noting that amyl nitrite decreased the amplitude of the last component of the first sound. (If this were originating in the right heart, it should have ben increased by the drug.)

These facts confirm clinical and experimental studies of van Bogaert *et al.* (196).

DELAYED FIRST HEART SOUND

A delayed first heart sound (prolonged Q-I interval) can be caused by several possible occurrences.

(1) Left BBB, right ventricular ectopic beats, or right ventricular pacing.
(2) Mitral stenosis.
(3) Left ventricular hypertrophy (moderate delay).

In case 1, delayed activation is responsible for the fact. In case 2, delayed closure of the mitral valve delays the rapid rise of pressure in the left ventricle. In case 3, no explanation has been given as yet.

Chapter 2

The Abnormal Second Heart Sound

UNDERSTANDING OF THE SECOND heart sound has greatly progressed since the early fifties. First, it was demonstrated that this sound includes two components, the *aortic* and the *pulmonary*, which separate more widely at the end of inspiration while they approximate or fuse in expiration (Leatham, 1954). Then, it was learned hat the best area of audibility (auscultation) and recordability (phonocardiography) is the third left interspace, where the "aortic" and the "pulmonary" areas are superimposed (Shah *et al.*, 1964). Finally, it was experimentally demonstrated that closure of the aortic and pulmonary valves is *followed*, after a certain interval, by the respective component of the second sound (MacCanon *et al.*, 1964).

Each component of the second sound is caused by vibrations of the closed valve, infundibulum, and vessel wall *at the time of the incisura*. The latter is caused by rebound of the last part of the stroke mass over the already closed semilunar valves producing a second pressure rise. Thus, there is acceleration followed by deceleration, which manifests itself as sound (Mori *et al.*, 1964).* Therefore, end of ventricular systole, valve closure, and sound are three events that occur in rapid succession. The intervals between them are etxremely brief but can increase under the influence of physiologic or pathological changes, especially in the case of the right heart. The amplitude of each component is

*The importance of vascular factors as cause of these vibrations is emphasized by clinical cases with only one valve (truncus) and two components of the second sound.

proportional to the peak value of the first derivative of the pressure gradient developing between vessel and ventricle at the beginning of diastole (Kusukawa *et al.*, 1966).

The widening of splitting that occurs in inspiration is not caused only by prolongation of right ventricular systoles. It is chiefly caused by a delay in the rebound of the pressure pulse of the pulmonary artery caused by the inspiratory dilatation of the pulmonary network (study in course of completion).

In recent years, numerous studies have been devoted to the second heart sound because of the clinical importance of its changes. As several doubts still are voiced about the hemodynamic causes of these changes, we shall examine the interplay of various factors that determine alterations of the second sound.

CHANGES IN AMPLITUDE

As mentioned above, the amplitude of the second sound is proportional to the peak value of the first derivative (dp/dt) of the pressure difference between vessel and ventricle (Kusukawa *et al.*, 1966).

High pressure in the aorta or pulmonary artery is usually accompanied by louder (auscultation) and larger (phonocardiography) aortic or pulmonary component. Thus, *systemic hypertension* and *coarctation of the aorta* usually cause a larger (and louder) aortic component (Fig. 14) while *pulmonary hypertension* causes a larger (and louder) pulmonary component (Fig. 15). It is noteworthy that subjects with *atherosclerosis* of the aortic wall also have a louder aortic component of the second sound even without hypertension (Fig. 16). This fact is due to the increased rigidity of the aortic wall, which causes vibrations of a higher frequency.

On the contrary, *valvular, sub-valvular,* or *supra-valvular stenosis of the aorta or pulmonary artery* causes a fainter (auscultation) and smaller (phonocardiography) component of the second sound. Thus, aortic stenosis causes a smaller aortic component (Fig. 17) while pulmonary stenosis causes a smaller pulmonary component (Fig. 18) *Peripheral stenosis* of the pulmonary branches causes either a normal or a smaller pul-

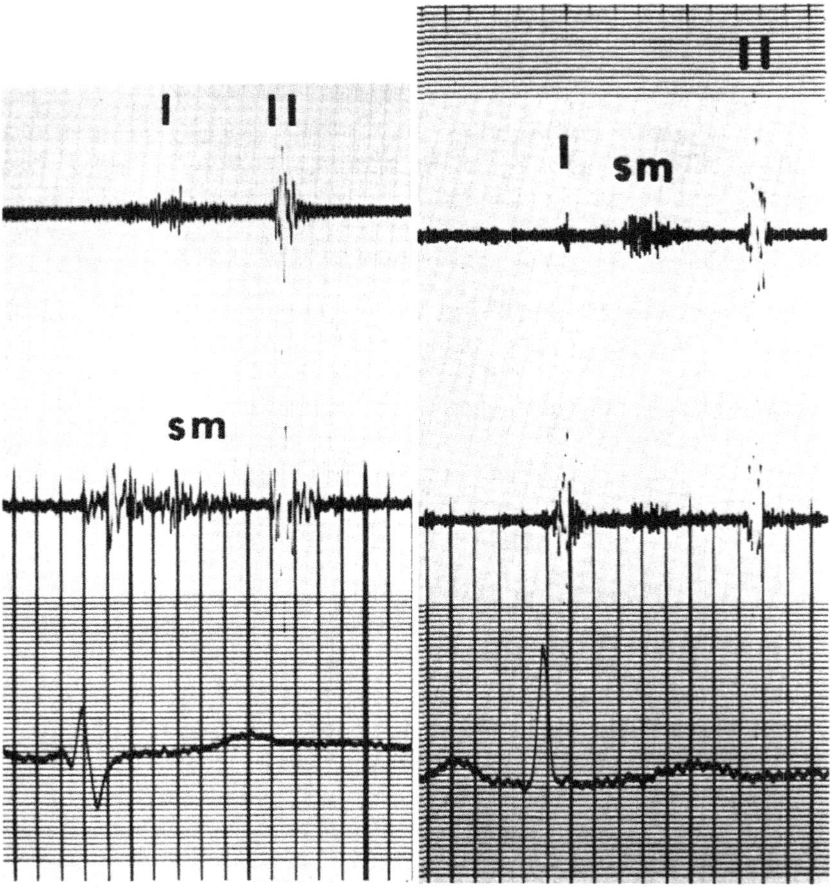

FIGURE 14. Two cases of systemic hypertension with a single second sound.
The upper tracings were recorded over the second right interspace; the center tracings, over the apex.
High pass filters = 400 Hz; slope 24 db/oct.

monary component because, even though systolic pressure is high in the main pulmonary artery of these cases, diastolic pressure is low so that a small gradient exists at the time of the incisura.

TIMING OF THE SOUND

As discussed in previous publications and in the first book of this series, end of ventricular systole and semilunar valve closure do not coincide with the respective component of the second sound.

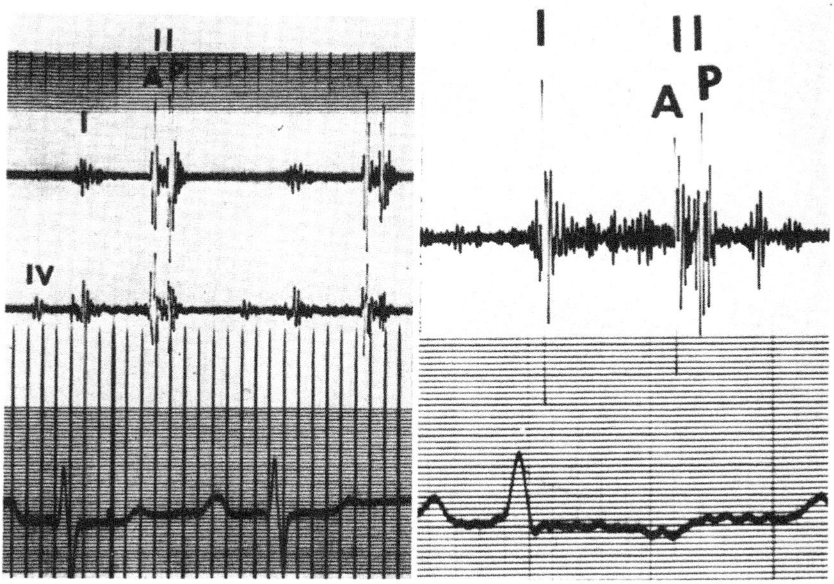

FIGURE 15. Two cases of pulmonary heart disease with severe pulmonary hypertension and wide splitting of the second sound.
(A) Scleroderma lesions of the lungs in a 20-year-old patient.
(B) Pulmonary fibrosis in a 38-year-old patient (opening sound; third sound).
In (A) the tracings were recorded over the second left i.c. (upper) and at the apex (center). In (B) they were recorded over the third left i.c.s. High pass filters = 400 Hz; slope 24 db/oct.

End of systole is indicated by the change in slope of the plateau-like pressure pulse of a ventricle (Fig. 19). Silent closure of the semilunar valve occurs a few milliseconds later (except in cases with valve stenosis—see below) as soon as the pressure of the ventricle falls below that of the artery (Fig. 19). The rebound of blood over the previously closed valve occurs some time later and the sound component grossly coincides with the incisura of the pulse, if this is recorded in the immediate vicinity of the valve (Fig. 19).

If we compare the pulse tracings of the two main arteries, we see that, on account of the lower resistance in the pulmonary system and greater compliance of the arterial walls, the rebound

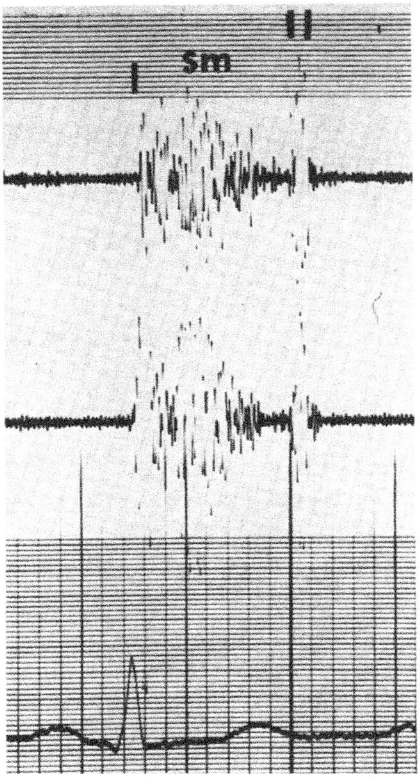

FIGURE 16. Case of aortic atherosclerosis and moderate systolic hypertension (170/90) in a 65-year-old man. The second sound is single and large.
Tracings recorded over the third right (upper) and the fourth left i.c.s. High pass filters = 200 Hz; slope 24 db/oct.

of the pulmonary pulse and the pulmonary component of the second sound are delayed over the respective events of the aorta (Fig. 20). End of systole is either simultaneous in the two ventricles or shows a minimal delay for one of them. This posible slight asynchronism is *not* revealed by the second sound, which is chiefly governed by the peripheral factors (Fig. 21).

Timing of a component depends upon two main factors: (a) time of closure of a semilunar valve; and (b) interval between this closure and the rebound of flow on the closed valve.

Time of closure may be altered by several factors as follows.

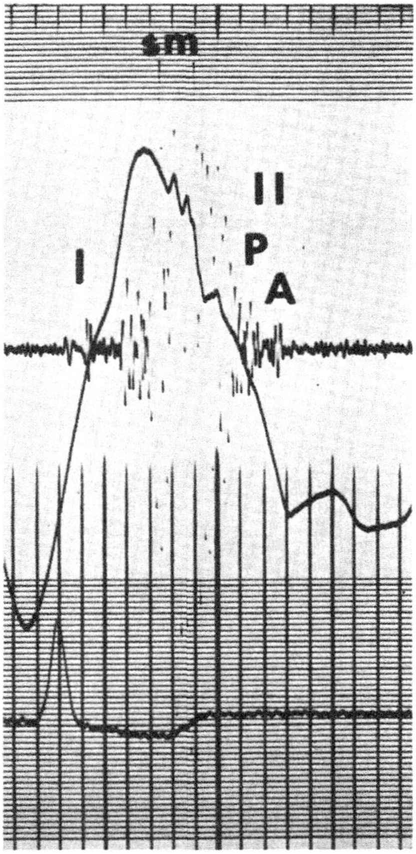

FIGURE 17. Severe aortic valvular stenosis in a 53-year-old man. Diamond-shaped systolic murmur and reverse splitting of the second sound.
Top = PCG over third right i.c.s.
High pass filter = 100 Hz; slope 12 db/oct.
Center = carotid tracing.

DELAYED ACTIVATION

If *activation of one ventricle is delayed,* systole starts later and ends later, so that the valve closes later (Fig. 22). While this mechanism is widely accepted for right bundle branch block (Fig. 23) and left ventricular ectopic beats (Fig. 24), it was denied in the case of left bundle branch block (Braunwald and Morrow, 1957). Our studies in right ventricular ectopic beats and right ventricular pacing (Fig. 25) (Luisada *et al.,* 1970) have revealed that, in these cases, left ventricular activation,

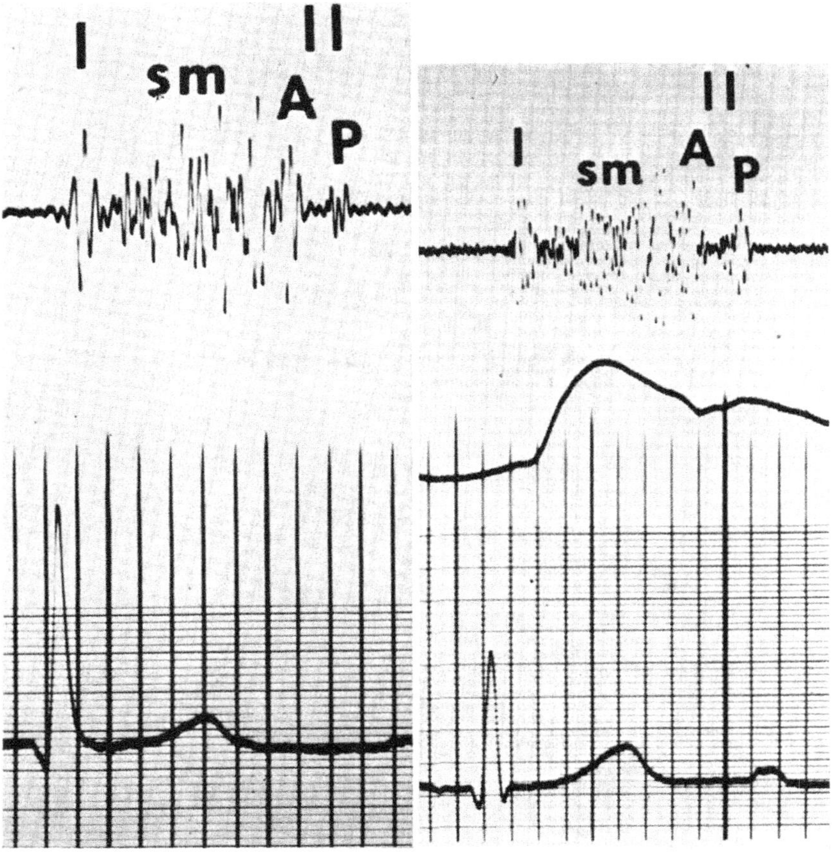

FIGURE 18. Two cases of medium-severe pulmonary valvular stenosis in children (3 and 4 years old, respectively). The tracings recorded over the second left i.c.s. demonstrate a wide splitting of the second sound.
(A) High pass filter at 100 Hz; slope 12 db/oct.
(B) Band pass filter at 100-700 Hz, slope 48 db/oct.
In (B) the identification of the components is aided by the carotid tracing.

instead of preceding that of the right ventricle, is simultaneous with it. Even in LBBB, 11 cases studied by us have shown the same abnormality, i.e., a later activation of the left ventricle with very close, simultaneous, or subsequent onset of pressure rise in the left ventricle in comparison with the right. On the other hand, prolongation of the tension period of the left ventricle is common in this condition as a result of left ventricular damage, and adds its effect by prolonging systole of this cham-

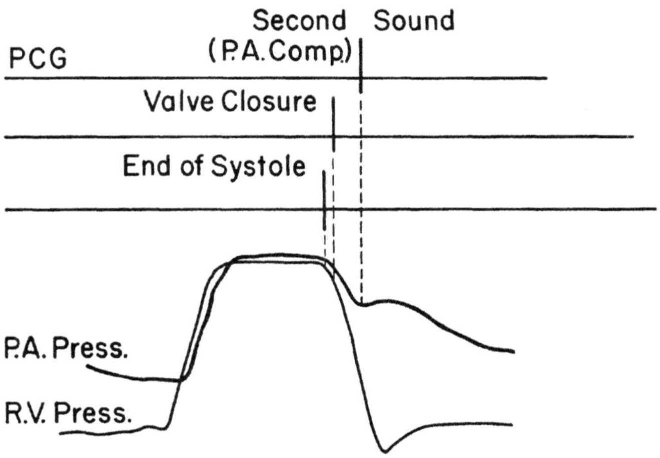

FIGURE 19. Scheme showing the relationship between end of right ventricular systole, pulmonary valve closure, and pulmonary component of second sound in a normal subject.

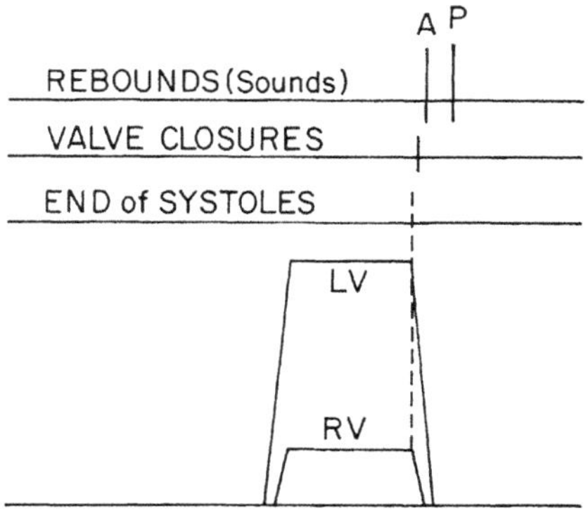

FIGURE 20. Scheme of the relationship between end of ventricular systoles, valve closures, and components of the second sound in a normal subject.

ber (Adolph *et al.*, 1969) and causing reverse splitting of the second sound (Fig. 26). In addition, peripheral factors are also involved (Luisada, Kumar and Pouget, 1972). It should be kept in mind that, while normal splitting increases or appears *in inspiration,* reverse splitting increases or appears *in expiration.*

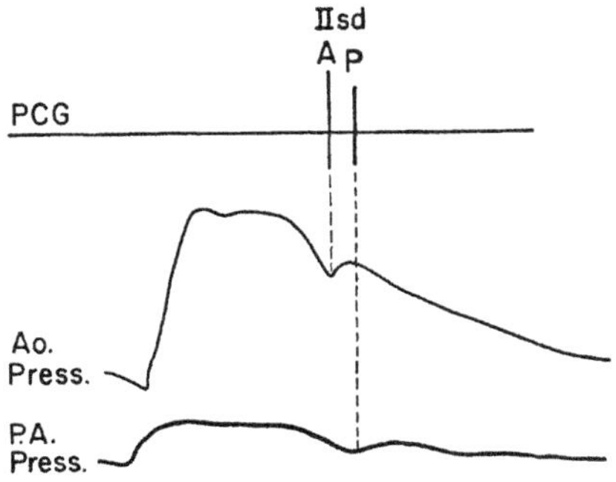

FIGURE 21. Scheme of the splitting of the second sound in a normal subject.
Note the relationship between the A and P components with the incisuras of the pulse tracings in the main arteries.

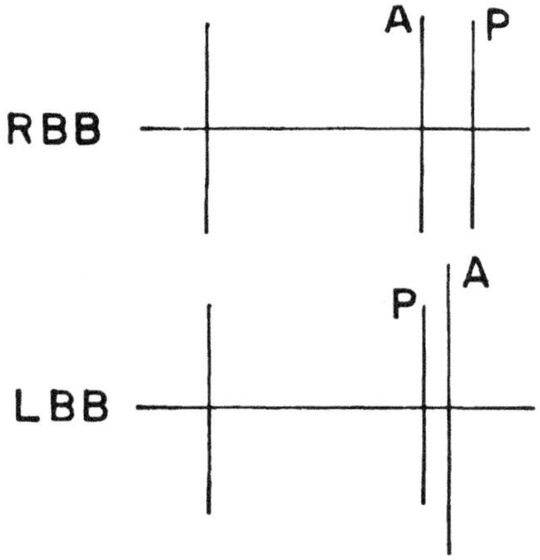

FIGURE 22. Scheme of normal but wide splitting in right bundle branch block and reverse splitting in left bundle branch block.

Changes in the Duration of Ejection

If the ejection period is prolonged because of diastolic overload, ventricular systole lasts longer and the valve will close later. This mechanism has been considered typical for the *right ventricle* in certain left-to-right shunts (ASD, partial anomalous venous return), and for the *left ventricle* in another left-to-right shunt (PDA). However, recent studies in our laboratory (Kumar and Luisada) have demonstrated that, in uncomplicated ASD, right ventricular systole ends at the same time as left ventricular

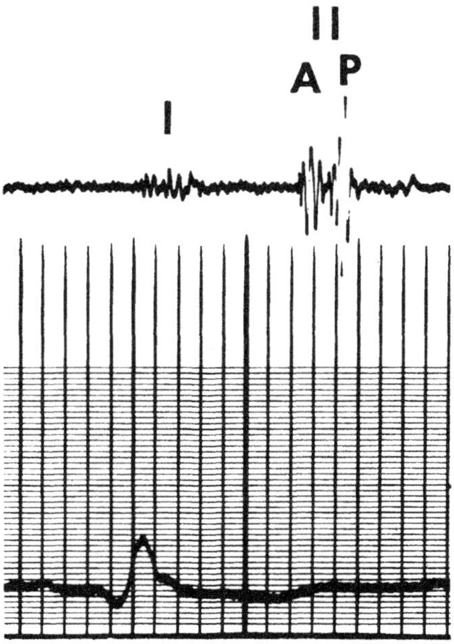

FIGURE 23. Right bundle branch block in a 72-year-old man. Wide splitting of normal type of the second sound.
Tracing recorded over the second left i.c.s.
High pass filter at 100 Hz; slope 12 db/oct.

systole. Therefore, it should be assumed that the delay of the pulmonic component of the second sound in this condition is caused *only by peripheral factors,* which increase the interval between pulmonary valve closure and sound component (see below).

Abbreviation of left ventricular ejection has been postulated in mitral insufficiency (Brigden and Leatham) and in ventricular septal defect (Blazek and Bliss). This would cause a wider splitting of the normal type. However, new studies are required.

CHANGES IN DURATION OF TENSION

If the isovolumic or tension period is *prolonged,* systole lasts longer (unless ejection is sharply decreased) and the valve will close later. While there was early recognition that diastolic overload of one ventricle is a common cause of splitting, the effect of systolic overload has been debated for a long time. Our observation that cases of chronic pulmonary disease often have

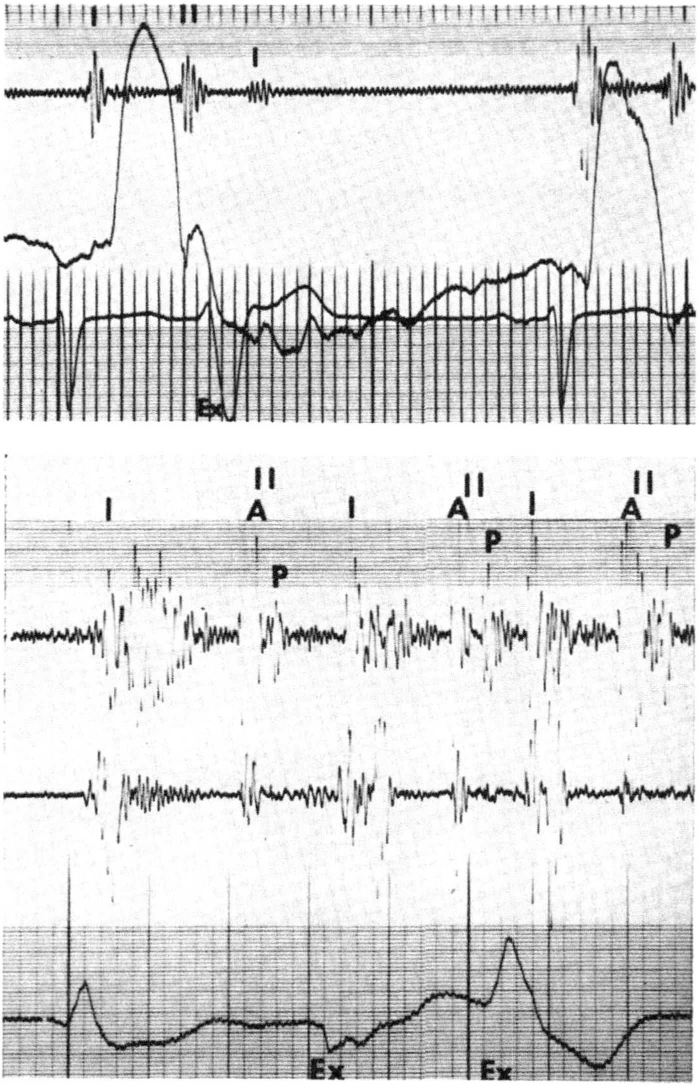

FIGURE 24. (A) Left ventricular ectopic beat.
Small first sound, no second sound.
From above = PCG, carotid tracing, ECG.
(B) Left ventricular ectopic beats.

Two LV ectopic beats from different foci cause a wide splitting of the second sound in a patient with atrial fibrillation and normal splitting in the usual beats (first complex).

a wide splitting of the second sound led us to state that systolic overload is a cause of prolongation of systole (Luisada and Shah, 1963). It is now accepted that *chronic* (Sutton and Harris) and even *acute* (Logue *et al.*) *pulmonary hypertension* is a common cause of wide splitting of the second sound. Wide inspiratory splitting of a normal type is present in cases of pulmonary hypertension without failure while wide and "fixed" splitting (inspiratory and expiratory) occurs in cases complicated by right heart failure (Bose). This phenomenon, attributed to prolongation of systole, has been recently studied in our laboratory. Animal experiments have shown that an early reflux of blood prevents for a brief time the closure of the pulmonic valve (Luisada *et al.*, 1972). Systemic hypertension has less

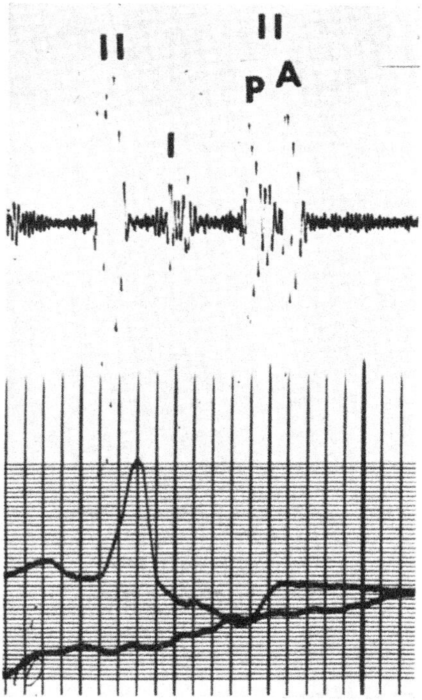

FIGURE 25. (A) Case of coronary heart disease. Right ventricular ectopic beat (ECG, lead II).
The second sound of this beat shows a wide, reverse type of splitting. The previous beat shows a single second sound. Velocity tracing recorded over the second left i.c.s. with band pass filter at 100-600 Hz; slope 48 db/oct.

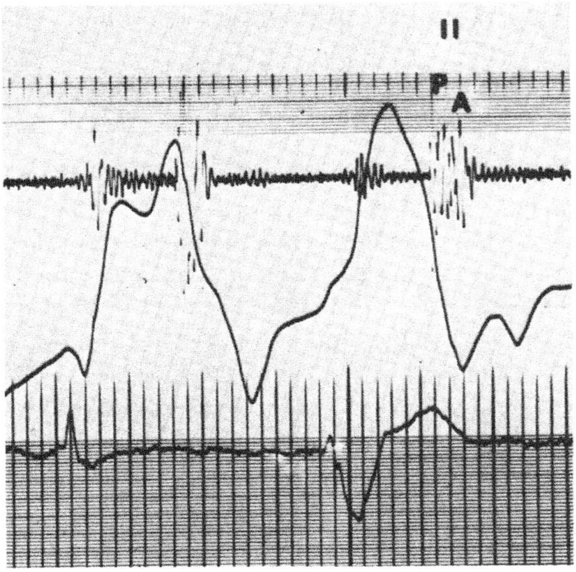

(B) Case of systemic and pulmonary hypertensions.
Upper tracing = PCG over second left i.c.s. velocity tracing, band pass filter at 50-300 Hz, slope 48 db/oct.
Center tracing = apex cardiogram.
Lower tracing = ECG, lead III.
The tracing shows a normal beat with single second sound and a *right ventricular ectopic beat* with reverse splitting of the second sound.

effect on left ventricular systole. However, severe hypertension causes the occurrence of a single second sound irrespective of age (Luisada and Argano), a fact that has been attributed to prolongation of left ventricular systole because of longer isovolumic phase with normal ejection phase. Reverse splitting can be observed only in severe systemic hypertension with left ventricular failure.

Delay of a component of the second sound is often observed in semilunar valve stenosis. However, splitting is seldom audible in this condition on account of the small magnitude of the delayed component. On the other hand, phonocardiography easily reveals the existence of this type of splitting.

In semilunar valve stenosis, several mechanisms contribute to the delay of the sound component. Even when the duration of systole is normal, closure of a valve is delayed because end of

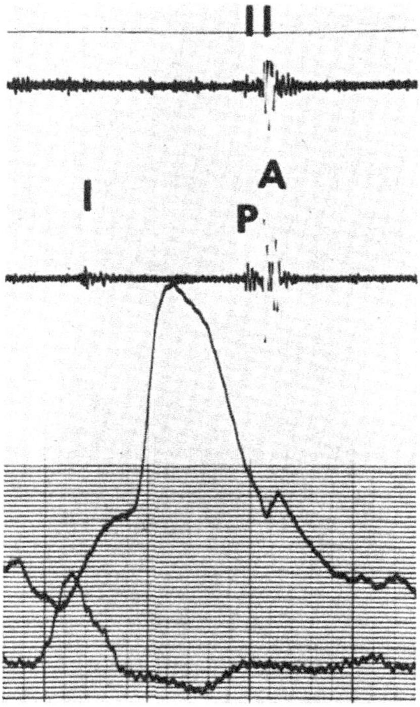

FIGURE 26. Left bundle branch block.
Reverse splitting of the second sound with larger and delayed aortic component.
Upper = PCG over second right i.c.s.
Center = PCG over apex.
High pass filters at 200 Hz; slope 24 db/oct.
The carotid tracing helps in recognizing the aortic component that coincides with the incisura.

systole is not immediately followed by valve closure, pressure in the vessel being much lower than in the ventricle (Fig. 27). In addition, peripheral factors contribute to further delay (see below).

Prolongation of the isovolumic period can be observed in two conditions:

(1) Ventricular damage (myocarditis, myocardiopathy, left BBB); and
(2) Ventricular failure, either right or left. It is still not known whether ventricular damage is a cause of splitting only through failure but this is likely.

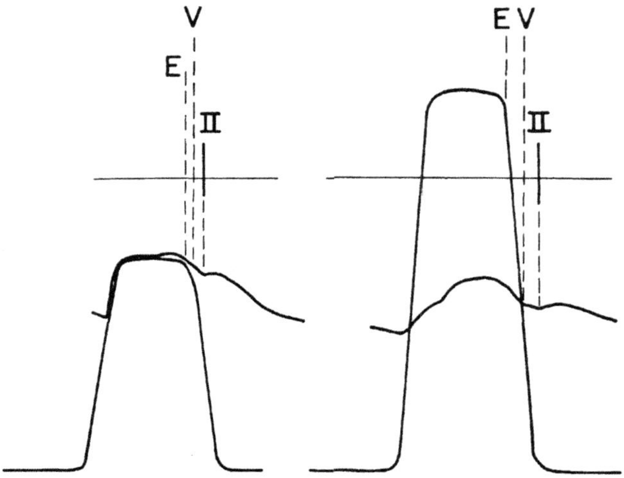

FIGURE 27. Scheme of dynamic events in semilunar valve stenosis. The lower level of arterial pressure causes a delay in semilunar valve closure.
E = end of ventricular systole.
V = valve closure.
II = components of second sound.

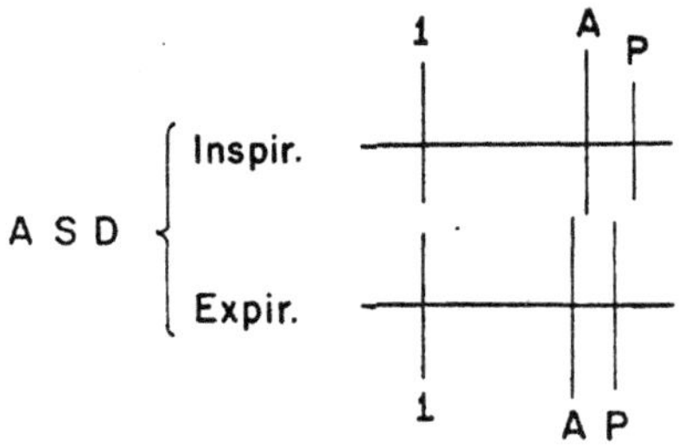

FIGURE 28. Scheme of behavior of the second sound in cases of atrial septal defect in inspiration and expiration.

Peripheral Effects

If one of the large vessels is dilated, the rebound over the closed valve occurs after a longer interval from valve closure (longer delay of both the rebound and the sound due to increased compliance of the vascular wall). This accounts for a delay of the sound component even though the end of ventricular systole is not delayed. This phenomenon is typically found in diastolic overload of the right ventricle (ASD, *secundum* type*; partial

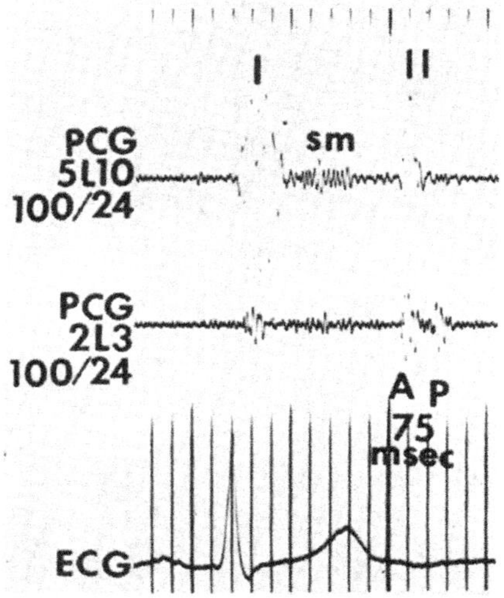

FIGURE 29. Case of atrial septal defect with marked splitting of the second sound, well recorded over second left i.c.s. (center tracing).

anomalous venous return) (Figs. 28 and 29). It is also typical in diastolic overload of the left ventricle (PDA). In certain other conditions (hypervolemia of pregnancy; renal failure) (Fig. 30), both the aorta and the pulmonary artery are dilated but the effect is greater for the pulmonary artery on account of its greater compliance. The importance of dilatation of the large vessels is demonstrated by experiments in dogs (Fig. 31). It is further shown by:

(1) Wide splitting of the normal type in idiopathic dilatation of the pulmonary artery;
(2) Frequent single second in severe dilatation of the aorta;
(3) Persistence of normal, but wide, splitting in ASD following repair of the defect, as long as the pulmonary artery is dilated.

Dilatation of the large vessels is an important contributory element in cases of semilunar valve stenosis (aortic, pulmonary) as well as subvalvular or supravalvular stenosis. In aortic

*In the *primum* type, mitral insufficiency is present; this may cause abbreviation of left ventricular systole with early closure of the aortic valve and early aortic component. In addition, dilatation of the pulmonary artery causes a delay of the pulmonary component like in the *secundum* type.

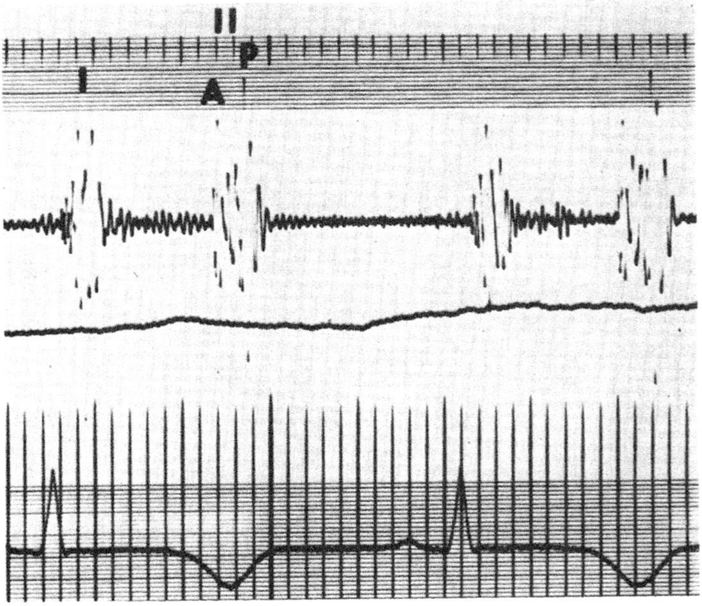

FIGURE 30. Case of urenia with marked splitting of second sound. Velocity tracing recorded over fourth left i.c.s.
High pass filter at 50 Hz; slope 48 db/oct.

stenosis, LV systole is often of normal duration (prolongation of the tension period is compensated by abbreviation of the ejection period) but the aortic rebound is still delayed. This delay seems to be related to peripheral factors (post-stenotic dilatation, slow arterial pulse). (Kumar and Luisada, 1971). In peripheral branch stenosis of either the pulmonary vessels (Perloff *et al.*) or the aortic vessels (Sakamoto *et al.*, 1968), both dilatation of the main vascular trunk and low pressure gradient at the time of valve closure (the diastolic pressure is normal or low in the artery) contribute to a delayed rebound. The delay of the sound component over closure of the valve is then greater than normal and the sound will be of either normal or lesser magnitude.

SLIGHT ABNORMALITIES OF THE SECOND SOUND

The normal second sound is made of two components, the *aortic* and the *pulmonary,* that follow each other in this order. There is a greater separation between the two at the end of inspiration while, in expiration, they either become closer or

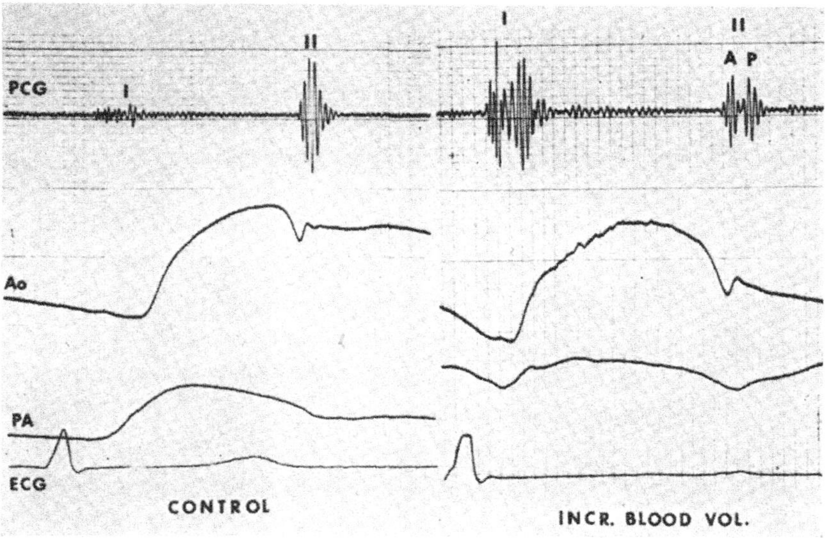

FIGURE 31. Experiment in a dog.
In a normal situation (control), there was a single second sound. After doubling the blood volume, a narrow but definite splitting of the second sound (20 msec.) has appeared.
Pressures in the aorta and pulmonary artery are recorded with catheter-tip manometers.
External phonocardiogram over third left i.c.s. recorded with G.E. equipment; high pass filter at 100 Hz; slope 48 db/oct.
The splitting may be decreased by the hypertension of the pulmonary artery.

fuse into a single sound. This inspiratory (wider) splitting of the second sound is a normal phenomenon easily appreciated by auscultation, and is usually more pronounced in children than in adults (Fig. 32). This splitting is usually best appreciated in the 3rd left i.c.s. close to the sternum. However, *normal subjects* may exhibit this splitting in the 2nd, 3rd, and 4th left i.c.s. along the left margin of the sternum and even in the 3rd or 4th right i.c.s. close to the sternum.

This phenomenon was explained in the past as being caused by changes in duration of right ventricular systole. Our studies in normal animals have shown that the right and left ventricular systoles have about the same duration in the two phases of respiration. Therefore, greater inspiratory volume of blood in the pulmonary tree during inspiration should be accepted as the

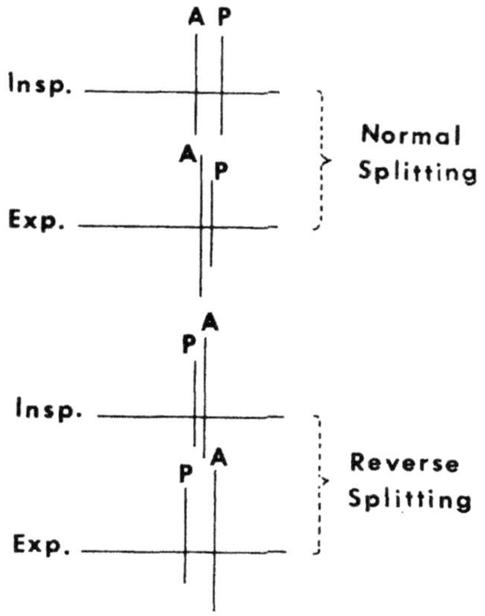

FIGURE 32. Scheme of the behavior of the A and P components with respiration in cases of normal or reverse splitting.

cause of the wider splitting by delaying the pulmonary pulse rebound and the pulmonary component of the second sound while an opposite, though lesser, change is shown by the aortic component.

The delay between the two components usually decreases with age and may disappear between 40 and 60 years of age. This fact can be explained by the greater pulmonary arteriolar resistance in mature adults, which would cause an earlier pulmonary pulse rebound and an earlier pulmonary component of the second sound.

The importance of vascular factors in causing the two components of the second sound is illustrated by the fact that often infants with *truncus arteriosus* and a single valve have a split second sound (Victorica *et al*). The two separate vascular systems (pulmonary and aortic), having different compliance,* cause two rebounds over the single valve.

Q-IIA Versus Q-IIP Intervals. The two intervals, Q-IIA (from onset of QRS to onset of aortic component of second sound) and the Q-IIP (from onset of QRS to onset of pulmonary component

*In these infants, in spite of the high pressure of the pulmonary arteries, the pulmonary vascular resistance is still much lower than the systemic.

of second sound), have been accepted in the past as indication of the duration of systole of the two ventricles (Shah and Slodki). However, they only give the timing of the sound, as discussed above. Formulas for correlating these intervals with the heart rate and the age of the individuals (from 20 to 50 years) have been developed, and they will help in determining whether a splitting is due to an earlier A component or a later P component.* More recent studies (Adolph and Fowler) have shown that, in about 60% of adults between 21 and 30 years of age, the second sound is split in inspiration and single in expiration. After age 50, the percentage falls to 34.6%.** In regard to older individuals, another study (Slodki *et al.*) has found that they all have prolongation of both Q-IIA and Q-IIP in comparison with younger individuals but that greater prolongation of one of these intervals causes the occurrence of either normal splitting (37%), single second sound (38%), or reverse splitting (25%). New formulas were then derived for old age.*** Expiratory splitting has been found in 11% of normal young subjects in the recumbent position but not when they are placed in the upright position. The same applies to expiratory splitting without heart disease in children (Ehlers *et al.*). Explanation of these changes through greater prolongation of systole of either ventricle is still speculative and will require documentation.

MULTIPLE COMPONENTS OF THE SECOND SOUND

The phonocardiogram of certain subjects, either normal or suffering from heart disease, may show a more complex configuration of the second sound. In them, auscultation would reveal either a "muffled" or a "prolonged" second sound.

The most frequent occurrence is that of finding a small, low frequency vibration at the apex that follows the pulmonic component and has a distance of 80-90 msec. from the aortic component. Less often there is a medium or high frequency vibration

*Shah and Slodki's formulas are as follows:

Q-IIA = 61.6 10.24 $\sqrt{RR}$

Q-IIP = 98.9 10.631 $\sqrt{R\text{-}R}$-0.860 Age

**Adolph and Fowler obtained the folowing formulas:

Q-IIA (in sec.) = 0.00182 HR (heart rate) 0.520 (S.D. ± 0.019)

Q-IIP (in sec.) = 0.00185 HR (heart rate) 0.529 (S.D. ± 0.022)

***Slodki et al. obtained the following formulas:

Q-IIA = 22.9 12.77 R-R

Q-IIP = 63.3 11.90 R-R

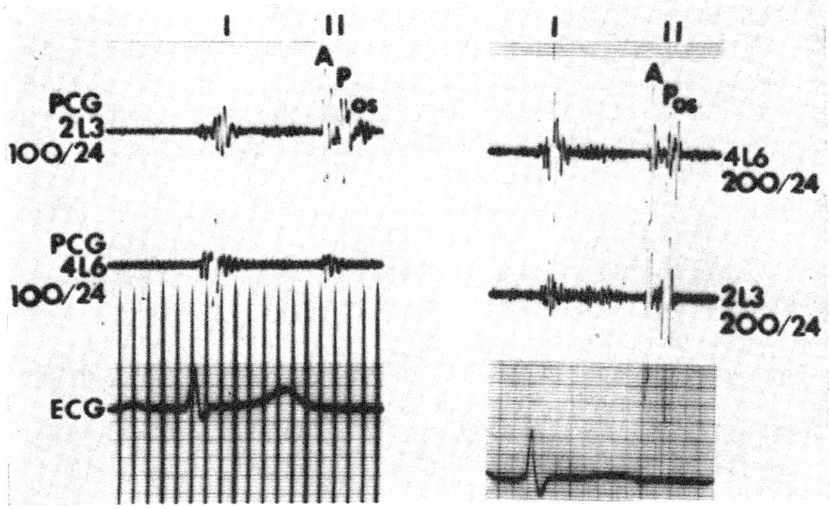

FIGURE 33. Physiologic opening sound (os) adding a third component to the second sound in two children with innocent murmurs (triplication of second sound).
At left, the component is of low frequency.
At right, the component is of high frequency.

in the second or third left interspace separated from the aortic component by the same interval (Luisada and Argano). As, in normal subjects, mtral valve opening follows aortic valve closure by an average of 85 msec., it seems logical to explain this additional vibration with the deceleration of left atrial blood that occurs at the completion of mitral valve opening.* Thus, the symbol *OS* (opening sound) should be given this vibration that is found in about 8 per cent of normal persons or subjects with innocent murmurs. The term *triplication of the second sound* seems appropriate for these cases (Fig. 33).

In other cases, the aortic component of the second sound is followed by *a second group of vibrations prior to the pulmonary component.* This additional group of vibrations (or component) is more easily studied whenever there is a reverse splitting of the second sound by left bundle branch block (Fig. 34). This additional component follows the delayed aortic component by

*This mechanism has been recently suggested for the opening snap of mitral stenosis, which has been attributed to "inertial vibrations of the entire left atrial system" (von Egidy and Dudeck).

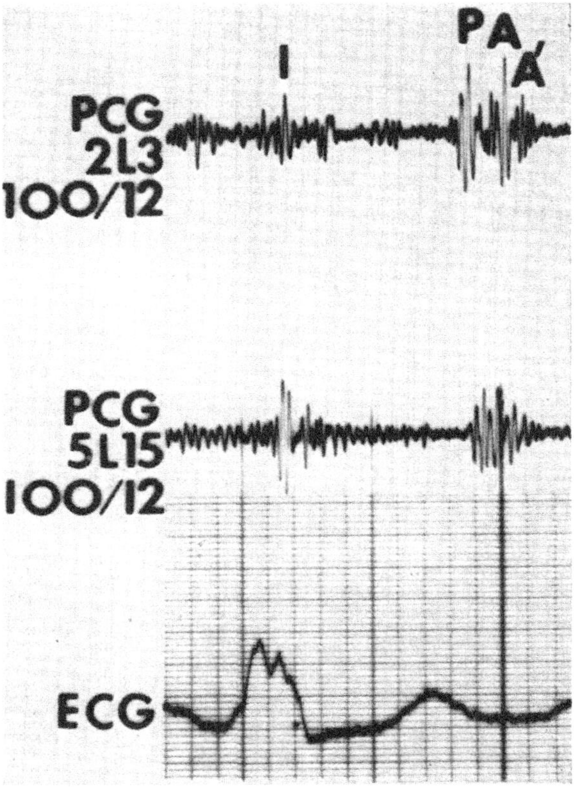

FIGURE 34. Double aortic component (A, A′) observed in a case with left bundle branch block and reverse splitting of the second sound. Triplication of the second sound.

from 15 to 35 msec. As opening of the mitral valve would occur much later, it seems reasonable to explain this component as due to *a second rebound of the aortic blood* causing vibrations of the aortic wall. A second rebound may be the result of a reflected wave from the periphery if the arteriolar resistance is high, and this seems to be the case in our patients, who were hypertensive.

Pathological triplication occurs of course in cases of mitral stenosis due to the fact that the opening snap follows the two normal components of the second sound.

The most unusual occurrence is *quadruplication of the second sound* (Fig. 35), probably caused by a double opening sound that occurs in normal persons with or without an innocent murmur. It is possible that increased cardiovascular dynamics plays a role in causing these opening sounds (one from the right heart, one from the left) in the absence of valvular lesions.

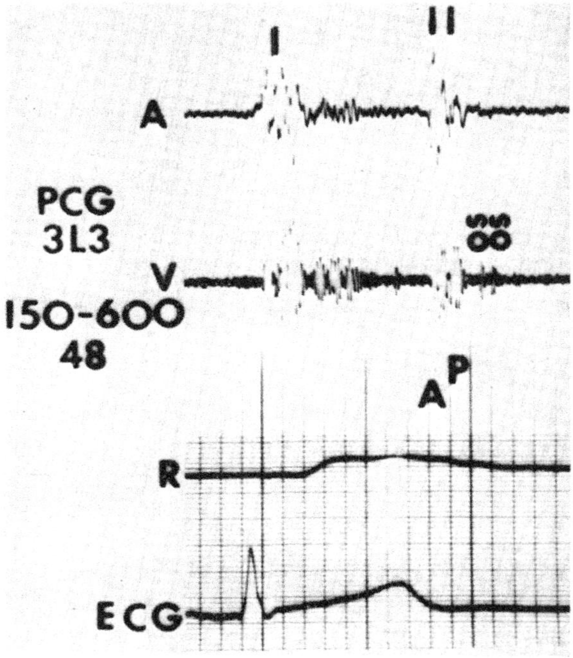

FIGURE 35. Double, small opening sound following a split second sound (quadruplication of second sound).
Twenty-two-year-old woman with pulmonary flow murmur.

WIDE SPLITTING OF NORMAL TYPE

Wider Inspiratory Splitting. In this type of splitting, the normal sequence of the two components is preserved. Two types can occur. In *one variety*, the splitting is wider than normal in inspiration but narrows in expiration to either close splitting or single sound. This occurs occasionally in normal children but is more common whenever either the pressure or the flow is increased in the right ventricle (with the exception of atrial septal defect and partial anomalous venous return). Thus it is found in ventricular septal defect, pulmonary hypertension (including pulmonary embolism), pulmonary stenosis, coarctation (Leatham and Segal), and has also been found in digitalized patients in congestive failure (Adolph and Fowler). It has also been demonstrated in cases of either acute or chronic pulmonary heart disease (Logue *et al.*)* and in cases of cardiomyopathy (Adolph and Fowler).

Fixed Splitting. In a *second variety,* splitting is wide in inspiration and persists in expiration. Even though minor respiratory variations are present, especially in cases with right bundle branch block), splitting persists throughout respiration and has been called *fixed splitting.* This variety includes patients who have either delayed activation of the right ventricle (RBBB, left ventricular ectopic beats), increased flow to the right heart causing dilatation of the pulmonary artery (ASD, partial anomalous venous return), or idiopathic dilatation of this vessel. In the first instance, the right ventricle contracts later than the left; in the others, dilatation of the pulmonary artery causes a delay of the pulmonary component. It is obvious that, in RBBB, splitting occurs both in inspiration and expiration; however, respiratory closing and widening still occur. In pulmonary venous or atrial shunts, a common blood reservoir usually prevents the occurrence of marked respiratory differences in the filling of the two ventricles.** (Figs. 28, 29).

In pulmonary stenosis, the splitting is usually fixed and the pulmonary component is small. Several factors contribute to this: one is a delayed occurrence of valve closure because the latter will take place later, during ventricular relaxation, for

*Recent experiments conducted in our laboratory have demonstrated a new mechanism for the wide splitting of the second sound in acute pulmonary embolism. When the late stage of the picture is reached, the right ventricle still ends its contraction at the same time as the left. However, its relaxation occurs very slowly so that pulmonary valve closure is markedly delayed. The only possible explanation it that, with severe dilatation of both the right ventricle and the pulmonary artery, end of ventricular contraction is followed by a regurgitant jet into the ventricle. When right ventricular pressure finally is approaching the diastolic level, the pulmonic valve closes and no further regurgitation takes place. Thus a murmur is recorded only between aortic and pulmonary components ending with the latter. Whether a similar mechanism also occurs in chronic pulmonary hypertension, we still do not know. Of course, in some of the experiments, right ventricular failure also takes place (with later ending of RV pressure) or a more conventional type of pulmonary insufficiency occurs (causing an early-diastolic murmur).

**Exceptions have been observed.

dynamic reasons; another is poststenotic dilatation of the artery causing a late incisura in the pulmonary artery.

Ventricular septal defect with transposition of the large vessels and large pulmonary flow causes wide splitting of the second sound (Flyer *et al.*).

It should be kept in mind that wide splitting of the normal type occurs in 37% of non-cardiac, apparently healthy old people. Wide splitting may occur in hypervolemia of pregnancy and is frequent in advanced renal failure; it disappears in the latter after removal of extracellular fluid by peritoneal dialysis (Gibson). This is most likely related to greater dilatation of the pulmonary artery than of the aorta due to its greater compliance. Of course right ventricular failure is a cause of wide and "fixed" splitting that persists in expiration (P. Harvey and Perloff).

A special acoustic and phonocardiographic abnormality has been described in *constrictive pericarditis* (Beck *et al.*): sudden appearance of wide splitting of the second sound at the acme of inspiration, associated with a marked weakening of the pulse

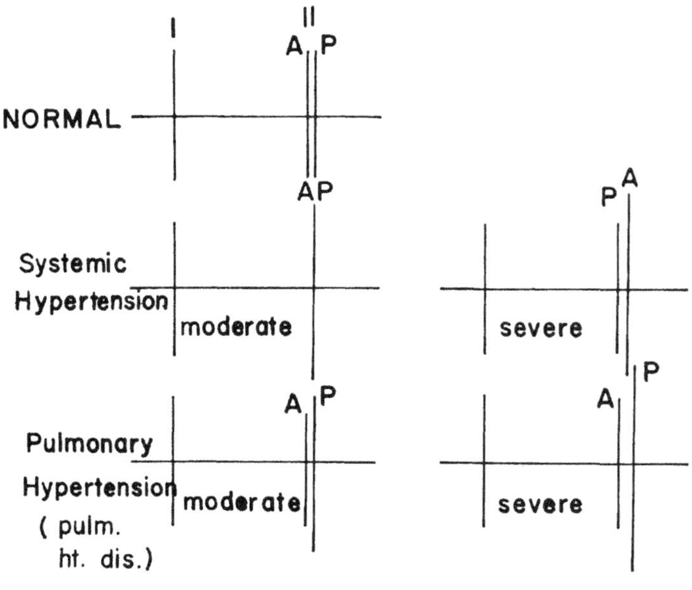

FIGURE 36. Behavior of the second sound in systemic and pulmonary hypertensions.

(*pulsus paradoxus*). This has been explained as due to sudden decrease of left ventricular filling causing a marked decrease in left ventricular output and shorter left ventricular systoles.

SINGLE SECOND SOUND

A single (or closely split) second sound that remains single in expiration can be related to various abnormalities. The most common occurrence is a prolongation of left ventricular systole, either as an isolated change or greater than prolongation of right ventricular systole.* This can be found in a percentage of *normal individuals* (more commonly above age 50) with 38% still found in old age (Slodki *et al.*, 1969). It is also commonly found in *systemic hypertension* irrespective of age (Luisada and Argano) (Fig. 36.)

Another cause of single second sound is present when there is either a *common arterial trunk* with a single quadricuspid valve** (Victorica *et al.*) or *pulmonary atresia* (Adolph and Fowler). In the first case, single second sound in the phonocardiogram is present only in some of the cases. The others have a normally split second sound that should be explained as the result of nonsimultaneous rebounds in the aorta and pulmonary arteries arising from the trunk, due to different compliance on account of different resistance in the two systems. Single second sound on auscultation is found in cases of *aortic stenosis, pulmonary stenosis, tetralogy of Fallot,* or *tricuspid atresia with shunts.* However, phonocardiograms often disclose *two groups of vibrations* in the second sound (the second component is not heard due to its small size and the auditory phenomenon of "masking").

Single second sound is typical of the *Eisenmenger syndrome.* Here the two ventricles contract simultaneously creating the same rise in pressure, and there is no valvular or vascular reason creating a difference in the time of the rebounds over the closed valves because of the high pulmonary vascular resistance.*** See footnote on next page.

*Recent studies on atrial septal defect, pulmonary embolism and aortic stenosis have shown how consistent is simultaneous ending of systoles of the two ventricles in spite of severe derangement of cardiovascular dynamics. Therefore, revision of this concept, based on exprimentation, seems imperative.

**In these cases, the valve can include from two to six leaflets.

In *severe aortic insufficiency,* a single second sound can be observed. This has been explained as the result of longer left ventricular ejection caused by the diastolic overload, so that the delayed aortic component would fuse with the pulmonary component Poggi *et al.*). However, further studies are needed because we have often observed splitting of the second sound in this condition.

WIDE SPLITTING OF THE REVERSE TYPE

A reverse type of splitting is found in most cases of *left bundle branch block* (Haber and Leatham),* in *right ventricular ectopic beats* (Oriol *et al.*), and in *right ventricular pacing.* Delayed activation of both ventricles is possible. However, left ventricular activation is usually *more delayed for the left than for the right ventricle.* The left contracts slightly after the right, contrary to what happens in normal subjects. In addition, the tension period of the left ventricle is often prolonged on account of less efficient contraction because of widespread damage of the left ventricular wall (LBBB, right ventricular pacing) or need for greater rise of pressure in order to open the aortic valve (ectopic beats). It is true that the ejection period becomes shorter but the total sum (tension plus ejection) may be slightly greater for the left than for the right ventricle. A further element, recognized in our study (Luisada, Kumar, and Pouget), is a peripheral element delaying the rebound in the aorta. (Fig. 32).

In *aortic insufficiency,* reverse splitting has been observed (Fishleder). This is possibly related to dilatation of the aorta plus left ventricular failure; the former would cause a delay of the aortic rebound; the latter might cause prolongation of left ventricular systole. (Fig. 37).

In *patent ductus arteriosus* with a large left-to-right shunt, there is a greatly increased flow through the left heart in comparison with the right. Splitting has been attributed to increased

***The difference between these cases and those with a common trunk resides in the fact that the latter, when observed in infancy, have a high pressure in the pulmonary arteries but a normal pulmonary arterial resistance.

*An interesting fact is *reverse splitting observed in RBBB complicating transposition of the great vessels.* The right ventricle is activated later but pumps its blood into the aorta (Zuberbuhler *et al.*).

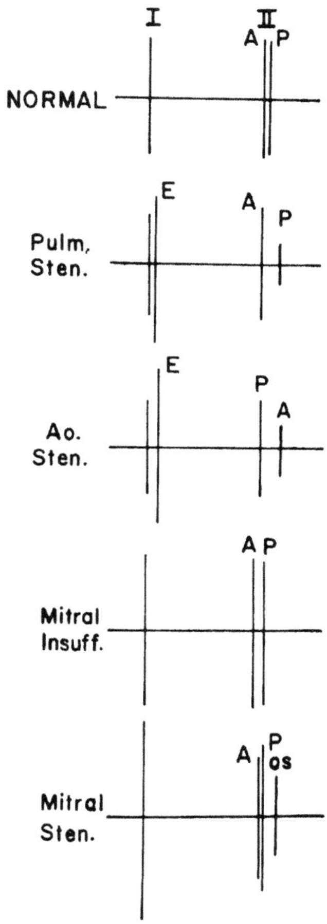

FIGURE 37. Behavior of the second sound in valve lesions.

duration of the ejection period of the left ventricle with delayed closure of the aortic valve (P. Harvey and Perloff). However, the importance of peripheral factors is such that delayed aortic rebound on account of dilatation of the aorta is more likely to be the responsible factor.

Reverse splitting may occur in *myocardial infarct* (Holldack *et al.*)*, *coronary heart disease* (Yurchak and Gorlin), either during attacks of angina pectoris or without pain (Agnew *et*

*It is interesting that some patients with infarct and some cases with cardiomyopathies have a reverse splitting that only appears in inspiration.

al.), and *cardiomyopathies* (Braunwald *et al.*, 1960). In infarct, the reverse splitting may disappear with recovery, though it may persist (Crampton); in angina, it is a transient phenomenon; in cardiomyopathies, when present, it is persistent. Reverse splitting can be observed in *systemic hypertension with left heart failure* (Gray), though this finding is not comomn.

Before considering that a reverse splitting is caused by coronary heart disease, one should keep in mind that reverse splitting occurs in 25% of *healthy senile subjects* without electrocardiographic or clinical evidence of such disease. This is probably due to senile alteration of the left ventricular wall (fibrosis).

In *aortic stenosis* (infundibular, valvular, or supravaluvular), left ventricular contraction may be somewhat prolonged. As prolongation of the tension period is often compensated by abbreviation of the ejection period, this alteration may not be significant. However, there are two causes for a delayed rebound: one is a delay in aortic valve closure, so that this will occur only during relaxation of the left ventricle (Fig. 27); the other is lower filling and post-stenotic dilatation of the aorta plus slower progression of the arterial pulse. As a result, the aortic component wil be small and delayed.*

An unusual occurrence has been found in the *Wolff-Parkinson-White syndrome type B* (anomalous bundle causing early activation of the right ventricle**). Here the P component occurs early while the A component occurs after a normal Q-IIA interval (Zuberbuhler *et al.*; March *et al.*).

Another rare occurrence is that of reverse splitting in cases of *transposition of the large vessels* (Zuberbuhler *et al.*, 1967). This is related to late activation of the right ventricule (from which the aorta is originating) whenever there is RBBB. A third unusual occurrence is that of reverse splitting in 44% of cases with "extensive aortitis" and narrowing of multiple arteries (*Takayasu's syndrome*) (Sakamoto *et al.*, 1968). This was explained as the result of left ventricular damage (fibrosis),

*Reverse splitting can be occasionally observed in patients with aortic stenosis that have received an aortic valve replacement. In these cases, the delayed valve click is extremely large. The delay is probably related to left ventricular failure.

**This phenomenon is not always present because the early activation may involve only part of the right ventricle.

equivalent to that encountered in chronic myocarditis. However, dilatation of the aorta may also be a factor delaying the rebound over the closed aortic valve.

Chapter 3

Abnormal Third and Fourth Sounds and the "Gallop Rhythms"

MOST OF THE POSSIBLE additional diastolic sounds were correctly evaluated by Potain at the turn of the century (1866 to 1900) on the basis of clinical auscultation. The addition of a diastolic sound gives rise to a cadence that has been called "gallop rhythm".

The varieties of gallop rhythm were correlated with the vibrations of the apex cardiogram (ultra-low frequency tracing) by Laubry and Pezzi in the twenties, and were further analyzed by means of phonocardiography by Wolferth and Margolies, as well as by Mozer and Duchosal, Routier and van Bogaert, and Orias and Braun-Menendez in the thirties. Numerous studies have been published since then, which did not change the basic clinical interpretation of these extra sounds but helped in clarifying some of their graphic details.

The pathologic rhythms, caused by the addition of one or two extra sounds to the two normal heart sounds, were first called *gallop rhythms*. Later the term *triple rhythm* was suggested by Evans whenever one extra sound was present, and the term *quadruple rhythm* was suggested by me whenever two extra sounds were present.

These rhythms can be explained as follows:

(1) *Triple rhythms* are caused by (a) accentuation of the third sound (diastolic or ventricular type); (b) accentuation of the fourth sound (presystolic or atrial type); or (c) accentuation of both the third and fourth sounds with fusion of them in a new sound (summation type).

(2) *Quadruple rhythms* are caused by the accentuation of both the third and fourth sounds, which remain distinct.

Ventricular Type of Triple Rhythm (Increase of the Third Sound)

The normal third sound is a low-pitched vibration that occurs in early diastole and is frequently audible in normal children and adolescents (Fig. 38). It is frequently found in the phonocardiogram of normal young people if the tracing is recorded in the low-frequency bands. It is recorded by the phonocardiograph in higher frequency bands, it is larger, and becomes audible, in several groups of clinical cases including mitral or tricuspid

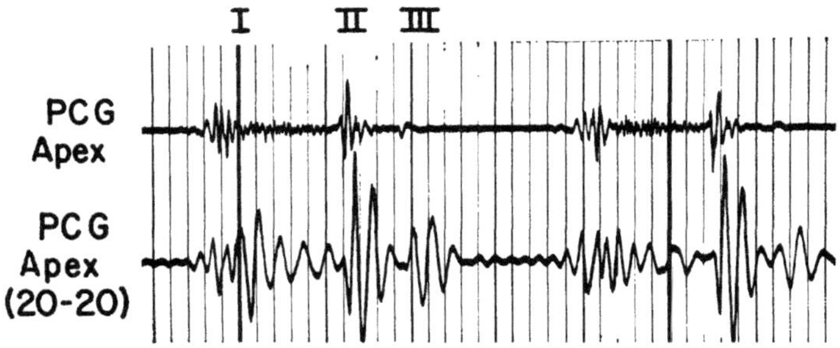

FIGURE 38. Third heart sound in a normal 12-year-old boy.
Above = stethoscopic tracing.
Below = filtered tracing in the 20 Hz range.

insufficiency, myocarditis, coronary heart disease, hypertensive heart disease with left ventricular strain or failure, constrictive pericarditis, and certain arrhythmias.

The normal third sound grossly coincides with the end of the phase of rapid, early-diastolic filling that follows the opening of the AV valves. Thus, this sound occurs at about the peak of the wave of rapid filling that is commonly present in the ultra-low frequency (Thayer, Luisada, 1953) or the low-frequency acceleration tracing (Rosa and Luisada) of the precordium.

Either *increased heart rate* or *increased filling* of either ventricle is frequently asociated with a loud third sound, though phenomena related to the *left* ventricle are the most important. Increased herat rate occurs in sinus or pathologic tachycardia. Left ventricular overload occurs in mitral insufficiency (Nixon, Bleifer *et al.*, 1960A) (Fig. 39); bilateral overload occurs in hyperthyroidism (Renfro *et al.*) and in pregnancy. A large sound is also frequently found in myocarditis (Taquini *et al.*; Luisada *et al.*, 1955, 1959; Harvey and Perloff) (Fig. 40) and in ischemic and hypertensive heart diseases (Parry and Mounsey), even though the third sound is less frequent than the fourth sound in the latter (Miles, Puchner *et al.*, 1960A; Luisada and Argano, 1970).

In conditions of *diastolic overload,* a greater mass of blood enters one or both ventricles, a moderate increase of the diastolic pressure having been accepted as the cause of the additional sound (Nixon). Similarly, a *relative* diastolic overload (normal filling of a dilated ventricle) may explain the large third sound that can be observed in cases of myocarditis, hypertensive heart disease, or coronary heart disease with either frank or latent failure (McKusick). (Fig. 39).

The third sound may decrease or disappear in the sitting or standing position (Parry and Mounsey; Ruskin and Gerner); it is frequently abolished by venous occlusion (Boyer, Warren and

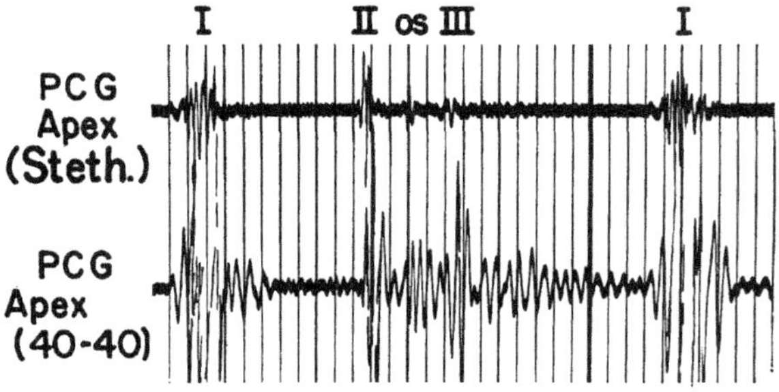

FIGURE 39. Small opening snap and large third sound followed by a mid-diastolic rumble in an 8-year-old boy with acute rheumatic fever. The amplitude and frequency range (40 Hz) are greater than in Figure 38.

Stead); and is increased by vasoconstrictor drugs (Ruskin and Gerner). This shows that it is deeply affected by factors that increase the ventricular volume (increase of the third sound) and vice versa.

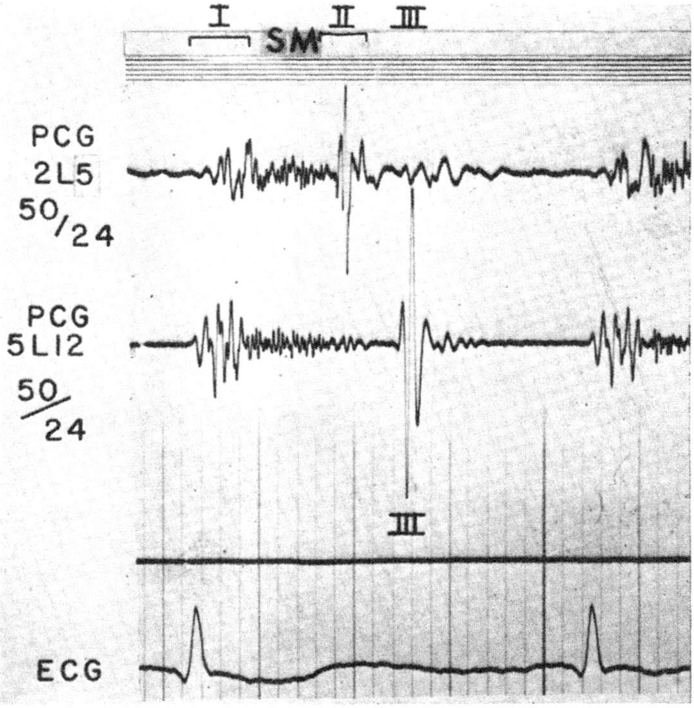

FIGURE 40. Gigantic third sound in a case of pure mitral insufficiency.

In a previous book, the conclusions of Arevalo *et al.* in our laboratory concerning the third heart sound were summarized. This sound cannot be explained by vibrations arising in the A-V valves and cannot be attributed to either blood flow or its arrest. Rather, a relationship was found between this sound and the process of ventricular relaxation. A possible source of energy is embodied in the blood volume present in the ventricles at the time of transition from active relaxation to passive distension. When the ventricular walls stop the adjustment for accommodating the volume of blood, vibrations arise in them. When the ventricular volume is increased because of greater filling or increased residual volume, the distending force is greater and the third sound is accentuated. These phenomena are more marked in the left ventricle than in the right, as proven by the greater effect of mitral insufficiency in comparison with tricuspid insufficiency or atrial septal defect as a cause of a large third sound. This concept explains the presence of a large third sound both in states of volume overload and in states of ventricular failure.

Right and Left Third Sounds. The phase of rapid filling may not be simultaneous in the two ventricles. If there is biventricular dilatation, one can record a third sound with different timing over two areas (Fig. 41). On the other hand, certain cases have a double third sound over the same area (Fig. 42). Both sounds could originate in the same ventricle, as was proven by intracardiac tracings within the right ventricle in one of our cases. It is more common, however, that they arise in the two ventricles.

Third Sound and Diastolic Rumble. The third sound may be unusually prolonged in conditions of severe ventricular overload (whether absolute or relative). Then a large third sound is ob-

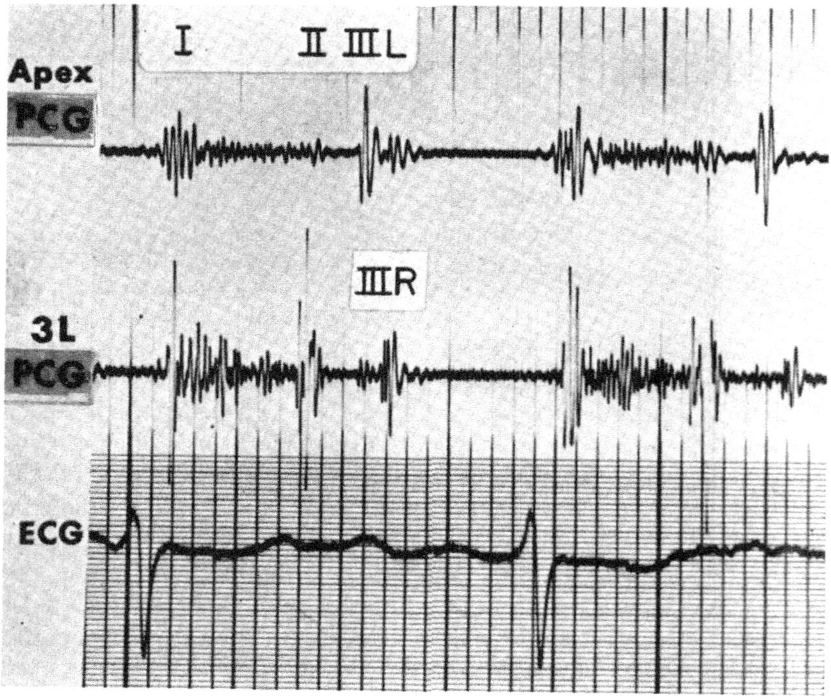

FIGURE 41. Tracings of a patient with mitral insufficiency and biventricular failure.
A double third sound is observed, the left chiefly at the apex; the right chiefly at the third left i.c.s.

served, and is followed by several aftervibrations (Fig. 39). This gives the auscultatory impression of a diastolic rumble and may be occasionally mistaken on auscultation with the rumble of mitral stenosis. These aftervibrations are often of a low frequency, as described by Weber, but can assume a higher frequency, as in cases described by Luisada **et al.** (1955), as well as by Zilli and Gamna (see also Chap. 7).

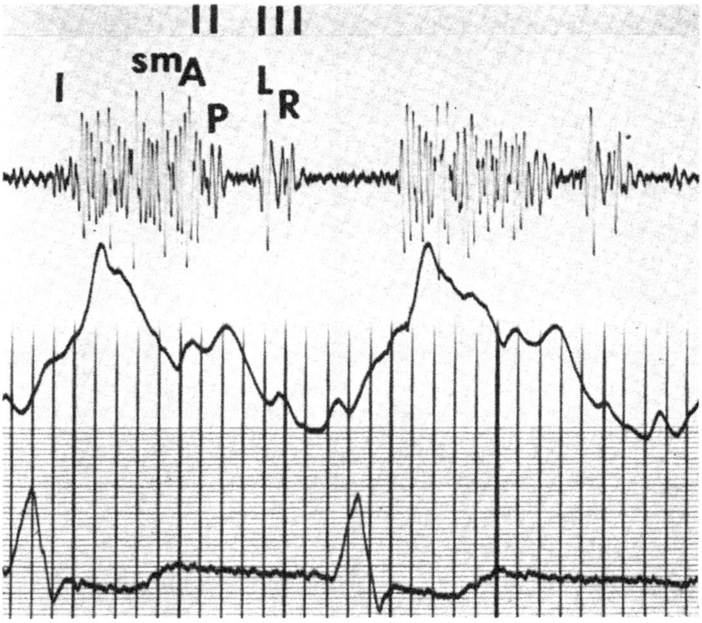

FIGURE 42. Double third sound recorded over the fourth i.c.s. in a case of cardiomyopathy. The incisura of the carotid tracing allows identification of the aortic component of the second sound.

Augmented Ventricular Gallop. When the phase of rapid filling is superimposed upon the phase of atrial contraction, the third sound increases in amplitude (Grayzel, 1959A). A similar phenomenon may be observed after a ventricular ectopic beat (Luisada, 1953), though through a different mechanism.

Third Sound in Constrictive Pericarditis. It is well known that a loud, snapping, and early third sound is frequently found in cases of constrictive or calcific pericarditis, more often in the latter. The name of *pericardial knock* or *pericardial click* was given to this sound to underline both its acoustic character and its cause (Chap. 14). It was described by Friedreich and was found by Mounsey (1955) in 18 out of 22 patients with constrictive pericarditis. The average time interval between the aortic component of the second sound and this diastolic sound is often short (average time is 0.10 sec.) but it increases to normal figures (average time is 0.17 sec.) after surgical decompression (Mounsey, 1955).

Simultaneous sound and pressure tracings from the right heart allowed Mounsey to ascertain that the diastolic sound coincided with the abrupt halting of right ventricular filling. The same was noted by us in 4 cases in which a similar coincidence was determined between the diastolic sound and the end of the diastolic dip.*

Atrial Type of Triple Rhythm

An atrial sound in presystole has been described in patients with delay of AV conduction, where greater separation between the atrial and the first sound makes auditory perception of the two sounds easier. The atrial sound is often audible in patients with *systolic overload* of one ventricle (aortic stenosis, systemic hypertension, pulmonary stenosis, pulmonary hypertension), as well as in patients with diffuse (myocarditis) or localized (ischemic heart disease) myocardial damage, often with latent or obvious failure (Figs. 43 and 44).

As a graphic phenomenon, the fourth sound can be found in the tracings of normal individuals, especially children; it is then a smal, low-pitched vibration. Whenever it has greater amplitude and a higher pitch, it is a pathologic phenomenon. The same authors who studied the third sound also studied the atrial sound.

In the early forties, the term *fourth sound* was suggested for the atrial sound. Even though this term is a misnomer (it gives the highest number to a phenomenon that actually occurs during the initiation of the cardiac cycle), it is in common usage and cannot be easily changed.

The atrial sound is heard and recorded more often than the third sound in patients with ischemic (Parry and Mounsey) or

*This coincidence was determined in three ways: by means of the apex or epigastric cardiogram; by means of the electrokymogram of the left ventricle; and by means of right heart catheterization.

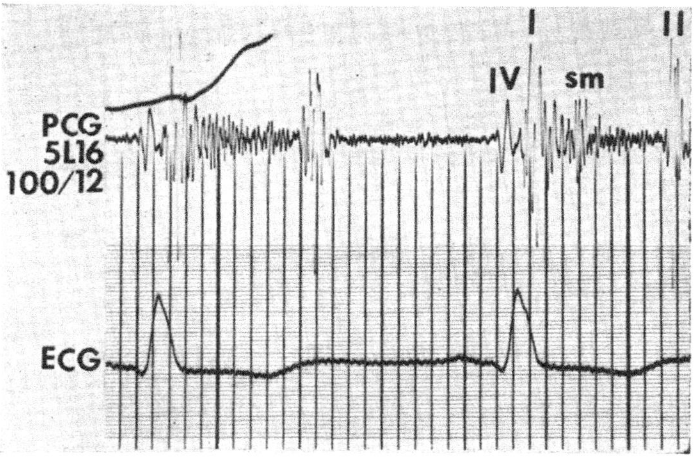

FIGURE 43. Large fourth sound in a case of systemic hypertension.

hypertensive heart diseases (Miles, Puchner *et al.*, 1960; Luisada and Argano, 1970), in aortic stenosis (Fig. 44B), and is common in myocarditis (Taquini *et al.*; Luisada, 1952, 1953; Harvey and Perloff), hyperthyroidism (Renfro *et al.*), and heart failure from any cause. The atrial sound is particularly increased in conditions of ventricular systolic overload, like in pulmonary stenosis (Dimond and Benchimol, 1960) or pulmonary hypertension (Sackner), and in aortic stenosis (Goldblatt *et al.*) or systemic hypertension. In the patients with aortic or pulmonary stenosis having a large fourth sound, the stenosis was severe and there was an increase of the end-diastolic pressure level of the overloaded ventricle. In hypertensive heart disease the atrial sound is usually found in patients with severe hypertension (Luisada and Argano, 1970), especially if they have a markedly elevated aortic diastolic pressure (Sakamoto *et al.*, 1960). These patients may also have an increase of left ventricular end-diastolic presure. A decreased compliance of the left ventricular wall, so that there is greater resistance to filling, has been considered as the probable cause of the additional sound. Atrial hypertrophy may result from the greater atrial effort.

The same maneuvers and positions that decrease the third sound also decrease the atrial sound through a decrease of venous return, whereas vasoconstrictor drugs often increase the

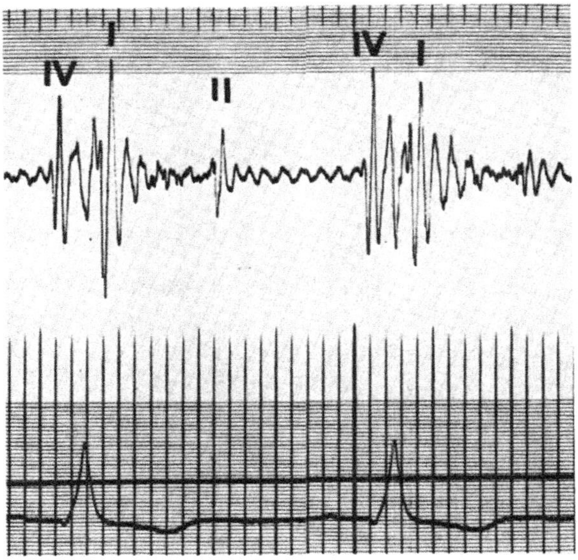

FIGURE 44. (A) Gigantic fourth sound in a 76-year-old patient with coronary and hypertensive heart disease. Displacement tracing over apex at 30-400 Hz.

atrial sound. This shows the connection between the atrial sound and an absolute or relative ventricular overload.

In patients with a large fourth sound (atrial gallop), the phonocardiogram in the medium low-frequency range (30 to 60 Hz) reveals either a diphasic or a triphasic wave in presystole. Occasionally, three or four small vibrations can be recorded, and sometimes *splitting of this atrial sound* can be observed 1952). The atrial sound may fuse with the first sound and may occur after the Q wave of the ECG, especially in children or adolescents with a short P-R interval, where the atrial sound may continue with the vibrations of the first sound.

A fourth sound may originate in either the right or the left heart. It has been stated that the fourth sound of the right heart has a typical behavior: it would become louder, and would become more widely separated from the subsequent first sound, in inspiration than in expiration (Kincaid-Smith and Barlow, 1959).

The *P-IV interval* (interval between the onset of the P wave of the ECG and that of the fourth sound) has been studied during various tests and maneuvers (Tabatznik). Ganglionic

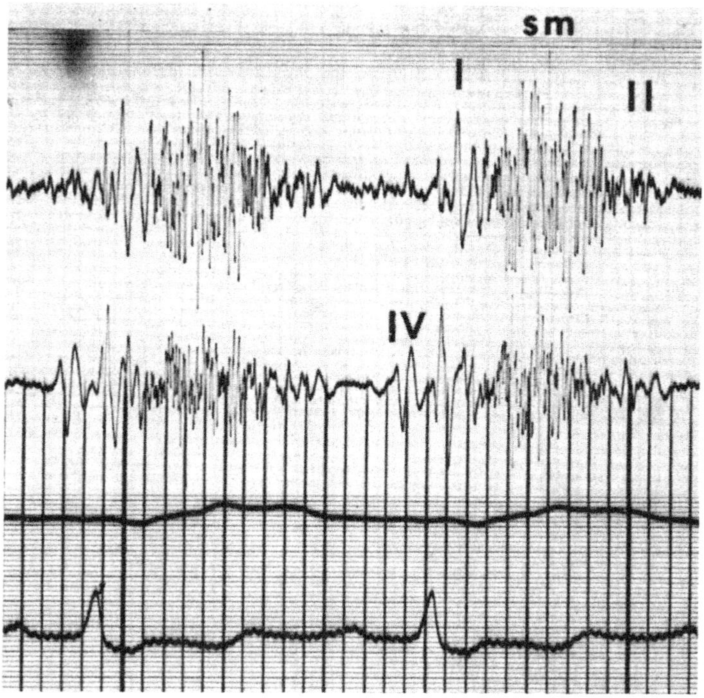

(B) Large fourth sound in a 73-year-old patient with aortic stenosis. The upper tracing is at the second left i.c.s., the central tracing is at the apex. Both are unfiltered tracings.

blocking agents, aminophylline or amyl nitrite lengthen this interval in hypertensive patients. Apparently a decrease of the diastolic pressure of he left ventricle is responsible for these changes. Carotid sinus massage may have a profound influence on the IV sound by (1) inducing atrial arrest, (2) prolonging the conduction time, or (3) causing bradycardia. Either of them would cause either disappearance (1) or a decrease (2,3) of the fourth sound.

The *cold pressor test* often increases the amplitude of the fourth sound, with or without a concomitant increase of blood pressure. Several explanations involving the venous return to the heart and the sympathetic stimualtion of the heart have been invoked for patients without secondary hypertension. Of course systemic vasoconstriction is involved when a pressure rise occurs, thus increasing the ventricular overload.

Atrial Sounds in Atrial Flutter and AV Block: The Augmented Atrial Sound. Long ago French cardiologists noted that, both in patients with atrial flutter and in those with AV block, it was possible to hear additional sounds, louder in the former than in the latter (*systoles en echo*). These sounds, which are recorded by the phonocardiograph, occur more commonly in early diastole. One of the possible explanations is that the atrial contraction is then superimposed over an otherwise silent phase of rapid passive filling. This phenomenon has been called *augmented atrial gallop* (Grayzel).

The phonocardiogram in AV block may record one low-pitched vibration, occasionally two (Lewis, T., Wolferth and Margolies, 1933, 1950) (Fig. 45), or even three (Calo, 1959). When two

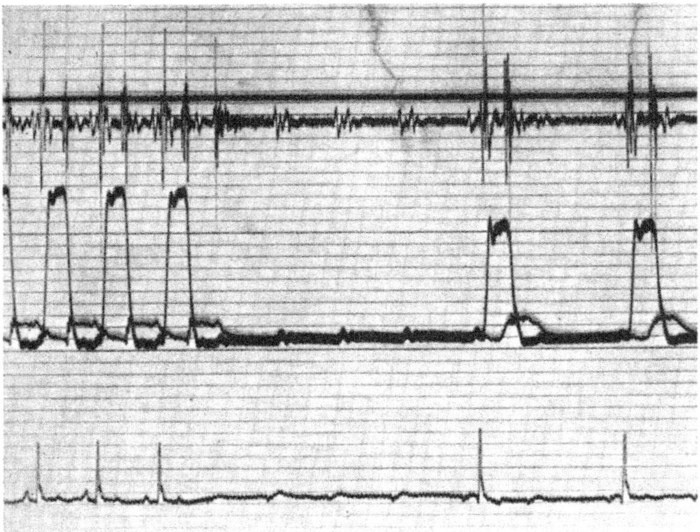

FIGURE 45. AV block induced in a dog by left vagal stimulation. Each atrial contraction causes two separate components of a IV sound. Tracing at 25 mm/sec. speed.
From above =
— respiration (apnea)
— phonocardiogram
— left atrial and left ventricular pressures
— electrocardiogram
The abnormal course of left atrial pressure in the sinus beats (at left) is caused by vagal stimulation. That in the junctional beats (at right) is caused by functional mitral insufficiency.

groups are recorded, the second has a higher frequency and has been explained as the result of tension of either the ventricular wall or the chordae and AV valves (Fig. 45). When three groups are recorded, the third has been explained as caused by an elastic rebound of the ventricular wall (Calo, 1959). These observations coincide with the intracardiac phonocardiograms of Muiesan *et al.* in our experimental laboratory. In these experiments, the first component, attributed to atrial contraction, was not recorded on the chest. Thus, the second component of Muiesan *et al.* would be equivalent in time to the first noted by clinical observers while their third component was equivalent to the second of others.

During atrial flutter, the phonocardiogram may reveal sounds of atrial origin, not only during ventricular diastole, but also during ventricular systole. These sounds have diagnostic value for patients in whom the electrocardiogram is not typical and simulates the pattern of atrial fibrillation (Chap. 11).

Increase of the fourth sound occurs most often in patients with systolic overload of one ventricle as a result of a more powerful atrial contraction. The sound that is heard and recorded over the overloaded ventricle (right or left) is the second component, which corresponds to the presystolic increase of pressure in the ventricle as a result of atrial contraction. In AV block, two components can be recorded corresponding to an increase of the two normal components of the atrial sound. Occasionally a third component may be recorded and may have a valvular origin.

Summation Type of Triple Rhythm—Quadruple Rhythm—Functional Diastolic Rumble

When diastole is short and both the third and fourth sounds are large, from one to three, often irregular, vibrations are recorded in middiastole. These are usually heard as a single sound, and the audible rhythm is then called the *summation type* of triple rhythm (Fig. 46). The same is true when diastole is normal but there is a marked delay of the AV conduction: the phase of early, rapid diastolic filling is either coincident with or continued by acceleration of filling connected with atrial contraction, and a summation type of triple rhythm occurs. On the other hand, if diastole is long and AV conduction is normal, both phases of diastolic filling (early, passive, rapid, late, presystolic) cause large sounds, and they are well separated: a *quadruple rhythm* then takes place (Luisada, 1953) (Fg. 47). In some of these patients, one can have the auditory impression of a diastolic rumble upon auscultation.

The factors causing a triple or quadruple rhythm can also cause a middiastolic murmur, a presystolic murmur, or both. The *middiastolic murmur* is most often of low frequency, usually starts with a large third sound, is frequently recorded over a wide area of the precordium, and is often of large amplitude in phonocardiograms (Luisada *et al.*, 1955, 1959). However, it may have higher frequency vibrations (Zilli and Gamma, Luisada and Di Bartolo, Harvey and Perloff), and may be prolonged into

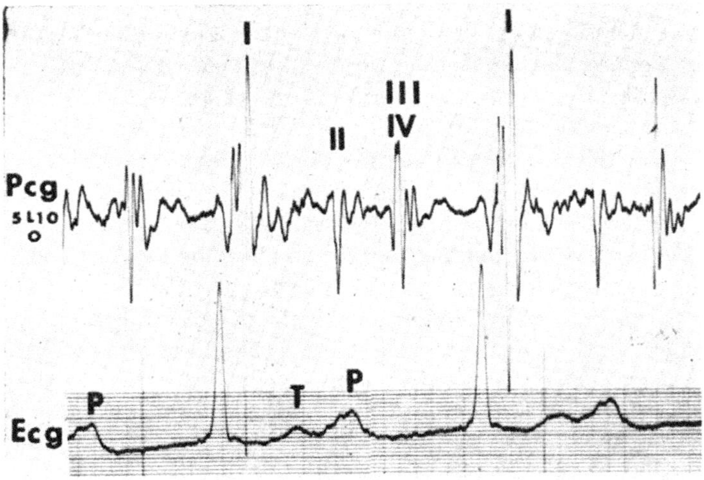

FIGURE 46. Summation type of triple rhythm in a patient with chronic myocarditis and prolonged conduction time (grade I AV block).

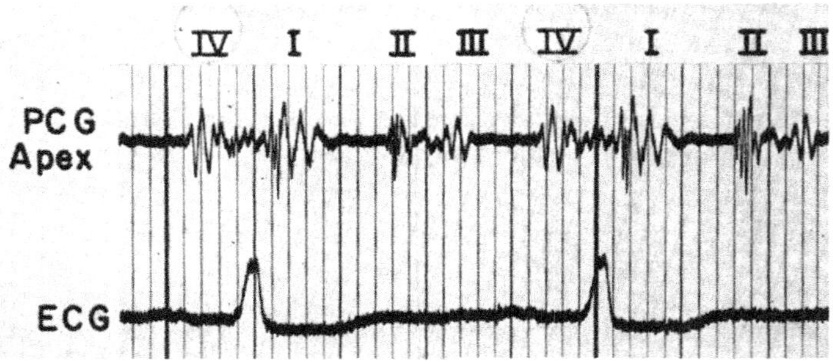

FIGURE 47. Quadruple rhythm with recording of both a third and a fourth sound during long diastoles.

late diastole becoming a *presystolic murmur*. Differentiation from the murmur caused by mitral stenosis is based partly on negative data (no opening snap, often no crescendo type of the presystolic murmur, no delay of the first sound) and partly on positive data (presystolic murmur often well separated from first sound, large third or atrial sound), which, however, have no absolute value. For example, a functional opening snap can occur in the absence of mitral or tricuspid valve lesions (see also Chap. 27).

GENERAL INTERPRETATION OF TRIPLE AND QUADRUPLE RHYTHMS

Triple and quadruple rhythms result from an increase of the third sound, the atrial sound, or both (whether separated or fused), often in the presence of ventricular overload. We are able to be more precise and to state the following:

(1) A triple rhythm of the ventricular (or early-diastolic) type is evidence of diastolic dilatation caused by either diastolic overload (with high output) or heart failure (increase of residual blood).

(2) A triple rhythm of the atrial (or presystolic type) is usually evidence of systolic overload of one side of the heart involving both the atrium and the ventricle; this overload may be relative (heart failure).

(3) A triple rhythm of the summation type is usually evidence of heart failure in the presence of tachycardia, whereas a quadruple rhythm is usually evidence of heart failure in the presence of normal heart rate or bradycardia.

The problem of separating patients with "absolute overload" from patients with "relative overload," and the separation of patients with "strain" from those with "failure" is difficult to solve on the basis of the phonocardiogram. It is a purely clinical problem because even normal diastolic pressure of a ventricle may still be present during failure if the compliance of the ventricle is altered.

Chapter 4

Systolic and Diastolic Clicks

SYSTOLIC SOUNDS

EJECTION CLICK

ONE TYPE OF SYSTOLIC sound is the so-called *ejection sound (ejection click, early systolic click)*. It was noted first by Potain and later by Lian and Welti (1937) as well as by Orias and Braun-Menendez. It was again described by Leatham (1954). When this vibration arises in the aorta, it represnts the accentuation of a normal vibration that can be found toward the end of the first sound. It coincides with a sudden change of slope of the aortic pressure curve and follows, after a brief interval, the opening of the aortic valve. In patients with *aortic stenosis* or *dilatation of the ascending aorta,* this vibration becomes larger, more high-pitched, and more delayed. The frequency of this vibration may be high, in the range of 200 to 400 Hz. It is best recorded over the third left interspace but it may be recorded over the entire left ventricular area, as well as over the entire aortic area (Shah *et al.*, 1964). The conditions associated with this click are aortic stenosis (valvular), aortic insufficiency, systemic hypertension, and hyperkinetic states.

A similar vibration can be heard and recorded in patients with *pulmonary stenosis* or *dilatation of the pulmonary artery.*

This vibration has no correlation with any vibration of the normal heart and only ocurs in abnormal conditions. Its frequency is similar to that of the aortic click. It is best recorded over the pulmonary area but it is also transmitted to the right ventricular area (Shah *et al.*, 1964).

The ejection sounds seem to originate in the walls of either of the great vessels resulting from eddies that form whenever there is a sudden change in caliber at the outlet of one ventricle, whether caused by valvular narrowing, arterial dilatation, or increased rapidity of flow. However, recent studies seem to associate the click to the end of opening of the stenosed valve, when the acceleration of flow is suddenly checked.*

MIDSYSTOLIC CLICK

Another snap can be heard and recorded. It is a *midsystolic* or *late-systolic snap or click*. It is of high frequency (200 to 400 Hz or more). It is often, though not always, recorded over the left ventricular area. In some cases, it is accentuated by deep inspiration. This late-systolic click has a typical variability in regard to its timing and can occur earlier or later when there are changes in the patient's position or a different phase of respiration (Fig. 48 a,b). *Two clicks can be recorded.* A late-systolic murmur following the click, an early-systolic murmur preecding the click, or a few midsystolic vibrations interrupted by the click can be recorded (Fig. 48 c,d,e).

In the past, this click was attributed to tension of pleuropericardial adhesions. Following angiocardiographic demonstration of mitral regurgitation and a particular type of billowing of one mitral leaflet, this click is now customarily explained as the result of a dysfunction of the mitral apparatus. This seems to be related to papillary muscle damage (fibrosis) caused by either a previous rheumatic lesion, a myocardial infarction, prolonged coronary insufficientcy, or a congenital weakness of the muscle fibers of the mitral valve. However, it cannot be excluded that

*It is possible that two distinct varieties of ejection clicks may be recorded. One would occur earlier and would be found only in semilunar valve stenosis. It would represent accentuation of the normal *b* component of the first heart sound. The other would occur later and would be found in conditions with increased flow or marked dilatation of the artery. This would represent accentuation of the *c* component. This interpretation, however, is so far theoretical.

individual cases, especially young people, present this click as a result of pericardial or pleuropericardial adhesions. Even older patients with coronary heart disease may have adhesions because a transmural myocardial infarct is most frequently followed by localized (or even diffuse) pericarditis. Unusual cases of non-obstructive cardiomyopathy without valve lesion may present a late-systolic click (Mercer *et al.*)

A confirmation of this concept can be found in a study of Kerber *et al.* by means of echocardiography. While, in cases of *click plus murmur*, a billowing of a mitral leaflet was documented, cases with an *isolated click* failed to show any abnormal mitral valve motion.

A mid- or late-systolic click should not be confused with an early systolic click (ejection click). The different timing of the two renders this differentiation easy (Fig. 49).

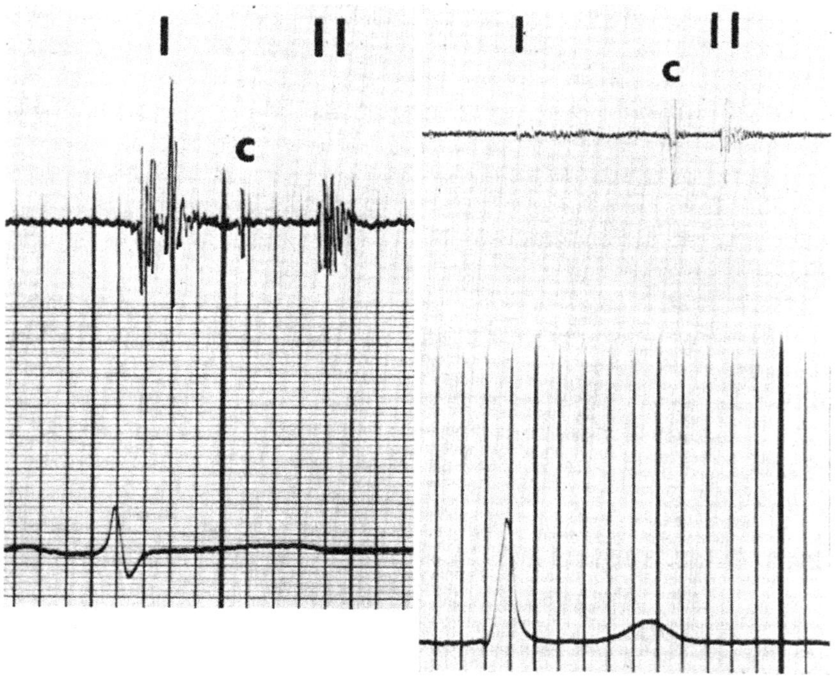

FIGURE 48. Examples of systolic clicks.
(a), (b) Single mid-systolic click in asymptomatic subjects between 20 and 30 years of age.

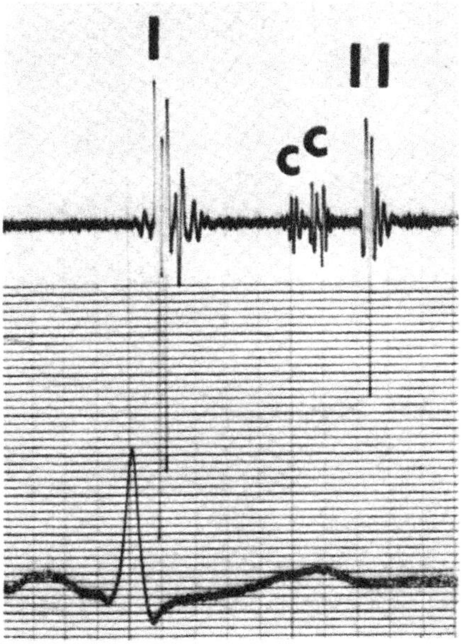

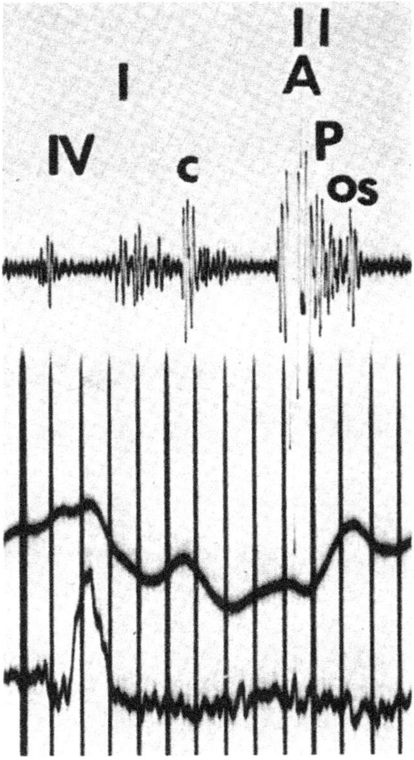

(c) Double systolic click in a 30-year-old asyptomatic subject.
(d) Single systolic click followed by a short systolic murmur.
Also present are a fourth sound and an opening snap. This 76-year-old patient had experienced a myocardial infarct several years previously.

DIASTOLIC SOUNDS

Opening Snap

The *opening snap* is a high-pitched sound that, both upon auscultation and in tracings recorded in the high-frequency range, may be as loud and as large, respectively, as the second sound. It is usualy best heard and recorded in the third and fourth left interspaces between apex and sternum. It may be larger in the second left interspace, however, and occasionally can be heard and recorded over the entire precordium.

The first description of the opening snap of the mitral valve was made by Bouillaud, but only in 1872 was this diastolic sound attributed to the mitral valve, in contrast with the previously

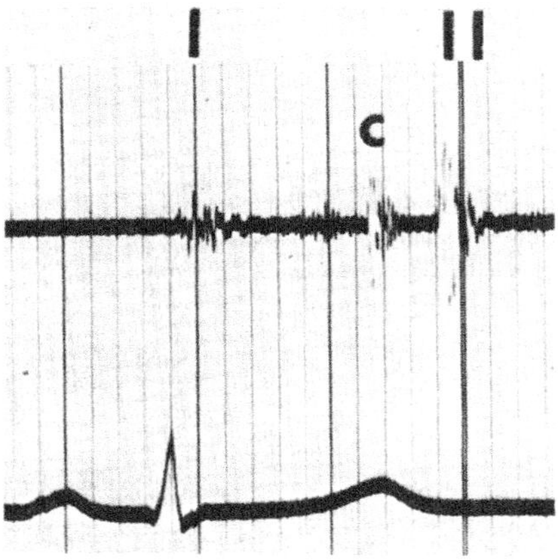

(e) Single systolic click preceded by a short systolic murmur in a 40-year-old asymptomatic subject.

admitted splitting of the second sound. The term "claquement d'ouverture de la mitrale" was first used by Potain, who believed that the sound was produced by the sudden stretching of the stenosed mitral valve. However, other mechanisms were suggested by Gallavardin (1905) and by Duchosal. Phonocardiographic studies of this vibration are numerous.

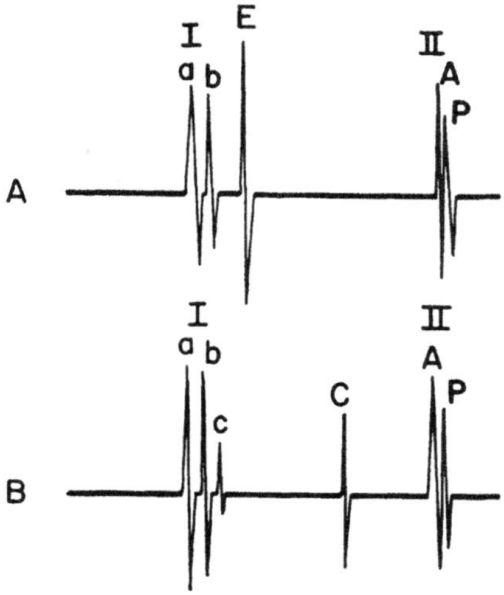

FIGURE 49. Differentiation of a mid-systolic click from an ejection sound based on their timing.

The typical opening snap consists of one or two rapid vibrations and lasts from 20 to 50 msec. Its frequency is between 50 and 300 Hz, the most common being in the 100 to 200 Hz range. Whenever more vibrations are noted, they represent the beginning of a diastolic murmur (Fig. 50).

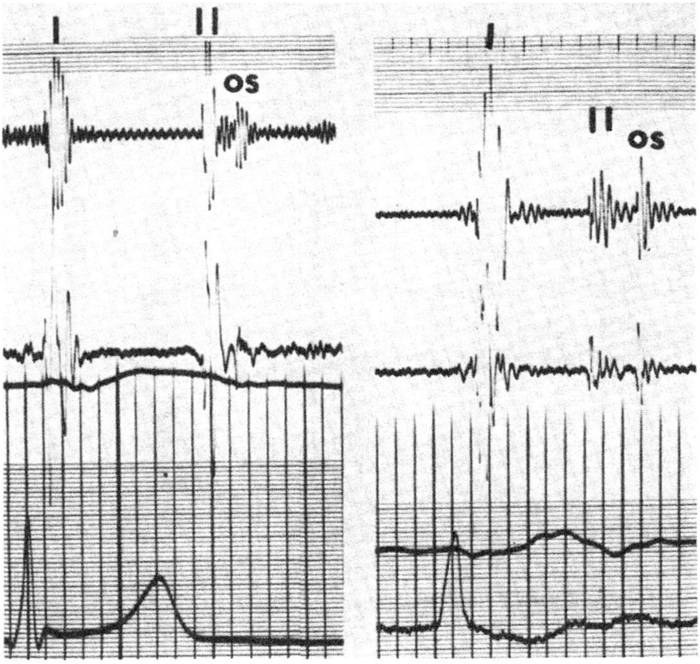

FIGURE 50. (A) Opening snap in a case of severe mitral stenosis before (left) and after (right) commissurotomy.

Epicardial Phonocardiograms (Magri *et al.*, 1958) revealed that the opening snap was best recorded over the left ventricular wall; in exceptional patients, it was recorded best over either the left atrium or the aortic root.

Thompson *et al.* (1969) have shown that the opening snap occurs soon after the crossing of the left ventricular and left atrial pressure, a point that marks the end of mitral valve opening. According to von Egidy and Dudeck, a sudden deceleration of the left atrial blood is responsible for this sound, which is grossly proportional in amplitude to the volume of the left atrium. Thus, it can be well understood why it persists even

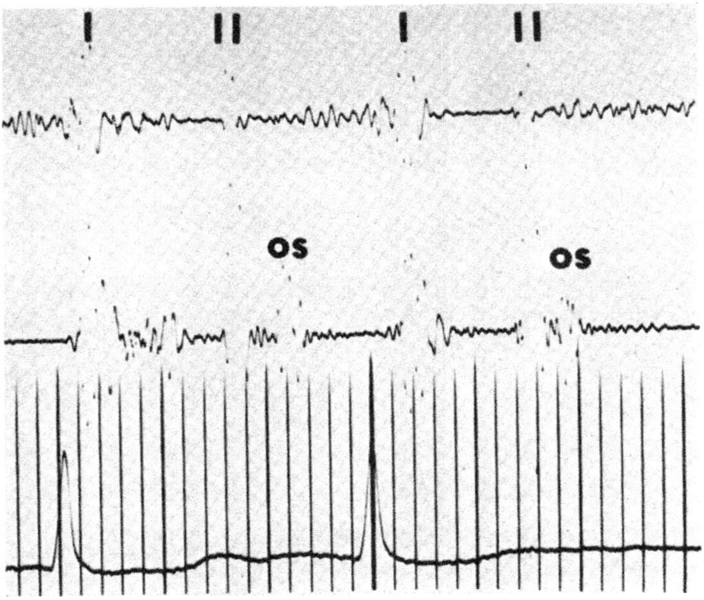

(B) Opening snap in a case of moderate-severe mitral stenosis with atrial fibrillation. Note the variable Q-I and II-os intervals. The first diastole has a rumble simulating a presystolic murmur.

after commissurotomy, the left atrial chamber being still larger than normal. Of course, a prosthetic mitral valve will give a large and late click upon completion of its opening. Whether this is due to blood vibrations, metallic vibrations of the cage, or both will depend on the residual pressure gradient and on the type of valve (Fig. 51).

A *tricuspid opening snap* has been described in tricuspid stenosis, a condition which is often associated with mitral valve lesion. Finally, an opening snap was described in *mitral insufficiency without stenosis, patent ductus arteriosus, ventricular septal defect,* and *atrial septal defect.** Occasionally, a normal phonocardiogram with a seeming triplication of the second sound can be seen, where obviously the third component is a *mitral opening sound* (Luisada, 1949 and 1953; Luisada and Argano, 1971). Sometimes a patient with triplication is sus-

*The fact that this snap is recorded in cases of ASD having right heart overload has been accepted as indicating a "functional tricuspid opening snap." However, demonstration that such snap originated in the right heart is still unavailable.

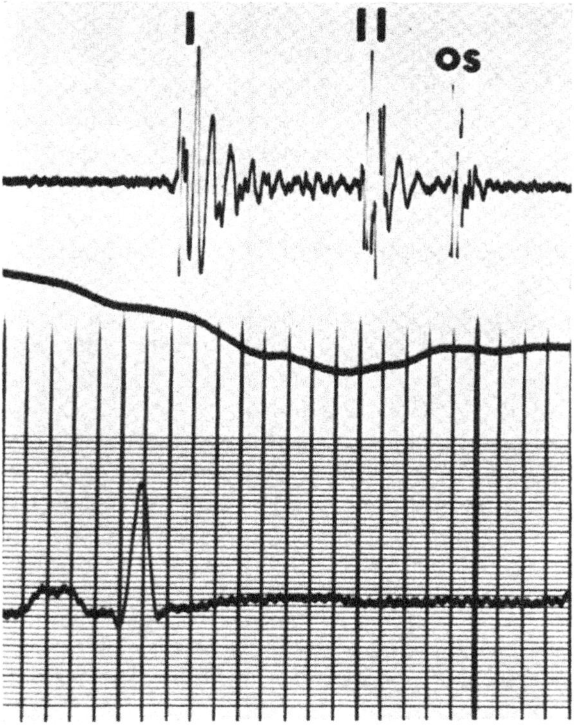

FIGURE 51. Mitral prosthetic valve clicks.
A rapid vibration precedes the main components of the first heart sound.
A delayed click indicates the opening of the valve.

pected of having a minimal mitral lesion but occasionally, this being excluded by right and left heart catheterization, one has to admit that the third component is a normal mitral opening sound.* In one of our cases, in which this triplication of the second sound was recorded, subsequent autopsy proved that all valves were normal.

In exceptional cases of mitral stenosis, an opening snap that has a slightly different timing over various areas can be observed. Since the earlier opening snap is found at the apex and

*The fact that a sound vibration can occur in conjunction with the opening of a normal valve most likely indicates that the origin of this sound is not in the valve itself. A vibration of the blood and cardiac walls caused by abrupt deceleration of the blood would seem the most logical explanation.

the later opening snap is found over the third left intercostal space, it can be surmised that the former is a mitral opening snap and the latter is a tricuspid opening snap (Luisada and Krol). In two of our cases with two opening snaps, autopsy revealed a normal tricuspid valve so that the tricuspid snap could only be explained as a sound caused by altered dynamics. More recently, we have observed cases of mitral stenosis with severe, relative tricuspid insufficiency that presented a *double opening snap* over the same intercostal spaces. Here again, the functional nature of one of the snaps should be admitted. We have also observed a few cases where two opening snaps were recorded over the entire precordium. In the absence of autopsies of these cases, we could not exclude the association of tricuspid stenosis to mitral stenosis, even though catheterization had revealed only a small right atrial-right ventricular end-diastolic gradient of pressure. Even more unusual is the observation of a double opening snap in pure mitral insufficiency (Fig. 52).

DIFFERENTIAL DIAGNOSIS BETWEEN THE PULMONARY COMPONENT (IIP), THE OPENING SNAP (OS), AND THE THIRD SOUND (III)

The problem of differentiation between the pulmonary component, the opening snap, and the third sound may occasionally arise, despite the fact that, in the majority of cases, different time intervals separate these sounds from the aortic component of the second sound.

It is necessary, first, to clearly identify the aortic component of the second sound (IIA). Among other characteristics, this component is usually well recorded along a line that goes from the second right interspace, through Erb's point, to the apex. An important element is represented by comparing the maximal and minimal intervals separating IIA from IIP, IIA from OS, and IIA from III.

The interval between the two components of the second sound is usually shorter than that between IIA and OS. However, borderline cases definitely exist. On the other hand, the interval between IIA and OS is shorter than the interval between IIA and III, the only exception being found in constrictive pericarditis. The former interval (IIA-OS) is usually below 120 msec., the latter (IIA-III) is usually between 140 and 180 msec., except

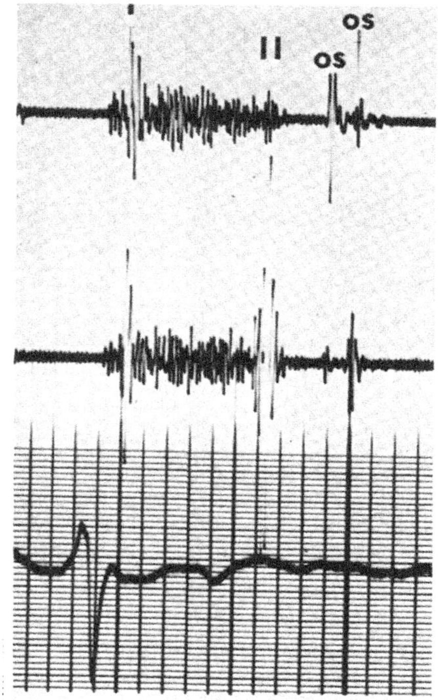

FIGURE 52. Double opening snap in a patient with pure mitral insufficiency (catheterization).

in children. As far as the IIA-OS interval is concerned, it can vary between 50 msec. (rare) and 120 msec. (rare) the average cases prior to commissurotomy being in the 60-90 msec. range.

In regard to the tricuspid opening snap, it occurred on an average of 100 msec. after the aortic component both in the two cases of Kossmann (organic stenosis) and in the two cases of Aravanis (1963) (relative stenosis). In six patients with large atrial septal defects studied by Luisada *et al.* (1964) by means of intracardiac phonocardiography, the IIA-OS interval varied between 90 and 120 msec. Thus, the interval is usually in the same range as that separating the mitral snap from the second sound. Differentiation between the tricuspid snap and the mitral snap can be made at times because the tricuspid snap is often greater in inspiration and increases after the administration of amyl nitrite, whereas the mitral snap has the opposite behavior.

In addition to this interval, attention should be paid to the area in which the extra sound is heard and recorded best. The third sound originating in the *left ventricle* is usually recorded best over the area of this ventricle (the older term was "mitral area"). The third sound originating in the *right ventricle* is usually recorded best over the xiphoid and at both sides of the sternum in the fourth interspace. A *mitral snap* is usually recorded best in the third and fourth left interspaces (occasionally in the second), while a *tricuspid snap* is recoded best in the same area as the right ventricular third sound (right ventricular area). The *pulmonary component* of the second sound is recorded best in the third left interspace.

Pulmonary Protodiastolic Click

A click or snap of the pulmonary artery has been described in one case of idiopathic dilatation of the pulmonary artery by Fishleder *et al.* (1963). This click followed the pulmonary component of the second sound by 90 to 170 msec. and seemed to coincide with the peak of the dicrotic wave of the pulmonary artery (right heart catheterization). It seemed to be the result of vibration of the wall of the extremely dilated pulmonary artery. A similar auscultatory phenomenon, caused by vibration of the wall of the extremely dilated pulmonary artery, was described by Kjellberg *et al.* in a case of supravalvular pulmonary stenosis, and was attributed to the vibration of the diaphragm causing the narrowing.

Chapter 5

Friction Rubs

ACCORDING TO LIAN, FRICTION rubs are groups of vibrations that last from 40 to 60 msec. and have frequencies of from 100 to 150 Hz. The phonocardiographic picture of "dry" pericarditis was first described by Cossio. He noted the frequent existence of vibrations in three phases: systole, early or middiastole, and presystole. The last two would be usually separated by an interval.

We have studied friction rubs since 1939. We soon noted that only two groups of vibrations (systole, diastole) or even one (systole) (Luisada, 1953) may occur, and we described the typical graphic picture consisting of vibrations of various frequencies but with a predominance of those of high frequency (100-400 Hz). Holldack described the pattern of the presystolic rub, which, in contrast with the presystolic murmur of mitral stenosis, does *not* have a crescendo appearance; he also described the diastolic rub, which, in contrast with the murmur of aortic insufficiency, does *not* have a decrescendo appearance. In a subsequent publication (Luisada, 1965) we have described the possibility of larger groups of vibrations within the main body of the rub, which may give on auscultation the impression of a "gallop." Spodick found that, in 18% of his cases, the rub occurred only in one phase; in 24%, in two phases; and in 58%, in three phases.

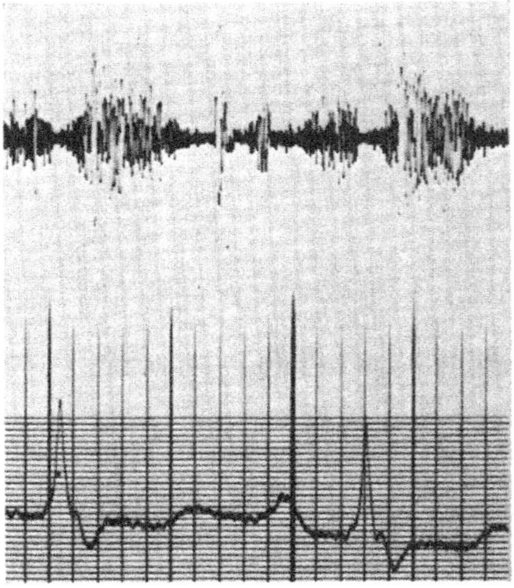

FIGURE 53. Case of pericarditis with three groups of vibrations (systolic rub, early-diastolic rub, presystolic rub).

Recently, we have studied 13 cases of friction rubs caused by pericardial inflammation of different etiology including surgical itervention (3 cases), Dressler's syndrome (1 case), and tuberculosis (1 case) ; 5 cases had presumably a viral pericarditis, and 1 case had a cardiomyopathy of obscure origin acompanied by pericarditis. In this series, we found evidence of rubs in either one phase, two phases, or three phase. The typical picture of rubs in three phases (systolic, early diastolic and presystolic) was found only seldom (Figs. 53 and 54). Two groups (systolic and presystolic) were found in 3 cases. Two groups (systolic and early diastolic) were found in 1 case. Two groups (systolic and pandiastolic) were found in 6 cases (Fig. 55). One case had only a systolic rub, and one case had only a presystolic rub (Luisada and Weiss).

Therefore, the original description by Cossio should be modified because, in many cases, the early-diastolic and presystolic rubs may fuse resulting in a single *pandiastolic rub*. This, associated with the *systolic*, is actually the most numerous group. In order of frequency, the second group is represented by cases with both a *presystolic* and a *systolic rub*. A few cases present either a systolic and an early diastolic rub, or a single rub in systole or presystole.

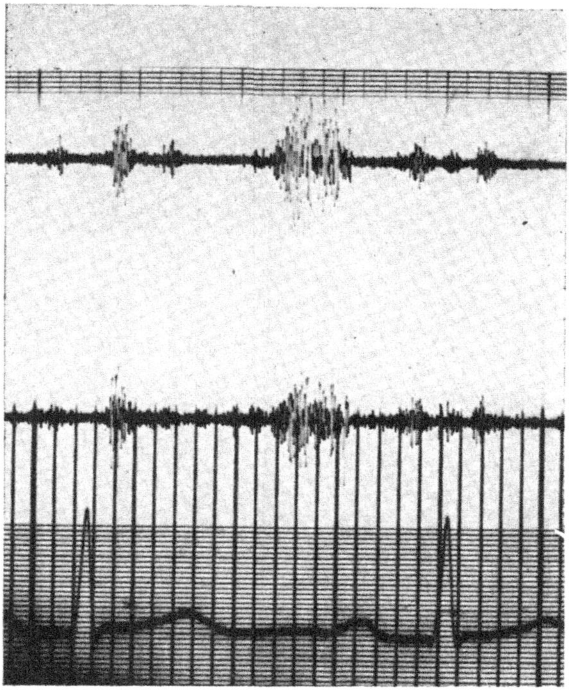

FIGURE 54. Case of pericarditis with three groups of vibrations in which the early diastolic rub has a large amplitude.

It should be noted hat most cases with a presystolic rub show an abrupt ending of the group of vibrations, which is typical of this condition (Figs. 53 and 54). It sometimes consists of separate, large components. The systolic rub usually consists of multiple vibrations. The early-diastolic group may have a much larger amplitude than the others (Fig. 54).

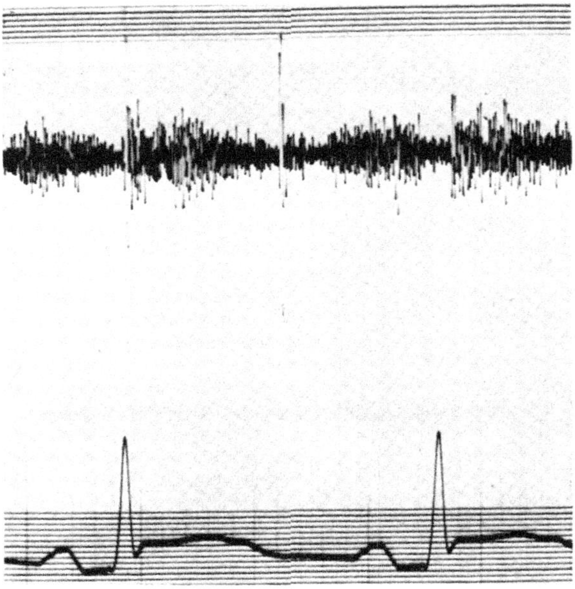

FIGURE 55. Case of pericarditis with two groups of vibrations (systolic rub, pandiastolic rub).

Chapter 6

Dynamic Correlations of Cardiovascular Murmurs

MURMURS HEARD IN PATIENTS with heart disease have been explained in various ways. The most common mechanism is believed to be *disturbed flow* in jets (Rushmer and Morgan). Blood flowing at high velocity through a narrow orifice or channel (stenotic valve, ventricular septal defect, patent ductus or other arterial or arteriovenous fistula) emerges as a jet into a wider channel or chamber. Eddies form at the interface with the surrounding stagnant or slowly moving blood, causing the downstream wake of the jet and vascular surroundings to oscillate audibly. The *impact of a jet* on the opposite wall may be a more direct mechanism causing noise. *Vortex shedding* around an obstruction may set up recurring pressure patterns, producing aeolian tones or musical murmurs. However, the downstream aperiodic fluctuating jet wake mechanism appears to be responsible for the non-musical noise of most murmurs.

Whatever the mechanism, there is no question that the speed of the blood is one of the most important factors of the intensity of the murmur, and thus of *the phase* of the cardiac cycle in which it occurs.

Studies performed with simultaneous pressure tracings during cardiac catheterization in man have revealed that *gradients of pressure* develop across either stenotic vessels or shunts. The gradient developed across a valve also applies in cases with valve incompetence. It is obvious that, the larger the gradient, the greater will be the velocity of flow, and thus the murmur will tend to be louder.

Marked decrease in cardiac output caused by heart failure would decrease the gradient. Also extreme narrowing of an orifice would limit the size of the jet. This explains the frequently observed decrease of murmurs in the advanced stages of heart disease.

The relationship between pressure gradients and murmurs in various types of heart disease will be discussed below.

Atrioventricular Valve Stenosis

Mitral Stenosis. It is typical of this condition to create a diastolic gradient of pressure across the mitral valve. When the stenosis is moderate, the gradient is present only in early diastole and presystole (Fig. 56A). On the other hand, when the stenosis is severe, the gradient is present throughout all of diastole (Fig. 56B). The typical diastolic rumble caused by this gradient is, therefore, present in either early and late-diastole (presystole) or throughout diastole. It starts typically after completion of the opening of the mitral valve, i.e., after the opening snap. It should be kept in mind that the gradient decreases when cardiac output is low, and this accounts for the leser magnitude of the murmur when either heart failure or extremely severe narrowing of the mitral orifice takes place.

The first heart sound is delayed due to the fact that mitral valve closure occurs later, when left ventricular pressure has risen above the elevated left atrial pressure. Some authors, in addition, have reported depressed left ventricular function in mitral stenosis.

The *opening snap* of the mitral valve occurs after a certain interval from the aortic component of the second sound that grossly measures the isovolumic relaxation period. This period is shorter in mitral stenosis but not in left ventricular failure for hemodynamic reasons (Fig. 57) (the gradient does not increase in failure).

Tricuspid Stenosis. A dynamic condition similar to that of mitral stenosis occurs in the right heart of these patients (Fig. 58). The main difference is that the presystolic murmur is

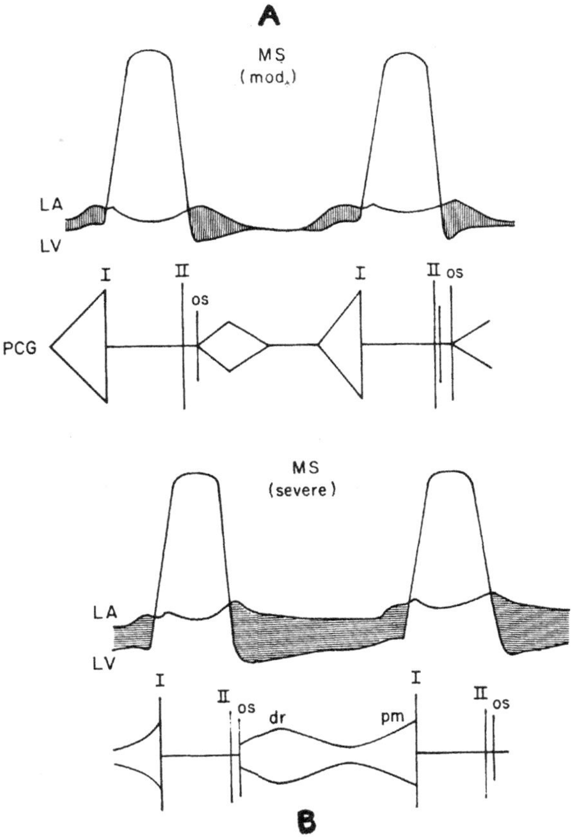

FIGURE 56. Scheme of the diastolic gradients of pressure across the mitral valve and of the shape of murmur recorded in phonocardiograms (PCG).
(A) Mild mitral stenosis. (B) Severe mitral stenosis.
OS = opening snap. I, II = heart sounds.
LA = left atrial pressure. LV = left ventricular pressure.
dr = diastolic rumble. pm = presystolic murmur.

usually more widely separated from the first sound. Also, the first heart sound, which is only caused by left ventricular dynamics, is not delayed as in mitral stenosis.

It should be kept in mind that, as extremely high venous pressures are incompatible with survival in tricuspid stenosis, the gradient of pressure found in this condition is always moderate, and this accounts for the low pitch and the rumbling character of the murmur.

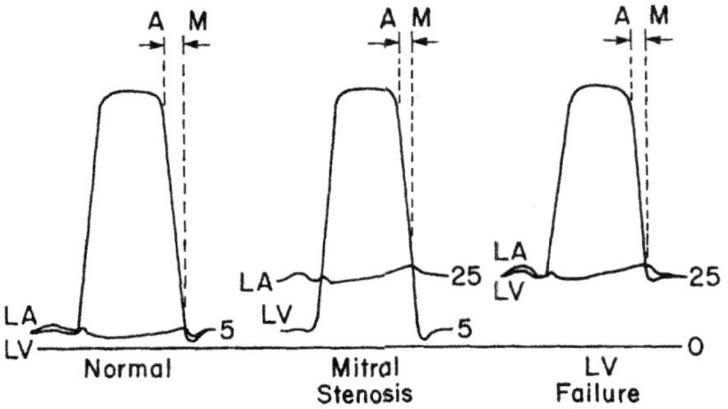

FIGURE 57. Scheme of the pressures in the left ventricle and left atrium in:
(A) A normal subject.
(B) A patient with mitral stenosis.
(C) A patient with left ventricular failure.
A = aortic valve closure.
M = mitral valve opening.
The interval between the two valve events of the left heart (isovolumic relaxation period) is shorter in mitral stenosis, not in left ventricular failure.

SEMILUNAR VALVE STENOSIS

Aortic Stenosis. In this condition, when severe, a very high gradient of pressure develops across the aortic valve (Fig. 59). The aortic pulse reaches eithr a normal or a slightly reduced level of systolic pressure but the pulse rise is slower; it is marked by low-frequency vibrations (shudder). The murmur is crescendo-decrescendo (diamond-shaped) and has a peak which

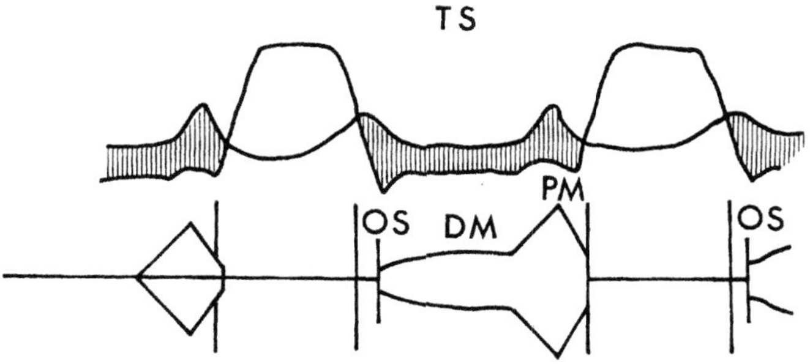

FIGURE 58. Scheme of the diastolic gradient of pressure across the tricuspid valve and of the shape of murmur recorded in severe tricuspid stenosis.

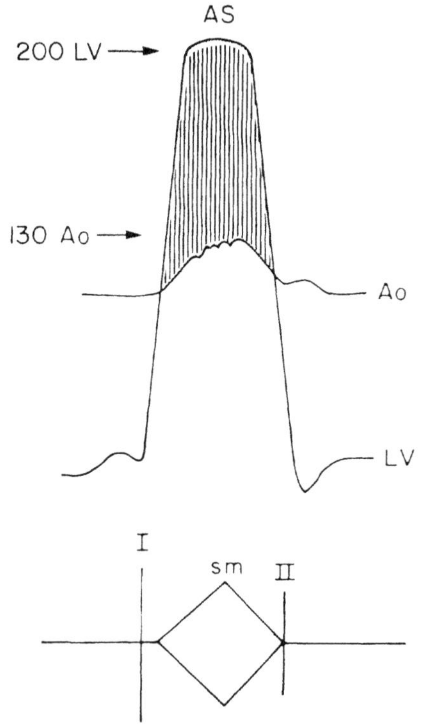

FIGURE 59. Scheme of the systolic gradient of pressure across the aortic valve and of the shape of murmur recorded in severe aortic stenosis.

is more delayed in severe stenosis. It is obvious that, as the murmur starts after the opening of the aortic valve, its onset occurs some time after the first sound and is usually marked by an "ejection sound" or "early-systolic click" unless the stenosis is extremely severe.

Pulmonary Stenosis. Severe pulmonary stenosis creates a condition similar to that of aortic stenosis as applied to the pulmonic valve (Fig. 60). The peak of the murmur tends to occur even later than that of aortic stenosis.

The Second Sound in Semilunar Valve Stenosis. It has been known for many years that the aortic component of the second sound is small and delayed in aortic stenosis (reverse splitting) and that the pulmonic component of the second sound is small and delayed (wide splitting) in pulmonic stenosis. This fact was attributed to prolongation of systole of the stressed ventricle with late closure of the valve. It was also thought that the murmur would terminate only with the delayed component, thus "riding over" the other component of the sound.

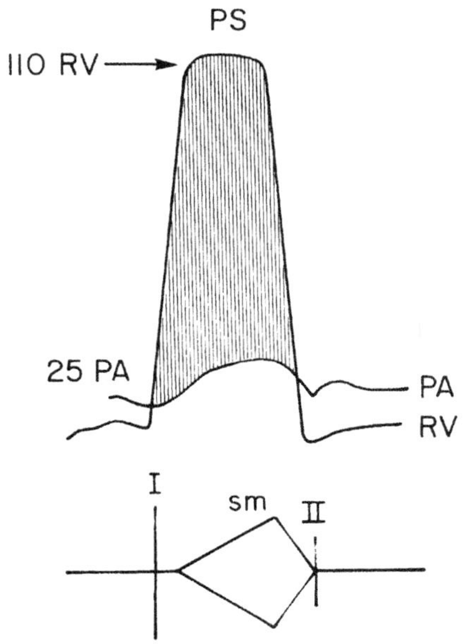

FIGURE 60. Scheme of the systolic gradient of pressure across the pulmonic valve and of the shape of murmur recorded in severe pulmonic stenosis.

Recent studies from our laboratory have shown that this concept is not correct. End of ventricular systole in a normal heart is followed after a short interval by semilunar valve closure. However, even in normal hearts, the sound component is slightly delayed over valve closure (especially the pulmonary component) because it is caused by the rebound of arterial blood over the closed valve, and this coincides with the incisura of the respective pulse. The momentum of the ejected blood can account for part of this brief interval.

When there is semilunar valve stenosis in the absence of failure, the systole of the stressed ventricle has a duration that is similar to, or even shorter than, that of the normal ventricle. However, valve closure takes place much later on account of

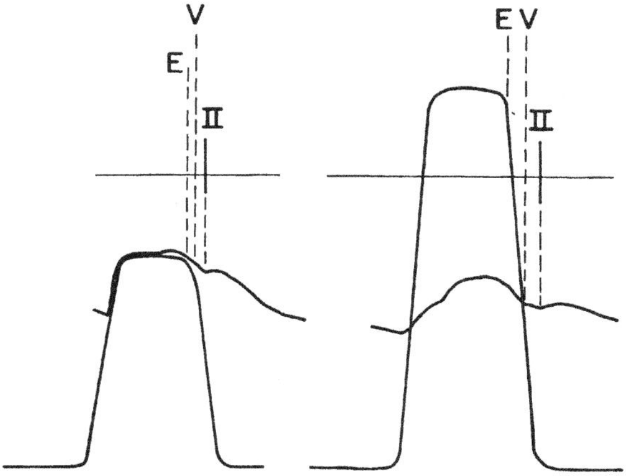

FIGURE 61. Scheme of the dynamic events in a normal heart and in severe aortic stenosis.
E = end of systole.
V = aortic valve closure.
II = second sound.
sm = systolic murmur.
Note the delay of aortic valve closure due to the relationship between ventricular and aortic pressures.

the difference in pressure between ventricle and vessel (Fig. 61) (Kumar and Luisada, 1971A). This delay accounts for most of the delay of the sound component. However, in some cases there is also a further delay caused by a delayed rebound that is due to the lower tension and the dilatation of the vascular wall requiring a longer time to reach its elastic limit. This would also cause a slower transmission of the pulse wave.

VALVULAR INSUFFICIENCY

Mitral Insufficiency. In this condition, the regurgitant jet across the mitral orifice usually causes a triangular wave of pressure in the left atrium with a late peak (Fig. 62). As the jet occurs during all or most of systole, we should accept that gradual elastic distension of the atrium prevents a marked rise of atrial pressure until greater tension of the wall is reached. This was actually demonstrated in our laboratory by comparing left atrial pressures with left atrial electrokymograms in experimental animals (Haring *et al.*, 1956). After the end of ventricular systole, a large forward flow crosses the mitral

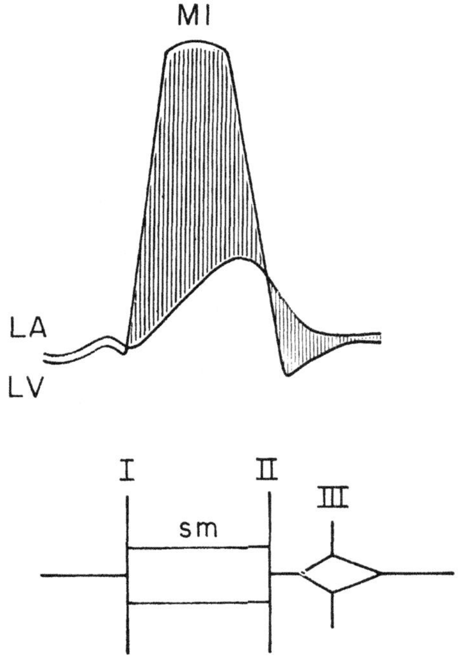

FIGURE 62. Scheme of left ventricular and left atrial pressures and scheme of the murmurs recorded in severe mitral insufficiency.
LA = left atrium.
LV = left ventricle.

valve and accounts for the frequently found early-diastolic gradient betwen left atrium and left ventricle. This large flow also explains the large third sound and the early or mid-diastolic rumble that are common features of this conditon. Other types of murmurs (early-systolic, late-systolic) are explained by either muscular dysfunction (dilatation of the mitral ring, stretching of the papillary muscles) or only minimal insufficiency (leak at one of the commissures) caused by minimal deformity of the valve apparatus. Another possibility is alteration of the intrinsic musculature of the mitral valve (Montiel).

Tricuspid Insufficiency. A regurgitant jet into the right atrium is typical of this condition. The pressure of the right atrium often shows a squarish systolic pressure rise and an early-diastolic gradient. The murmur is pansystolic and similar to that of mitral insufficiency (Fig. 63).

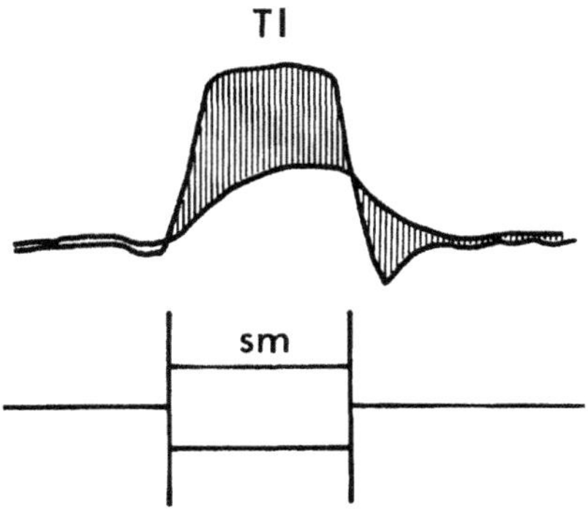

FIGURE 63. Scheme of right ventricular and right atrial pressure as well as of the murmur recorded in severe tricuspid insufficiency.

Aortic Insufficiency. The pressure pulse of the aorta has a low incisura in comparison with normal subjects. The phonocardiogram frequently shows an ejection sound and a diamond-shaped systolic murmur with an early peak (flow murmur). The typical feature, however, is the diastolic murmur (Fig. 64). This murmur may start *before* the aortic component of the second sound in severe insufficiency because the regurgitant jet then starts at the beginning of ventricular relaxation, which precedes the sound caused by the aortic rebound. The murmur may be

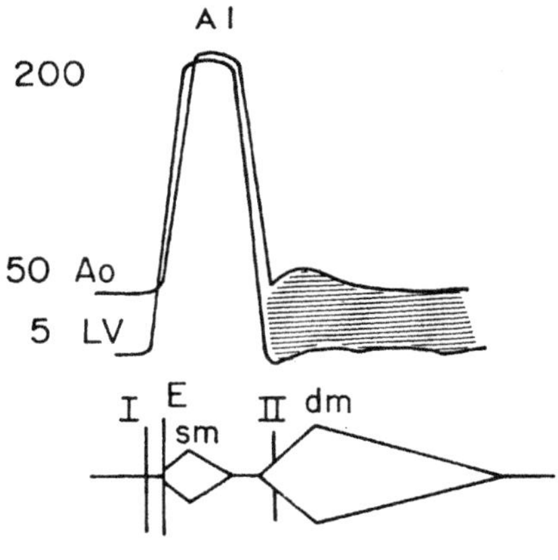

FIGURE 64. Scheme of left ventricular and aortic pressures and of the murmur in aortic insufficiency.

diamond-shaped on account of the gradient of pressure. This is maximal in early diastole due to the casual coincidence between the aortic dicrotic wave and the peak of left ventricular rapid filling. However, in some cases, the murmur may be decrescendo in character and encompass all of diastole (pandiastolic murmur). The diastolic rumble (Austin Flint murmur) that can be observed in patients with very large hearts is a series of low frequency vibrations, which are partly related to the regurgitation, partly to the meeting of two streams of blood in the left ventricle (from the atrium and from the aorta), and possibly—in some cases—to displacement of the anterior mitral leaflet by the regurgitant aortic jet producing a functional narrowing of the mitral orifice.

Pulmonary Insufficiency. In this condition, there is an early diastolic murmur that is usually of lower frequency and shorter duration than that of aortic insufficiency. The low pressure gradient between pulmonary artery and right ventricle accounts for these characteristics of the murmur.

Ventricular Septal Defect. In this condition, an opening between the two ventricles creates a shunt flow in systole between left and right ventricles that follows a constant course on account of the constant gradient between the two ventricles. Thus, the typical murmur is pansystolic and its pattern resembles that of mitral insufficiency (Fig. 65). When *pulmonary hypertension* develops, the flow gradient decreases and shows a gradual rising and falling. Then the murmur, still pansystolic, assumes a crescendo-decrescendo pattern. Dilatation of the pulmonary artery probably contributes to this pattern by adding a vascular flow murmur to that of the defect.

Atrial Septal Defect. In this type of shunt, the blood flows from the left to the right atrium during the entire cardiac cycle due to lesser compliance of the left atrial wall. As the effect of inspiratory increase on venous return is compensated by lesser flow through the shunt, the respiratory variations in the splitting of the second sound are practically absent in this condition.

The wider splitting of the second heart sound in this condition was attributed in the past to a prolongation of right ventricular systole. However, dynamic studies in our laboratory (Kumar and Luisada, 1971B) have shown that this does not occur. On the other hand, the increased pulmonary flow causes a dilatation

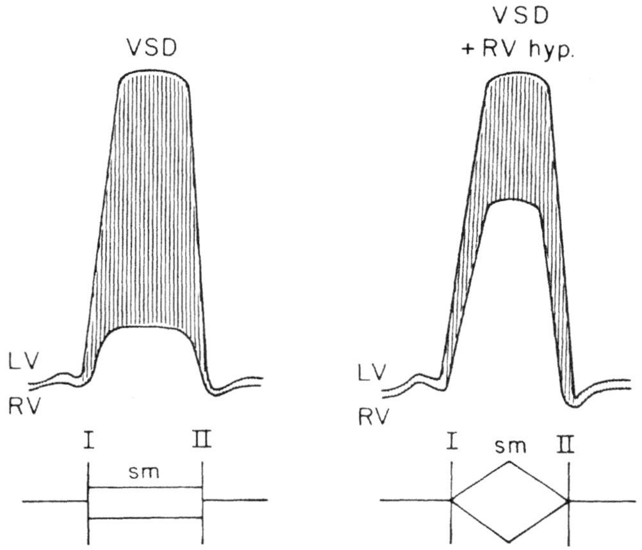

FIGURE 65. Scheme of right ventricular (RV) and left ventricular (LV) pressure and of the murmur recorded in ventricular septal defect.
At left = without pulmonary hypertension.
At right = with pulmonary hypertension.

of the pulmonary bed. This is followed by a delayed rebound of the pressure pulse over the closed pulmonic valve and a delayed pulmonary component of the second sound. The typical crescendo-decrescendo systolic murmur heard at the left of the sternum is a *flow murmur* that develops in the dilated pulmonary artery.

The above considerations are based on accepted findings at cardiac catheterization and clinical phonocardiograms. Several, still unexplained, facts should be kept in mind:

(1) In cases with *ventricular dilatation* distending either the mitral or the tricuspid ring, it is customary to observe a moderately large systolic murmur and a low pitched, mid-diastolic (and occasionally diastolic-presystolic) rumble. The exact dynamic cause of the rumble is still unknown as there is no greater diastolic gradient of pressure than the normal minimal difference between atrial and ventricular pressures. Thus, speculations include eddies forming within the dilated ventricular chamber and

stretching of the papillary muscles that would keep under tension the mitral or tricuspid leaflets. The latter might cause a minimal functional, mitral or tricuspid stenosis even though no pressure gradient is usually observed.

(2) When *the rapidity of circulation is increased or either the flow or the pressure is increased* in either the pulmonary or the systemic circulation, it is customary to observe either a pulmonary or an aortic flow murmur, the former more commonly than the latter. It is customary to explain these murmurs as caused by eddies which form in passing from a normal valve to a large vessel. However, it should be noted that they usually take place without any pressure gradient, the only exception being represented by cases of atrial septal defect with an enormous shunt, in which a gradient of 10-20 mm Hg can develop across the pulmonic valve. The reason for such a lack of gradient in the other cases is still under discussion

Part II

Clinical Phonocardiography

Chapter 7

Mitral Valve and Tricuspid Valve Defects

MITRAL STENOSIS

BOTH AUSCULTATION AND PHONOCARDIOGRAPHY have great importance for the clinical diagnosis of mitral stenosis.

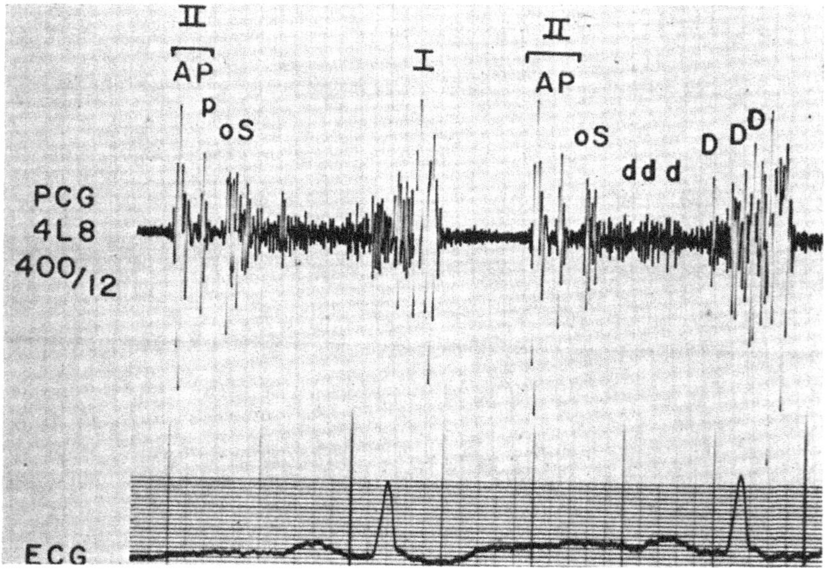

FIGURE 66. Typical phonocardiogram of severe mitral stenosis with sinus rhythm.

It is unfortunate that no proportion exists between the loudness (and magnitude) of a murmur and the severity of the process. On the one hand, there may be small, or even absent, murmurs in extremely severe stenosis, related to reduction in cardiac output or marked bradycardia (either decreased or slower flow). On the other, a loud presystolic murmur may be heard in mild mitral stenosis.

First Sound. The first sound is typically louder on auscultation and of increased magnitude in graphic tracings (Fig. 66). Delayed closure of the mitral valve, slow rise of left ventricular pressure prior to the closure, and a rapid rise after it are some of the factors of this increase (Kurz *et al.*). Others are a stiffer mitral apparatus that resists more to the left ventricular pressure rise, and a smaller left ventricle. Such an accentuation is often absent in patients with mild stenosis and in cases with a severely calcified mitral valve* (Dack *et al.*), as well as in cases with dilated left ventricle. Associated mitral insufficiency may be another factor of a decreased amplitude of the first sound.

In atrial fibrillation, the intensity of the first sound varies from cycle to cycle being loudest after a short diastole or when diastole is longer than 0.25 sec. (Rytand, Spiegl *et al.***). These facts are related to the arrhythmia and not to the stenosis. However, it should be noted that patients with mitral stenosis and atrial fibrillation exhibit a marked variability in the Q-I interval, far greater than in patients with fibrillation but no stenosis. This was pointed out by the author a long time ago (Luisada, 1941).

Leo and Hultgren have advocated the view that the first sound in severe mitral stenosis presents, first, a smaller group

*Cases with a severely calcified mitral valve are usually cases in an advanced stage of the disease. The lack of increase of the first sound may be related to myocardial damage of the left ventricular wall causing less rapid rise of pressure. If such is the case (and it has been documented by angiography—Heller and Carleton), the left ventricular cavity may be enlarged instead of reduced, thus adding an additional factor to the decrease of the sound.

**The intensity of the first sound varies in cases with atrial fibrillation from beat to beat, according to Ferrero et al.; if there is correlation between these changes and the length of the previous diastoles, then the pulmonary capillary pressure would be below 20 mm Hg. When this pressure is higher, the variability has no correlation with diastolic intervals.

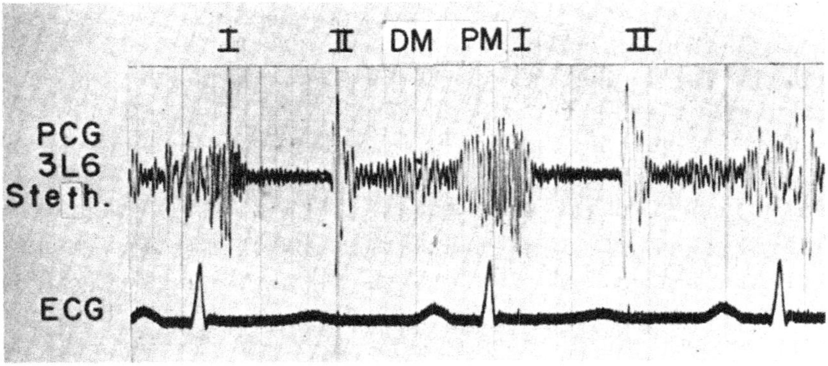

FIGURE 67. Typical phonocardiogram of severe mitral stenosis with sinus rhythm. No opening snap is present.

of vibrations due to the right ventricle (so-called tricuspid component), then a larger group due to the left ventricle (so-called mitral component). However, the left ventricular vibrations are the exclusive cause of the first heart sound in normal hearts. Theoretically the hypertensive right ventricle might be expected to cause sound vibrations but so far we have beeen unable to demonstrate them. Moreover, as right ventricular pressure still rises *after* the left, one would expect a single first sound.

An *ejection sound* of the pulmonary artery can be found in mitral stenosis (Leatham and Vogelpoel, Kunos *et al.*, Leo and Hultgren, Dack, *et al.*). This sound is found in patients with marked dilatation of the pulmonary artery, is caused by the sudden vibration of the wall of this artery at the beginning of ejection, and is equivalent to the aortic ejection sound that is found in conditions causing aortic dilatation. The mean interval between the largest component of the first sound and this ejection sound was found to be 40 msec. by Leo and Hultgren.

Q-I Interval. The Q-I interval is typically prolonged in severe mitral stenosis. The diagnostic significance of this, however, is impaired by several facts:

(1) There is still uncertainty about which vibration of the first sound should be considered for measurement in mitral stenosis. Most likely the largest and latest group of vibrations is the most significant. If there is a large presystolic murmur, however, measurement my be difficult.

(2) The Q-I, interval is different according to the filter used. Studies on this problem have ben effected, and special Tables have been prepared in our laboratory (Aravanis *et al.*, 1971).

(3) The Q-I interval is prolonged in several other cardiac diseases involving ventricular hypertrophy.

The Q-I interval has a typical variability in patients with atrial fibrillation (as first shown by the author in 1941), having an inverse relation with the length of the previous diastole. Apparently, when diastole is very short, incomplete emptying of the left atrium delays the closure of the mitral valve because of the higher left atrial pressure. This will delay the rapid rise of pressure in the left ventricle and thus the first sound.

Second Sound. The pulmonary component of the second sound is abnormally loud and is recorded as a large vibration, which may be greater than the aortic component. Splitting of the second sound may be normal (Fig. 66), less evident than normal (Fig. 67), or occasionally wider than normal.

The factors of this variable behavior are as follows:

(1) The factor of age should be considered (older people have a narrower splitting).

(2) Pulmonary hypertension may delay the pulmonary component by causing dilatation of the pulmonary vessels.

(3) Right ventricular failure would prolong right ventricular ejection delaying the pulmonary component.

(4) Decreased cardiac output would abbreviate both right and left ventricular ejections (no change in the degree of splitting).

In the study of Sutton *et al.*, 1968, splitting of the second sound was about normal in mitral stenosis and was not affected by the level of the pulmonary artery pressure.

For the above reasons, a split second sound cannot be considered as a typical sign of mitral stenosis. Actually, the classic statement about "a split second sound" was a reference to the easily heard sequence of the second sound followed by the opening snap. It is important, however, to identify the two components of the second sound in the phonocardiogram, because this will permit us to identify with certainty an opening snap, which may be close to the second sound in severe stenosis (triplication of the second sound.)

Opening Snap of the Mitral Valve. The opening snap is an important part of the auscultatory picture of mitral stenosis and was discussed in Chapter 4.

The opening snap is usually evidence of mitral stenosis. However, the value of this fact, which has practical significance, is decreased by the following considerations:

(1) A fair percentage of patients with mitral insufficiency have an opening snap.

(2) The opening snap may disappear in a patient with extremely tight mitral stenosis having rigid and calcified leaflets (Holldack and Bayer, Wynn) (Fig. 67). Moreover, in a certain number of patients, the snap is visible in the phonocardiogram, though it is inaudible because it is too close to the pulmonary component of the second sound.

(3) Patients with conditions simulating mitral stenosis (ASD in failure, pulmonary heart disease in failure) may have an opening snap, possibly occurring in the right heart.

On the oher hand, there are patients with barely audible and recordable diastolic murmurs, in whom the opening snap is prominent and has great diagnostic significance. The best location for hearing and recording an opening snap is often not the apex, as commonly stated. In our experience the best area corresponds to the third or fourth left interspace half-way between the apex and the margin of the sternum. Even in cases of wide transmission of this sound over the percordium, this area is the area of maximum loudness. This corresponds to the findings of Shah *et al.* (1963A) in regard to the closing sound of a mitral prosthetic valve, and was demonstrated by Sainani in our laboratory in regard to the usual type of snap.

The vibrational frequency of the opening snap is high and may approach that of certin components of the aortic second sound. Schlitter and Schoelmerich evaluated it at about 75 Hz. In our experience, it may be as high as 400 Hz. The opening snap may be the loudest sound of the cardiac cycle (Dack *et al.*) but this is exceptional. The opening snap consists of 1 to 4 vibrations according to the degree of filtration used.

The opening snap is a vibration of the left atrial system when the blood is decelerated by the end of opening of the stenosed mitral valve, and the volume of the atrium is of paramount importance in the determination of its fundamental frequency (von Egidy and Dudeck).

Inhalation of *amyl nitrite* usually decreases the amplitude of the mitral opening snap.

Intracardiac phonocardiograms show well the mitral opening snap in both the left atrium and ventricle. They may show it with the catheter placed in the wedge position, in the pulmonary artery, or in the right ventricle (Luisada and Liu; Luisada *et al.*, 1964).

II-OS Interval. The II-OS interval is inversely proportional to the gradient of pressure across the mitral valve, as shown by Wells (1954), Kelly, Haring *et al.* (1956), and many others. Therefore, the interval is longer (8 to 12 msec.) in mild mitral stenosis, and vice versa. In spite of early belief, however, the relationship between the II-OS interval and left atrial pressure is only approximate and not exact. The reason for this is obvious. In severe mitral stenosis, the crossing point between left atrial and left ventricular pressure occurs earlier and, therefore, the opening snap is closer to the aortic component of the second sound (Fig. 57). This is not true in left ventricular failure, where both the left ventricular end-diastolic pressure and the left atrial pressure are raised (Fig. 57).

In exceptional cases, one can observe both a mitral and a tricuspid opening snap. The former occurs earlier and is recorded best at the apex; the latter occurs later and is recorded best over the third left intercostal space. In some cases, both can be recorded in succession over the 4th left i.c.s.

Diastolic Rumble. The dynamic correlations between diastolic murmur and hemodynamic alterations have been outlined in Chapter 6. In patients with sinus rhythm, this murmur has two distinct phases of greater loudness: (1) an early-diastolic phase coinciding with a slow and prolonged phase of filling (so-called mid-diastolic rumble); and (2) a presystolic (or atriosystolic) phase coinciding with the powerful contraction of the hypertrophied left atrium (presystolic accentuation) (Fig. 65).

The presystolic murmur has been explained as the result of approximation of the mitral leaflets prior to ventricular contraction; this would narrow the mitral opening and cause turbulence.

In atrial fibrillation, the early and middiastolic rumble is the only evidence of the stenosis. It is a murmur of low pitch (15 to 80 Hz) and small amplitude and may have a decrescendo quality, so as to fade in middiastole if diastole is long. On the contrary, it may be larger and have a crescendo quality (simulating a presystolic murmur) when diastole is short (Fig. 68).

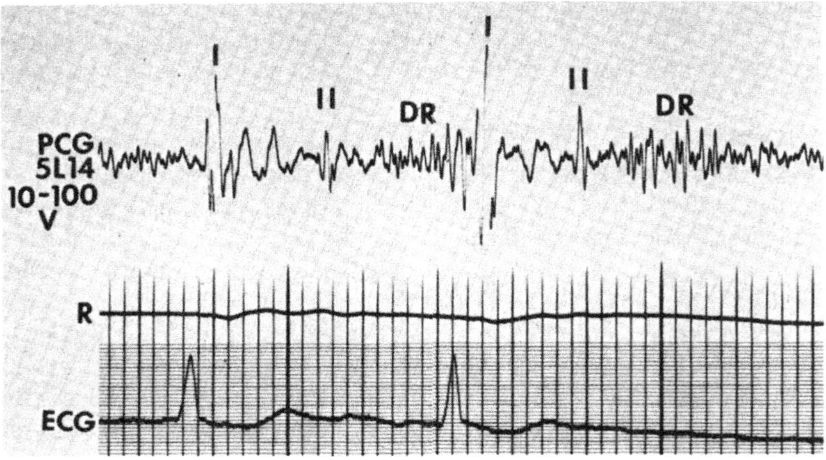

FIGURE 68. Phonocardiogram of a patient with mitral stenosis and atrial fibrillation. The first short diastole causes an accentuation of the rumble before the following first sound that simulates a presystolic murmur.

Because of its low frequency, this murmur is heard best with an open bell applied gently against the skin. Unusual cases of rheumatic heart disease (thickened chorda) may present a high-pitched, musical diastolic murmur (Schrire and Vogelpoel, 1961). Other cases may have a softer, high pitched murmur, though the cause of his fact stil has not been recognized (Mehta and Luisada) (Fig. 69).

The real presystolic murmur obviously occurs only when sinus rhythm is present. It has a higher pitch than the previous phase of diastolic rumble. Its frequency components range from 30 to 200 Hz and may occasionaly reach 400 Hz. The vibrations of the murmur terminate with the delayed first sound. Patients with mild stenosis may have a small presystolic murmur having few vibrations between the R peak of the electrocardiogram and the first sound. In such cases, one might erroneously interpret the murmur as an abnormal initiation of the first sound, a fact that is occasionally encountered in subjects with a normal heart.

The middiastolic and presystolic murmurs are often heard best between the apex and the left sternal border in the fourth left interspace, in contrast with statements to the contrary (apical murmur). This is well explained by the direction of the flow passing through the narrow mitral valve, which moves from an upper posterior area to a lower anterior area and hits the medio-inferior portion of the left ventricle. However, patients in whom the apex is formed by the right ventricle may have the murmur displaced more laterally.

The diastolic-presystolic murmur is increased by tachycardia. Thus, it is accentuated by the left lateral decubius, exercise, often by amyl nitrite, and by isoproterenol; it is decreased by pressor amines, which cause a reflex bradycardia (Fig. 71).

Certain patients present a presystolic murmur without diastolic rumble. In our experience, these patients have a minimal or moderate stenosis (Fig. 70).

In patients with Grade 1 AV block, the presystolic murmur is

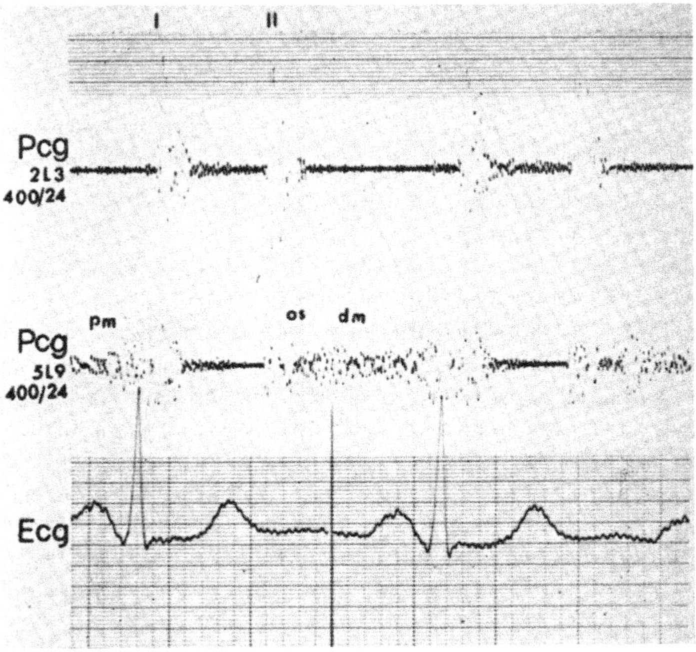

FIGURE 69. Phonocardiogram of mitral stenosis with sinus rhythm and a high frequency diastolic-presystolic murmur that follows an opening snap.

well separated from the first sound and may have a crescendo-decrescendo configuration.

Systolic Murmurs. A systolic murmur in decrescendo is frequently recorded over the apex (Figs. 69 and 70). It is likely

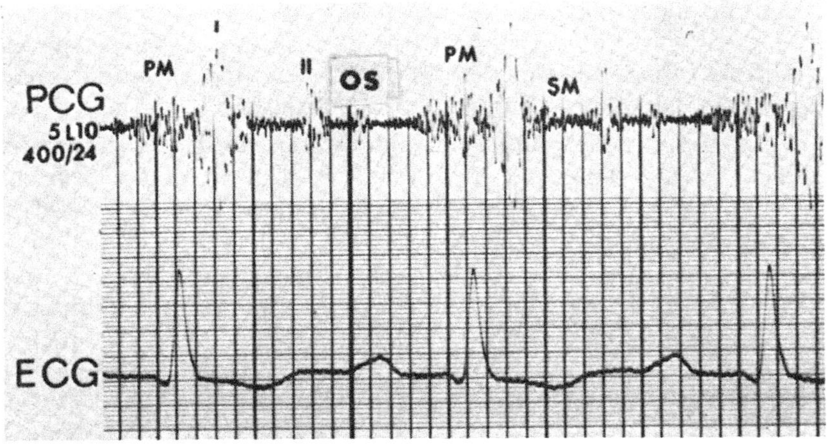

FIGURE 70. Patient with moderate mitral stenosis. There is a late opening snap and a presystolic murmur.

to be caused by a minor degree of mitral or tricuspid insufficiency, the latter being particularly common in cases with severe narrowing of the mitral valve. A diamond-shaped systolic murmur is frequently recorded over the pulmonary area. It is a flow murmur related to dilatation of the pulmonary artery, secondary to pulmonary hypertension.

Blowing Diastolic Murmur of Pulmonary Insufficiency (Graham Steell Murmur). A blowing, decrescendo diastolic murmur is sometimes heard over the second and third left interspaces. It is recorded as a series of high-pitched vibrations with gradually decreasing intensity and short duration. However, many cases with a reported G. Steell murmur were shown through angiocardiography to have aortic insufficiency. Of course both aortic and pulmonary insufficiency may be present.

Third Sound. A third sound is usually considered evidence of only a minor degree of mitral stenosis. Connolly and Wood did not find a third sound unless there was a significant degree of mitral regurgitation.

Certain patients with mitral stenosis do present a third sound in the phonocardiogram, as noted by Mounsey (1953), Ongley *et al.* (1955), and Auinger. Although it is probably true that the majority of such patients have only a mild form of stenosis, this is not always so.

The following facts should be considered in regard to the third sound:

(1) A third sound can be superinmposed on the vibrations of the diastolic rumble and thus may be inaudible, as well as poorly observable in phonocardiograms.

(2) The interpretation of the third sound has changed, and we can admit its occurrence even without a large wave of rapid filling if the left ventricle is distended and has a high diastolic pressure. This can occur (a) if there is an associated aortic insufficiency: (b) if mitral insufficiency and stenosis are both present; (c) if there is systemic hypertension or coronary heart disease; or (d) if there is left ventricular failure. The latter is more often caused by rheumatic carditis but may occur as a result of fibrosis from unknown origin (Heller and Carleton).

(3) There can be a right ventricular third sound. This can be ascertained by scanning the precordium and observing that the third sound is larger over the xiphoid and the right fourth interspace near the sternum in contrast with the apex.

Fourth Sound. The same considerations advanced for the third sound also apply to the fourth sound, which theoretically should be absent in mitral stenosis. Occasionally, a fourth sound can be identified within the presystolic murmur, even in cases of severe stenosis. Either a right ventricular fourth sound or a distinct vibration that can still occur because of mild stenosis should be considered. Scanning of the precordium may supply the answer. Mild stenosis plus a Grade 1 AV block were accepted as causes of the fourth sound in a case of Adams. Severe stenosis in pregnancy was present in one of our cases.

Low-frequency Tracing. Since Weitz, and Luisada (1953), the apex cardiogram of mitral stenosis has been considered to have certain typical aspects, such as a small presystolic wave IV, deep point O coinciding with the opening snap, and a slow rise during rapid filling with a small wave III. It is unfortunate that the variability of this tracing does not allow us to assign a real diagnostic value to these changes. However, Warembourg and Ducloux found it useful for an evaluation of the severity of the stenosis.

FUNCTIONAL TESTS

During *respiration,* the mitral murmur is greater in expiration and smaller in inspiration.

Amyl nitrite increases the murmur of mitral stenosis but only in the second phase (increased flow from the lungs to the left atrium) or if there is severe tachycardia (Fig. 71). *Pressor amines* usually decrease the mitral diastolic murmur because of the reflex bradycardia which favors a slower flow through the mitral orifice (Fig. 71).

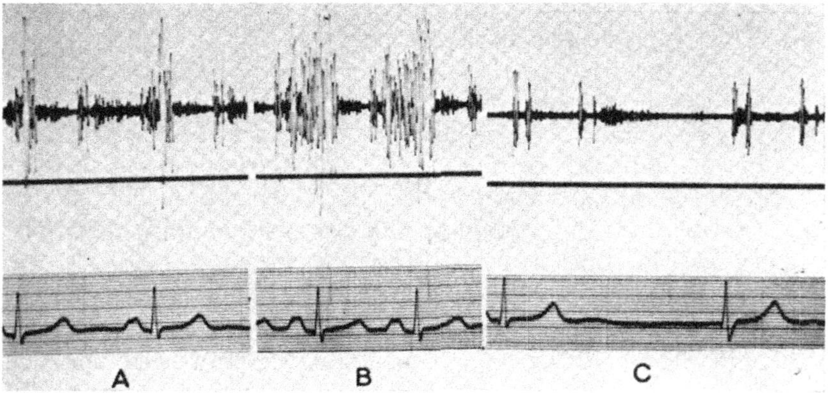

FIGURE 71. Functional tests in a case of mitral stenosis with sinus rhythm.
(A) Control.
(B) Effect of inhalation of amyl nitrite. Tachycardia. Increase of the diastolic murmur.
(C) Effect of i.v. injection of methoxamine. Bradycardia. Decrease of the diastolic murmur.

RELATIVE MITRAL STENOSIS

The possibility of apical diastolic and presystolic rumbles caused by a functional mechanism, i.e., without mitral stenosis, has been known for a long time.

Clinical reports first described functional murmurs in aortic regurgitation (Austin Flint murmur), pericardial or hypertensive heart disease (P.D. White), rheumatic heart disease without mitral stenosis (Bland *et al.*, 1936; Gibson, 1954; Coombs, Bramwell, 1943; Hurst), and myocarditis (Weinstein and Lev). Subsequently, phonocardiographic studies both by our group and by others documented such murmurs in rheumatic, coronary, hypertensive, and congenital heart diseases (Ravin and Darley;

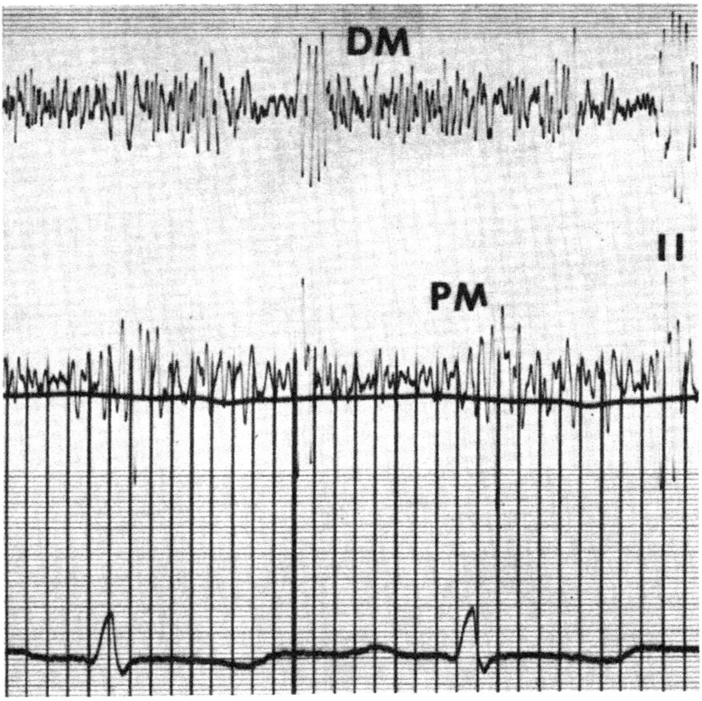

FIGURE 72. Diastolic-presystolic murmur *(Austin Flint murmur)* in a case of rheumatic aortic insufficiency. Upper tracing was recorded over the second right i.c.s., middle tracing was recorded at the apex. Both were taken using a 50 Hz, high pass filter with a 12 db/oct. slope.

Fishleder and Medrano; Likoff *et al.*). Several studies also documented the middiastolic rumble of patients with pure mitral insufficiency.

Five previous reports from our group (Luisada and Perez Montes; Luisada *et al.*, 1955, 1959, 1971; Zilli and Gamna) described numerous such cases, proved the functional nature of the murmurs by autopsy or through disappearance of the mur-

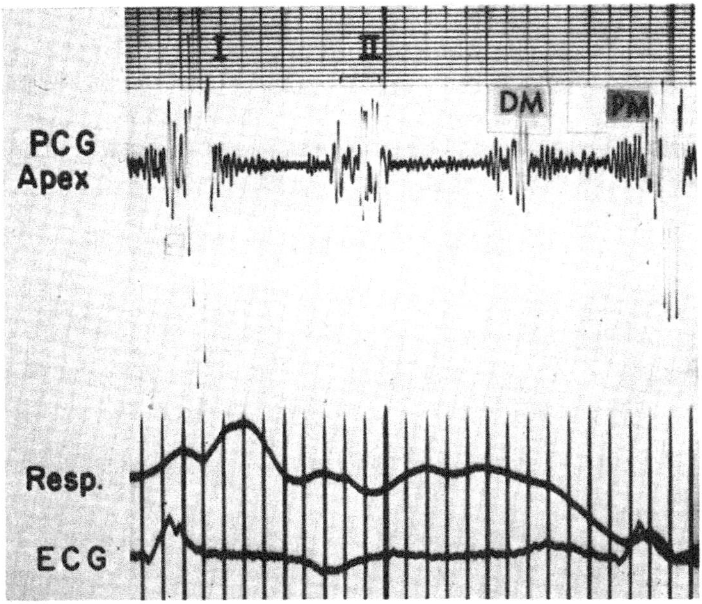

FIGURE 73. (A) Phonocardiogram recorded at the apex in a 70-year-old man with coronary heart disease and a history of old infarct. The tracing reveals a reverse splitting of the second sound, a late middiastolic murmur, and a presystolic murmur.

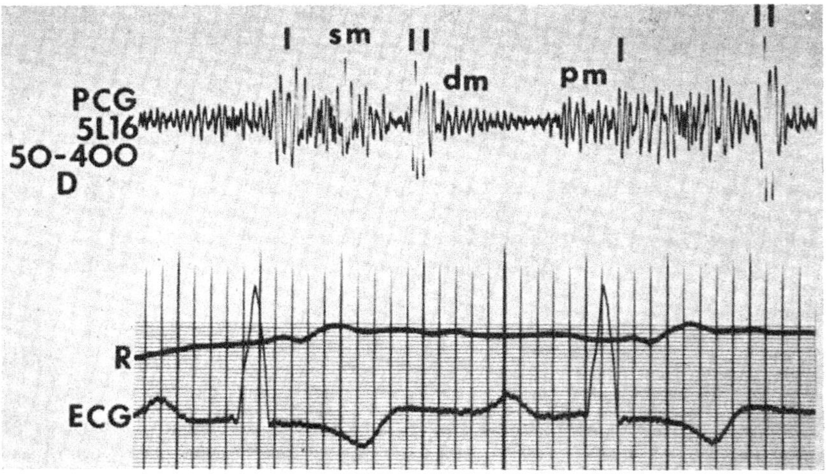

(B) Phonocardiogram of another patient with a history of infarct and both a diastolic and a presystolic murmur.

mur, and described the graphic characteristics of the functional murmur (Figs. 72-75). In spite of such impressive evidence, the classic teaching that a diastolic or presystolic rumble is always organic in nature unfortunately still prevails. our studies revealed not only the possibility of a middiastolic rumble but also that of a presystolic murmur (Figs. 72-75). This functional diastolic murmur can be found in acute rheumatic fever (myocarditis) (Carey Coombs murmur) as well as in other forms of acute or chronic myocarditis. It can be found in aortic insufficiency (Austin Flint murmur). It can be found in cardiomyopathies and in cases of coronary heart disease (mostly after infarcts), as well as in other patients with heart failure.

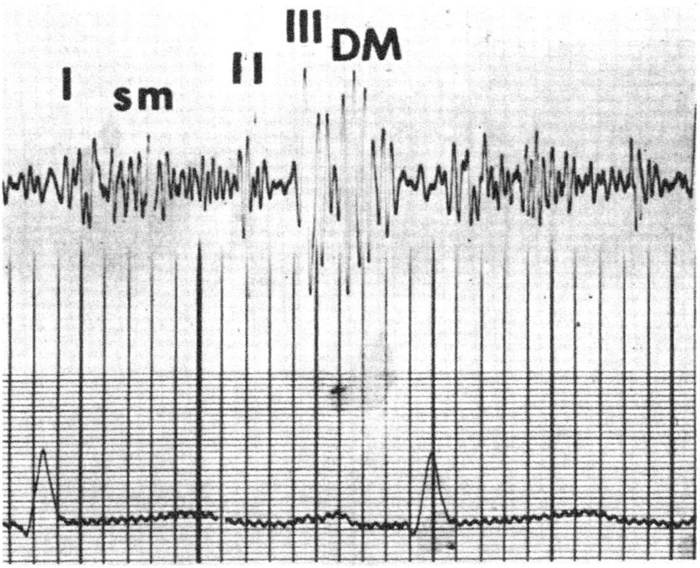

FIGURE 74. Tracing of a 16-year-old girl with acute rheumatic fever. Large third sound and gigantic presystolic murmur (not in crescendo) at the apex. Unfiltered tracing.

In several cases, certain graphic characteristics give a clue for the diagnosis (Chap. 27).

(1) The murmur often starts in middiastole with a large third sound or can be considered as a prolongation of the latter; if it is a presystolic murmur, the murmur is *not* in crescendo and is often separated from the first sound.

(2) The murmur has often a large amplitude in contrast with that of the average case of mitral stenosis.
(3) The murmur can be recorded over a wide area of the myocardium as a result of ventricular dilatation.
(4) The opening snap is absent in the majority of cases with functional diastolic rumbles.

However, these characteristics are not always present. *Intracardiac phonocardiography* helps by documenting the fact that the murmur is found in one of the ventricles; should this be the right ventricle, obviously mitral stenosis would be less likely.

Collateral data, supplied by electrocardiography (normal ECG, left ventricular hypertrophy, left bundle branch block) or

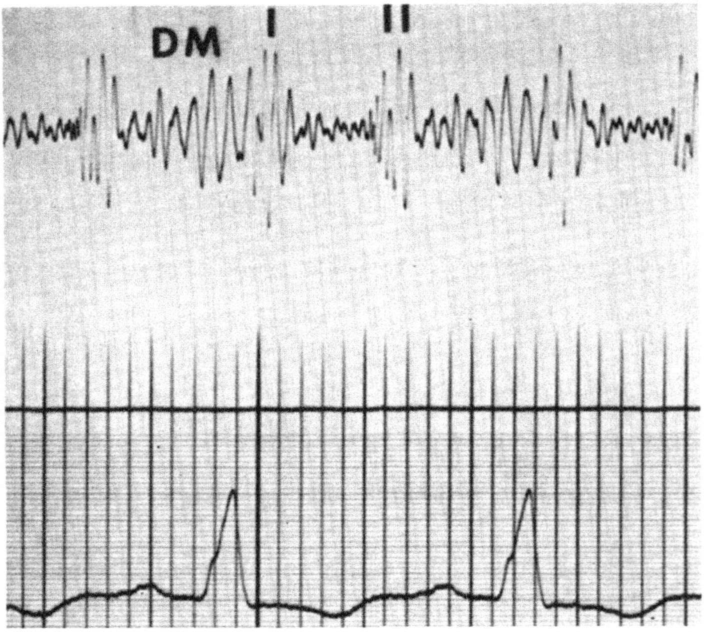

FIGURE 75. Tracing recorded over the third left i.c.s. in a 42-year-old patient with alcoholic cardiomyopathy.
Large diastolic-presystolic murmur of low frequency.
Displacement tracing at 20-400 Hz. filtration (filter slopes 48 db/oct.).

catheterization (absence of a gradient across the mitral valve) are at times decisive in excluding an organic mitral stenosis.

LEFT ATRIAL MYXOMA

Systolic and diastolic murmurs, as well as an opening snap, have been recorded in this condition. Several differential points should be kept in mind:

(1) The Q-I interval is usually borderline.
(2) The murmurs may vary from day to day or by changing the patient's position.
(3) The vibrations of the diastolic murmur are often present during isometric relaxation, i.e., before the opening snap.

TRICUSPID STENOSIS

Tricuspid stenosis is associated to mitral stenosis in about 15 per cent of the cases of rheumatic valvular lesions.

Many of the auscultatory and graphic characteristics of tricuspid stenosis are liable to be confused with those caused by

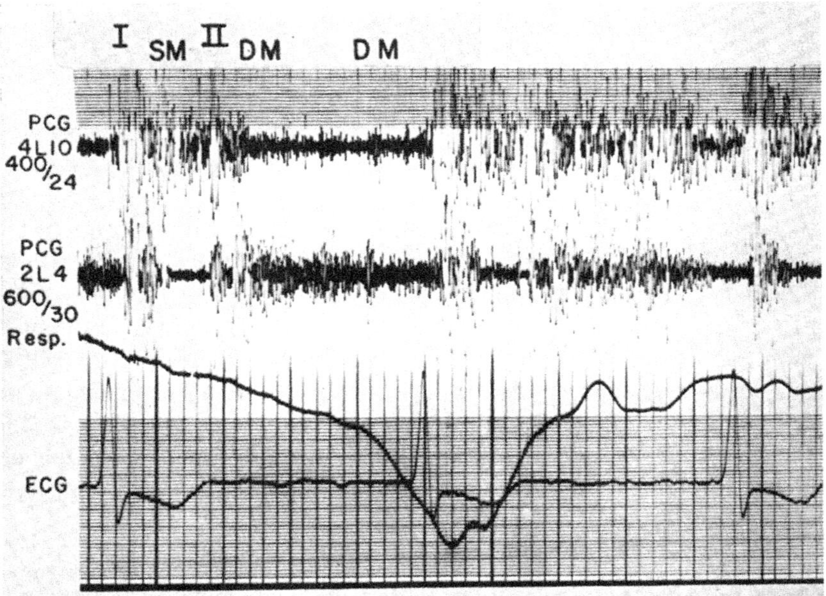

FIGURE 76. Two cases of rheumatic heart disease with mitral and tricuspid lesions. The phonocardiograms over the fourth (A) and third (B) interspaces at the left of the sternum demonstrate an increase of the murmurs during inspiration.
In (A) the diastolic murmur is especially increased.

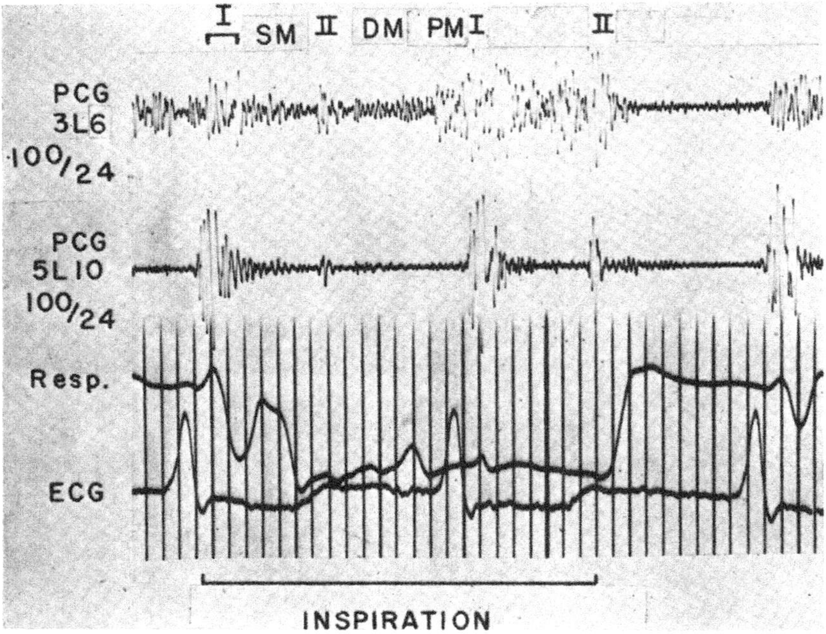

In B) both the systolic and diastolic murmurs are increased.

the lesion of the mitral valve. However, the murmur of tricuspid stenosis is frequently of a higher pitch (Rivero Carvallo, 1950; McKusick, 1958; Bousvaros and Stubington) and has been confused at times with that of aortic insufficiency. Tricuspid stenosis *per se* does not cause a late, loud, snapping first sound and does not prolong the Q-I interval because the first sound originates in the left ventricle, and the gradient across the mitral valve is not increased by tricuspid stenosis. However, such data may be present because of a concomitant mitral stenosis.

Tricuspid stenosis gives rise to a *diastolic rumble* and a *presystolic murmur*, which are heard and recorded best over the third left, fourth left, or fourth right interspace, or over the xiphoid; thus, they may be confused in cetrain cases with similar murmurs caused by mitral stenosis (Fig. 77). The presystolic murmur, being related to right atrial contraction (which occurs slightly before the left atrial), is better separated from the first sound, has no crescendo type, and may have a crescendo-

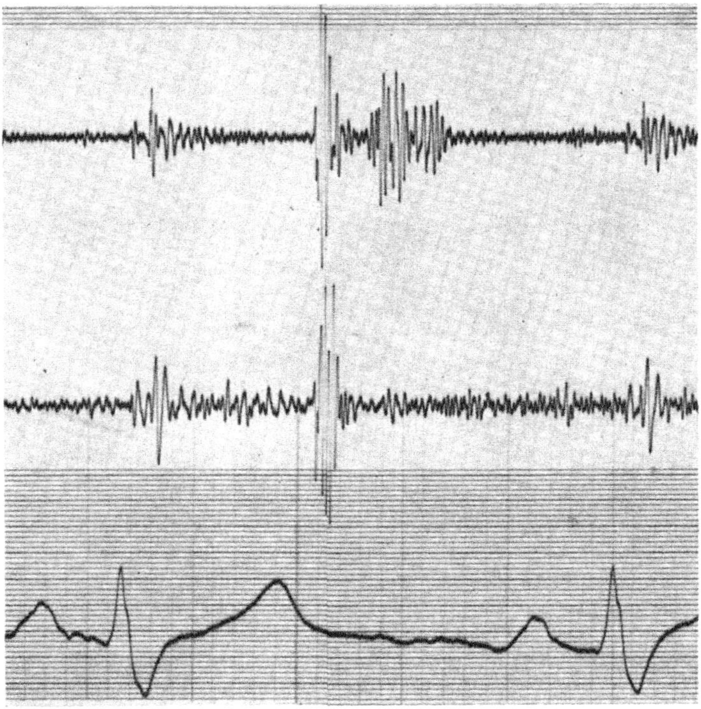

FIGURE 77. Case of mitral and tricuspid stenosis.
Above—PCG over the third left i.c.s.
Center—PCG over the second right i.c.s.
Note the low-frequency, large murmur in early diastole (upper tracing), well separated from the second heart sound.
High pass filter at 100 Hz (slope 24 db/oct.).

decrescendo configuration, even in the presence of a normal AV conduction (Bousvaros and Stubington).

An *opening snap of the tricuspid valve* has been documented by Rivero Carvallo, 1950, Kossmann, and Bousvaros and Stubington. The opening snap occurred 30 msec. after the onset of the drop of the right atrial curve in one of Kossmann's patients. It took place from 97 to 129 msec. after the aortic component of the second sound in his first case; and from 84 to 97 msec. in his second case. It was heard best over the xiphoid in one patient, and over the right precordium in the other patient.

Both he diastolic-presystolic murmur and the opening snap can be differentiated from similar murmurs and sounds of the left heart by means of functional tests.

1. *Inspiration and inspiratory apnea* increase the abnormal sounds of tricuspid stenosis and decrease those of mitral stenosis (Fig. 76) (Carral, 1949; Rivero Carvallo, 1950).
2. *Amyl nitrite* markedly increases the abnormal sounds of tricuspid stenosis in an early phase; it increases (if it causes marked tachycardia) those of mitral stenosis only in a later phase (Romeo *et al.*).

RELATIVE TRICUSPID STENOSIS

Relative or functional tricuspid stenosis occurs in conditions associated with increased flow through the tricuspid valve (atrial septal defect, partial anomalous pulmonary venous return) or right ventricular dilatation connected with hypertension and failure (pulmonary heart disease, mitral stenosis). Occasional evidence of relative stenosis can be found in patients with systolic overload of the right ventricle (pulmonary stenosis, primary pulmonary hypertension), obstruction to right ventricular filling (constrictive pericarditis, calcified endomyocarditis), or displacement toward the right of the ventricular septum (Bernheim syndrome), as reported by Lenzi *et al.*

The same tests used for recognition of organic stenosis also apply to relative stenosis (Rivero Carvallo *et al.*). Therefore, although these tests help in the differential diagnosis between tricuspid and mitral diseases, they do not help in recognizing whether a tricuspid murmur is caused by valvular narrowing or ventricular dilatation. On the other hand, the same graphic criteria outlined for recognition of a murmur caused by relative stenosis of the mitral valve also apply to that of the tricuspid valve.

MITRAL INSUFFICIENCY

Mitral insufficiency is most often caused by rheumatic endocarditis. However, it may be the result of bacterial endocarditis, rupture of a chorda or papillary muscle (myocardial infarction), fibrosis of chordae or a papillary muscle (rheumatic, postinfarctual, or resulting from coronary insufficiency), endocardial fibroelastosis or another congenital malformation,* or endomyo-

*Congenital anomalies causing mitral insufficiency are found in common atrioventricular canal, foramen primum, corrected transposition with Ebstein's anomaly of the left AV valve, or isolated anomalous chordal insertion.

cardial fibrosis. A possible explanation for cases with a "billowing" of the mitral valve is alteration or poor development of the intrinsic musculature of the valve leaflets.

First Sound. The first sound is typically decreased and faint in experimental mitral insufficiency and in some clinical cases.* However, a normal first sound may occur and even a loud first sound has been exceptionally documented in our laboratory. The typical weakening of the first sound can be explained by the fact that, when the mitral valve is unable to completely close the orifice, the rise of pressure will necessarily be slower in the left ventricle. This fact, which has been documented both in animals and man by van Bogaert *et al.*, is followed by softening of the first sound because the loudness of this sound is proportional

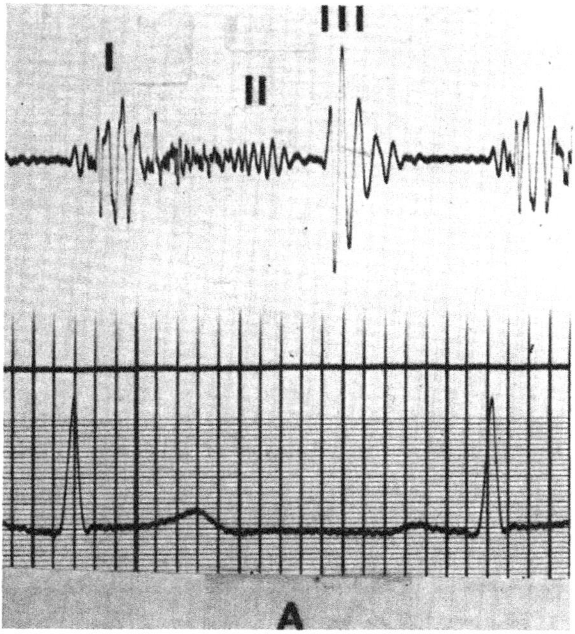

FIGURE 78. Pure mitral insufficiency in a 16-year-old patient. (A) Displacement tracing with band pass filter at 30-300 Hz (slopes 48 db/oct.) revealing the low frequency vibrations. Gigantic third sound followed by a short, functional, mid-diastolic rumble.

*The first sound was decreased in 29 out of 33 patients of Perloff and Harvey (1962).

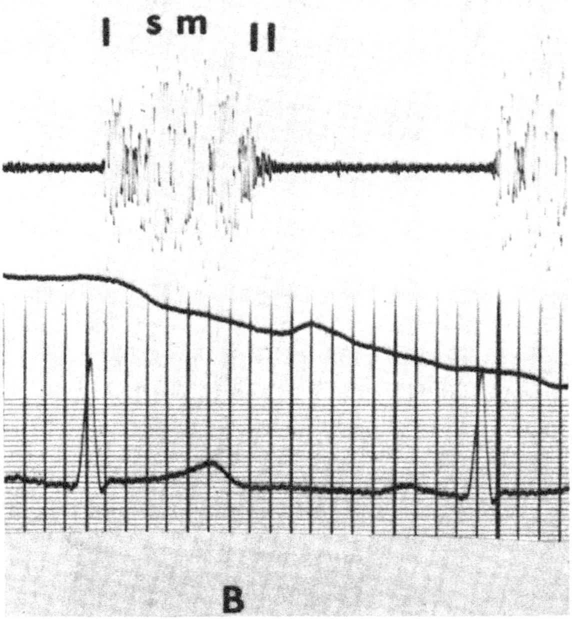

(B) Velocity tracing with band pass filter at 150-800 Hz (slopes 48 db/ oct.) revealing the high frequency pansystolic murmur.

to the rapidity of the rise of presure in the left ventricle, as demonstrated by records of the first derivative of left ventricular pressure. Normal or louder first sounds result from compensatory hypertrophy of the left ventricle.

Systolic Murmur. The systolic murmur may be from Grade 1 to Grade 5 but is usually Grade 3 or 4. It is described clinically as blowing, soft, or moderately harsh. It may be musical. It is exceptionally accompanied by a thrill. It is well heard over the left ventricular area, this being the most accessible, but it should be heard even better over the left atrium. As this chamber is located deep within the chest, the atrial murmur is audible (1) above and to the left of the apex; (2) if loud, at the left interscapulovertebral area; and (3) within the esophagus—all typical areas of radiation. It is occasionally audible along the left or right sternal borders, and sometimes at the base and neck. This unusual radiation has been documented by Bleifer *et al.* (1960) (who attributed it to the existence of a giant left atrium) and by Perloff and Harvey (1962). Deformity of the

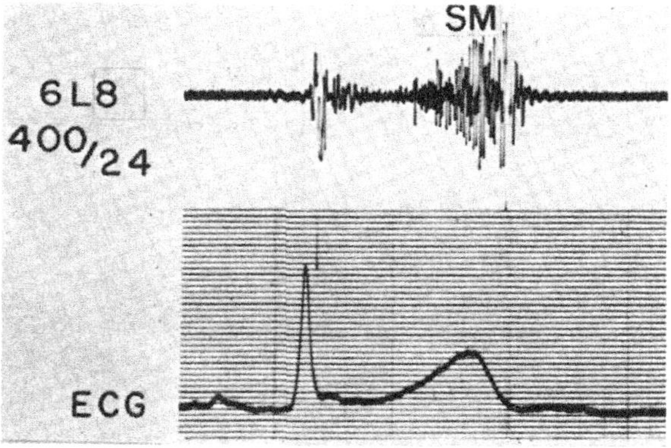

FIGURE 79. Pure mitral insufficiency (probably relative) in a 6-year-old child with acute rheumatic carditis and glomerulonephritis. Tracing at apex with high pass filter at 400 Hz (slope 24 db/oct.) revealing a systolic crescendo murmur lasting through the second sound.

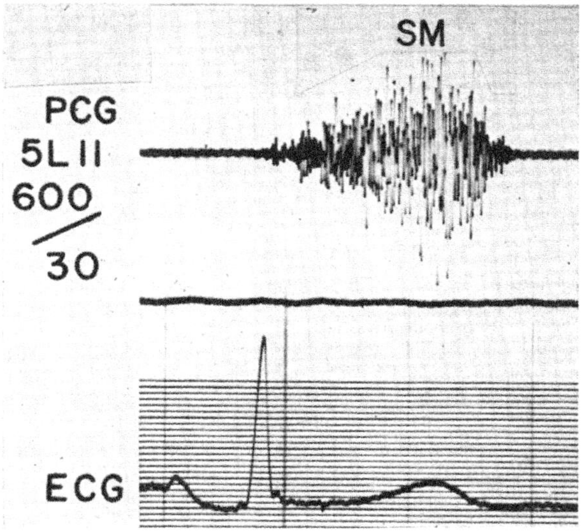

FIGURE 80. Pure mitral insufficiency in a 67-year-old man. Probable papillary muscle dysfunction. Tracing at apex with high pass filter at 600 Hz (slope 30 db/oct.) revealing a pansystolic crescendo murmur lasting through the second sound.

posterior leaflet tends to project the murmur toward the base* and to cause its changing to a more diamond-shaped configura-

*This would direct the regurgitant stream forward and medially against the atrial septum near the aortic wall.

tion. Deformity of the *anterior* leaflet tends to project the murmur more posteriorly and, through contact with the spine, even to cause transmission to the skull.

In the configuration of the murmur, there are four patterns that can be recognized.

(1) The most typical is a high-pitched, all-systolic (*pansystolic*) murmur with vibrations of similar amplitude throughout the phase (Fig. 78).

(2) A less comomn, but not unusual, pattern is a high-pitched systolic murmur in crescendo (Perloff and Harvey, 1962; Segal *et al.*) ; in such cases the crescendo phase may have a higher pitch than the rest.* The murmur may be late-systolic *in crescendo* (Figs. 79 and 80). This particular pattern sems to be caused by either billowing of one leaflet (congenital, bacterial, or rheumatic lesions) or damage to a papillary muscle (rheumatic, coronary). The studies of Burch *et al.*, Barlow and Bosman, Linhart and Taylor, Tavel *et al.*, Criley *et al.*, and Burch *et al.*, on the "papillary muscle syndrome" have been followed by many others.

(3) There may be a systolic murmur *in decrescendo.* This usually has a medium pitch and may occur only in the third left interspace; the decrescendo shape may be present only with low filtration.**

(4) The least common pattern is a *diamond-shaped murmur,* often recorded best at the base. This has been explained by Perloff and Harvey (1962) as the result of a lesion of the posterior mitral leaflet.

In general, the murmur has most of its vibrations between 50 and 300 Hz; however, certain cases have important components between 300 and 500 Hz.

The murmur caused by a *ruptured papillary muscle* is loud and harsh, and may be accompanied by a thrill; it may be maximal over the second right interspace; it has a predominance of medium low-frequency vibrations.

When one or more of the leaflets are calcified, the murmur

*It is now recognized that the late-systolic crescendo murmur is most often found in patients with mitral insufficiency resulting from papillary muscle dysfunction.

**Older tracings presented this configuration more often because of their inability to record high-frequency vibrations.

becomes either harsh or musical; in the last instance, regular vibrations can be observed in the tracing. A *late-systolic click* has been observed in some cases (Brigden and Leatham), and may precede the murmur by several years.

A late-systolic murmur in crescendo can be observed in other conditions, such as: (1) patent ductus arteriosus with pulmonary hypertension; (2) coarctation of the aorta; (3) coarctation (or stenosis) of the branches of the pulmonary artery; and (4) pleuropericardial adhesions.

Second Sound. Ususally, the pulmonary component of the second sound is normal or moderately increased. *Wide splitting* is common (Brigden and Leatham; Perloff and Harvey, 1958; Bleifer *et al.*, 1960). This has been explained as the result of a shorter Q-IIA interval in the presence of a normal Q-IIP interval. Splitting is variable but does not disappear in expiration. Current theories attribute this fact to an earlier ending of left ventricular systole but hemodynamic studies are required before accepting this interpretation.

Early Diastolic Murmur. A soft, blowing, early diastolic murmur, best audible over the left ventricular area, has been observed in some patients and is more often recorded than heard (Brigden and Leatham) (Fig. 80). This murmur is actually a prolongation of the systolic murmur beyond the second sound*, and was found by Perloff and Harvey (1962) in 21 per cent of their patients. Such a prolongation is easily understandable because atrial and ventricular pressures may not become equalized until after the closure of the aortic valve (Wiggers and Feil), so that the expansion of the left trium may continue for a short time after the second sound. This was also proved by simultaneous phonocardiograms and electrokymograms (Luisada and Fleischner), as well as by simultaneous phonocardiograms and left atrial presure tracings (Luisada and Liu, 1958).

Third Sound. An audible, loud third sound was found by Sprague and White and by Nixon in the majority of their patients, and in about two-thirds of their patients by Perloff and Harvey (1962). We have found this *gallop sound* frequently, though probably in no more than half of our cases. It is audible and recorded best over the left ventricular area. It is well re-

*Obviously, a concomitant aortic insufficiency should be excluded before attributing such a peculiar quality to this murmur.

corded in the medium low frequency bands but can be recorded at 20 and 400 Hz and can assume a gigantic magnitude (Fig 78).

Fourth Sound. A few patients of Perloff and Harvey (1962) presented a loud fourth sound. However, in 3 of their 4 cases it was attributed to the right ventricle as a result of pulmonary hypertension. A fourth sound is found more frequently in cases of mitral insufficiency caused by papillary or chordal lesions than in others, due to the basic myocardial disease.

Diastolic Rumble. A short diastolic rumble is often found over the left ventricular area (Bleifer *et al.*, 1960; Nixon, Perloff and Harvey, 1962). The last authors found it in about one-third of heir cases, whereas it was present in about one-fourth of our patients. It follows the third sound and terminates in middiastole. We have found this rumble as a common occurrence in cases with severe mitral insufficiency.

Opening Snap. An opening snap is present in certain cases. Perloff and Harvey (1962) found this snap in 4 out of 33 cases (the IIA-OS interval was 0.06 to 0.10 sec.); Nixon found it in the majority of his patients; and Bloemer *et al.* found it in 4 out of 31 cases. It can be differentiated from the pulmonary component by observing a seeming triplication of the second sound over the second and third left interspaces. It can be differentiated from the third sound by recording simultaneous tracings at the base and apex and by the fact that it occurs earlier than a third sound. However, it may coexist with the latter. The OS is better recorded in the second and third left interspaces, whereas the third sound is recorded best in the fifth and sixth left interspaces.

It is obvious that, whenever an opening snap, a third sound, and a diastolic rumble are present in a patient with a blowing systolic murmur, the clinical diagnosis will tend to be "double mitral defect." The phonocardiogram is of great help in excluding mitral stenosis by showing the typical systolic murmur, the gallop sound, and the late onset of the diastolic murmur. Cardiac catheterization will further prove that there is a pure mitral insufficiency. However, in some cases, a minor fusion of a commissure may contribute to the auscultatory findings.

Intracardiac Phonocardiography

The murmur of mitral insufficiency is recorded best in the cavity of the left atrium, particularly if the pickup (or the tip of the catheter) is near the mitral valve (Luisada and Liu, 1958; Luisada *et al.*, 1964). The murmur most often is diamond-shaped or spindle-shaped, even in patients that have a long, plateau-like or pansystolic murmur in the external phonocardiogram. As high-frequency pickups reproduce more often a plateau-like murmur, it is possible that the filtering properties of the chest wall tend to reduce the vibrations in the central part of the murmur. On the other hand, it is well known that only high-frequency tracings reproduce well the pansystolic murmur. Intracardiac phonocardiograms of the left ventricle show only a small murmur.

Esophageal Phonocardiography

The esophageal phonocardiogram, recorded at the left atrial level, demonstrates the systolic murmur as well as or better than the external phonocardiogram. In cases where there is doubt whether the murmur is caused by mitral or tricuspid insufficiency this tracing may be important because only the murmur of mitral insufficiency is well recorded through the esophagus.

Apex Cardiogram

The apex cardiogram has a fairly typical pattern in pure mitral insufficiency. As described by Weitz and later by me (1953), and as confirmed by Benchimol and Dimond, the pattern is as follows: there is a large, tall, conical, positive wave during systole, a deep point O, and a rapid and tall wave III. This pattern is the resut of left ventricular hypertrophy, relatively short ejection, and increased diastolic filling of the left ventricle. Warembourg and Ducloux found this tracing useful in their evaluation of the severity of mitral insufficiency.

Right-sided Cardiogram

In patients with severe mitral insufficiency and a giant left atrium, a low frequency tracing recorded over the 3rd and 4th *right* interspaces may reveal a large, plateau-like, systolic pulsation that is similar to a ventricular pressure tracing except for a late termination of the rise. This is evidence of the systolic expansion of the enlarged left atrium caused by the regurgitant jet.

Functional Tests

An *amyl nitirate test* causes a decrease of the systolic murmur and of the third sound. *A pressor amine test* causes an increase of the systolic murmur and of the third sound (Fig. 81). Braunwald *et al.* (1958), first, showed that pressor amines increase the severity of the regurgitation, and this established the value of this test for the study of mitral insufficiency. According to Perloff and Harvey (1962), pressor amines frequently cause a

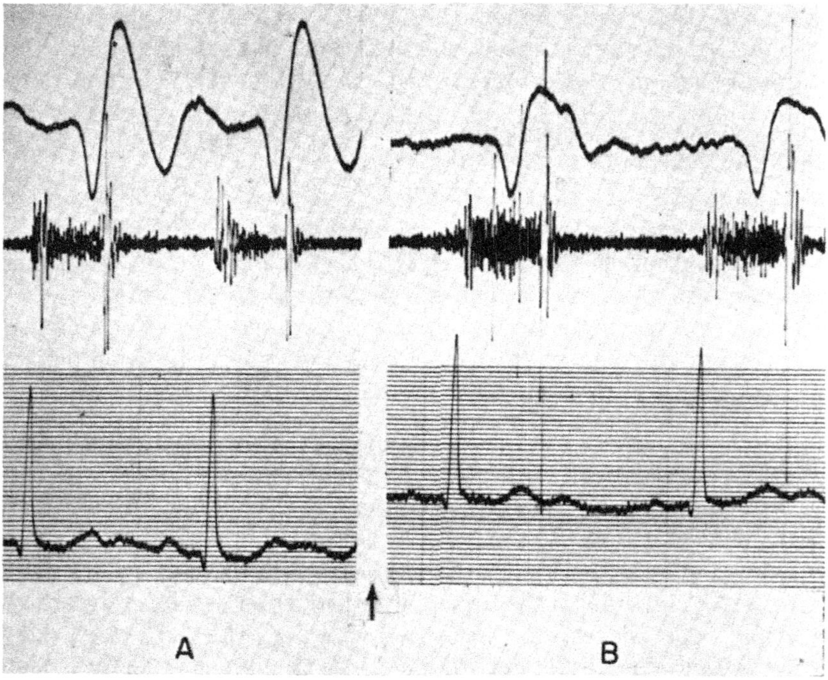

FIGURE 81. Tracings recorded at the apex in a patient with mitral insufficiency. A pressor amine test (Aramine) increases the pansystolic murmur.

Above—digital pulse
Center—PCG
Below—ECG

At left—control
At right—after drug administration

prolongation of the murmur into the isovolumic relaxation period, a fact that we have also occasionally noted.

RELATIVE MITRAL INSUFFICIENCY

Relative mitral insufficiency usually produces a small or moderately loud murmur, which frequently is of a decrescendo type. However, severe mitral insufficiency caused by marked dilatation of the mitral orifice or by a functional anomaly of a papillary muscle may cause a Grade 2 or 3 murmur, having either a pansystolic or a crescendo configuration. In one patient, we observed a regurgitant wave of 45 mm Hg in the left atrium following a myocardial infarct. The phonocardiogram showed systolic, early diastolic, and presystolic murmurs. Autopsy revealed the absence of valvular lesions.

TRICUSPID INSUFFICIENCY

Tricuspid insufficiency usually causes a pansystolic murmur that is audible over the third left, third right, or fourth right interspace. In cases with marked right ventricular dilatation (mitral stenosis, pulmonary heart disease), it can be heard even at the apex. The murmur of tricuspid insufficiency may last through the aortic component to the pulmonary component of the second sound. This fact will be noticeable, however, only in cases of wide splitting of the second sound, and may not occur.

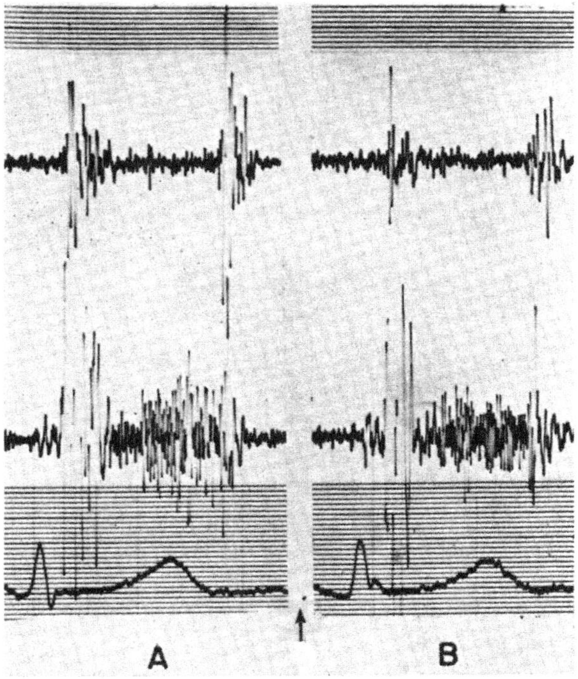

FIGURE 82. Pansystolic murmur of tricuspid insufficiency. The murmur is larger in inspiration (A) than in expiration (B).

In general, the murmur is similar to that of mitral insufficiency. The murmur may be hard to differentiate from other murmurs because of the frequent association of mitral and tricuspid defects or the existence of a trivalvular lesion. However, the murmur of tricuspid insufficiency is often *increased by inspiration or inspiratory apnea* (Fig. 82) in contrast with that of mitral insufficiency (Rivero Carvallo, 1946, Aceves and Carral) unless there is severe right heart failure.

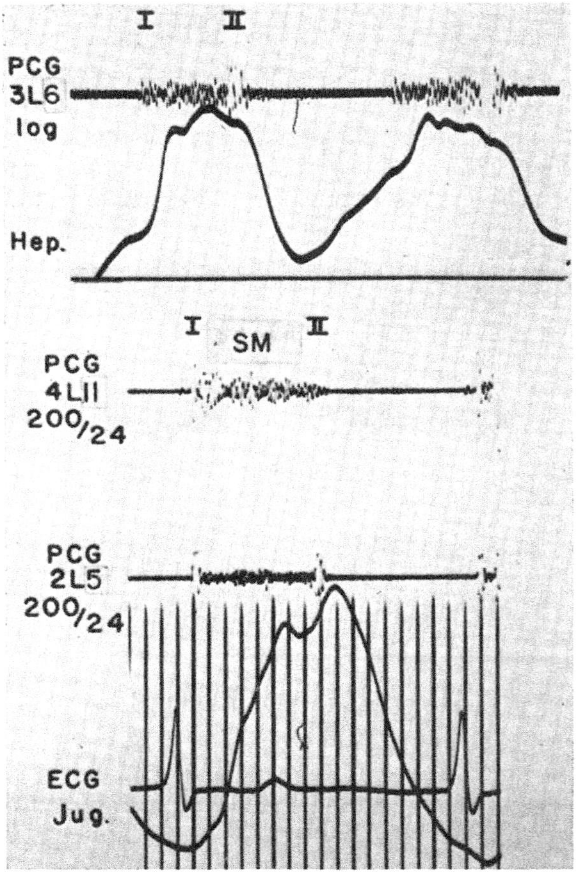

FIGURE 83. (A) Pansystolic murmur of tricuspid insufficiency. Evidence of the regurgitation into the right atrial-venous system is supplied by both the hepatic (Hep) and the jugular (Jug) tracings.

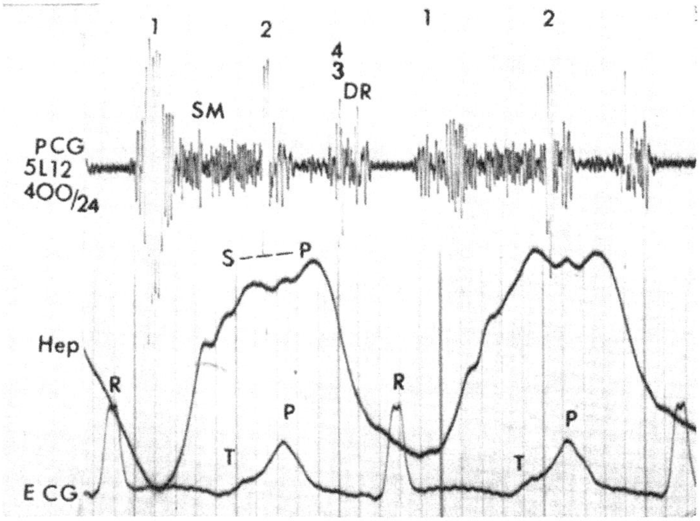

(B) Pansystolic murmur and functional diastolic murmur in a patient with mitral and tricuspid insufficiency. Typical systolic plateau in the hepatic tracing (Hep).

Drug Tests

Amyl nitrite often causes a sharp and early increase of the pansystolic murmur 20 to 30 sec. after the inhalation.

Venous Tracings

Both the *jugular* and the *hepatic tracing* show a typical, plateau-like wave during systole, which is evidence of the systolic regurgitation into the right atrium. This plateau often ends *after* the second sound (V-wave) and drops slowly in early diastole (Fig. 83).

Chapter 8

Aortic Valve and Pulmonary Valve Defects

AORTIC STENOSIS

AORTIC STENOSIS CAN BE caused by narrowing of the ascending aorta (supravalvular stenosis), of the aortic valve (valvular stenosis), or of the outflow tract of the left ventricle (subvalvular stenosis). Whereas supravalvular stenosis is usually congenital and valvular stenosis is usually acquired,* subvalvular stenosis can be caused by either a congenital abnormality or an acquired (muscular hypertrophy) process.**

As a result of the lesion, the majority of patients present an abbreviation of the ejection period and a prolongation of the isovolumic contraction period so that the total duration of systole is often unchanged. Systolic pressure is markedly elevated in the left ventricle, and a systolic gradient develops across the stenotic orifice or section. Left ventricular diastolic pressure is elevated in severe cases. This elevation is caused by ventricular hypertrophy and, in the late stages, by ventricular failure. The

*Cases of valvular aortic stenosis in infants or young children may be caused by endocarditis but often are truly congenital.

**Among the various causes of subvalvular stenosis, the possibility of an abnormal anterior leaflet of the mitral valve (congenital abnormality, rheumatic alteration) encroaching on the outflow tract of the left ventricle should be remembered (Bjoerk *et al.*, Moberg *et al.*). Subaortic muscular stenosis may have a familial type indicating a congenital predisposition.

former alters the pressure-volume relationship of the chamber; the latter increases the amount of residual blood.

Obviously, marked differences occur according to the site of obstruction. Changes of ventricular and aortic pressure, location of the gradient, and pulse contour differ in the various forms (Fig. 84) and explain some of the peculiarities of the murmurs.

As a result of the increased resistance to filling of the left ventricle, left atrial pressure rises, a compensatory hypertrophy of the left atrium develops, and a giant A wave may be seen on catheterization of the left atrium.

First Sound. The first sound may be normal or faint and may seem split (Fig. 85). The faintness and low pitch mentioned by several authors have been explained by prolongation of the isovolumic contraction period,* heart failure, or poor transmission. However, phonocardiograms, while showing a small first sound at the base, frequently show a normal first sound over the mid precordium and apex (Fig. 87).

Ejection Sound. Apparent splitting of the first sound is caused by the existence of a loud and large ejection sound,** i.e., of an increased and delayed third component of the first sound (Figs. 85 and 87). This pathologic sound is a group of large, rapid vibrations well recorded over the entire outflow tract of the left ventricle, from the third left interspace to the second right interspace. This group of vibrations is delayed over the beginning of the first sound as a result of the prolongation of the isovolumic tension period, which is in turn caused by the need to raise left ventricular pressure to a higher level than in normal conditions, and probably also by the slower rise of pressure in the aorta.

This loud and large ejection sound is not related to valve calcification but is partly related to poststenotic dilatation (Calo, 1962). Wood found it more frequently in mild-to-moderate stenosis than in severe stenosis. This fact is logical because, in severe stenosis, flow through the narrow section is represented

*A small first sound can be the result of a slower rise of left ventricular pressure. However, in aortic stenosis the prolongation of the isovolumic contraction period would be caused by the need to raise LV pressure to a higher level, and not by a slower rise.

**This vibration, described by Potain (1900), was further studied by Gallavardin (1924), Calo (1962), Wood (1958), and many others, and emphasized by Leatham (1954) and Leatham and Vogelpoel.

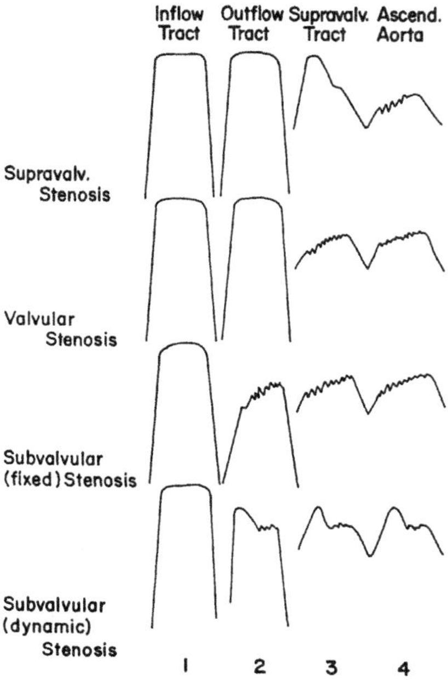

FIGURE 84. Scheme of changes of pressure and pulse contours and magnitude in the various forms of aortic stenosis. (Adapted from Fishleder, B. L. *et al.*: *Arch Inst Cardiol* Mexico, 1962.)

by a thin jet, and there is less impact on the aortic wall. However, according to Calo, occasional patients with severe stenosis do present this group of vibrations, a fact that we can confirm. The vibrational frequency of this group of vibrations is high, so that it can often be recorded with a filter at 300 Hz. It occurs from 60 to 80 msec. after the first component of the first sound and from 30 to 50 msec. after the onset of ejection (Leatham and Vogelpoel). *This ejection sound is less frequent in subaortic stenosis,* though occasionally it may occur even in this type of lesion (Fig. 89). It may be replaced in this condition by a *midsystolic click* (Benchimol *et al.*, 1963). It is also absent in older patients with calcified and immobile valve (Fig. 88).

Second Sound. The intensity of the second sound can vary according to the location (smaller sound in valvular stenosis), severity (smaller sound in severe stenosis), type of lesion, and presence or absence of systemic hypertension (larger in hypertensive cases). The severity of the lesion is undoubtedly the

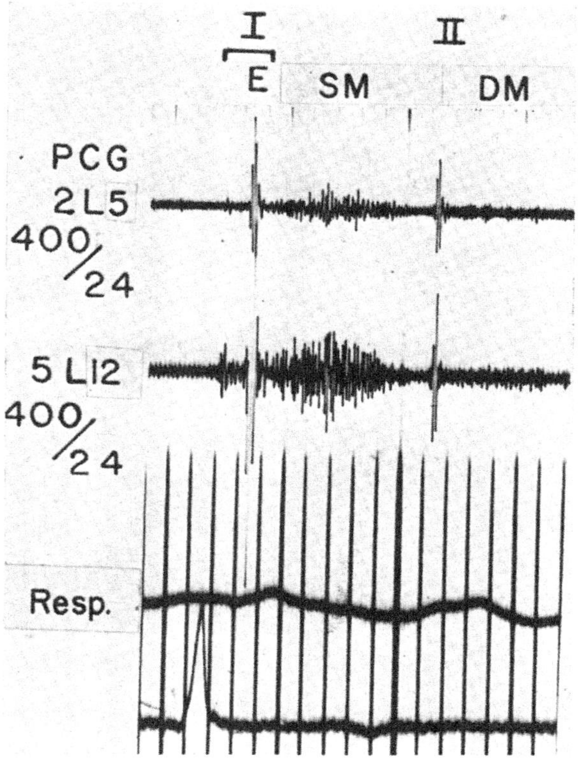

FIGURE 85. Tracings of a 35-year-old woman with rheumatic aortic valvular stenosis of moderate degree. There is an ejection sound at the base (upper tracing) and a diamond-shaped systolic murmur with early ending (middle tracing), especially large at the apex. The first sound at the apex seems split due to the presence of the ejection sound.

most important factor, and this is also revealed by a different type of splitting (Leatham and Vogelpoel, Kumar and Luisada, 1971).

When aortic stenosis is moderate, the aortic component of the second sound is slightly delayed and falls close to the pulmonary component, even in young patients (who normally have a wider splitting). When aortic stenosis is more marked, the aortic component occurs simultaneously with the pulmonary component in young people, and may occur slightly after this component in older patients (who normally would have a single second sound). When aortic stenosis is severe (gradient over

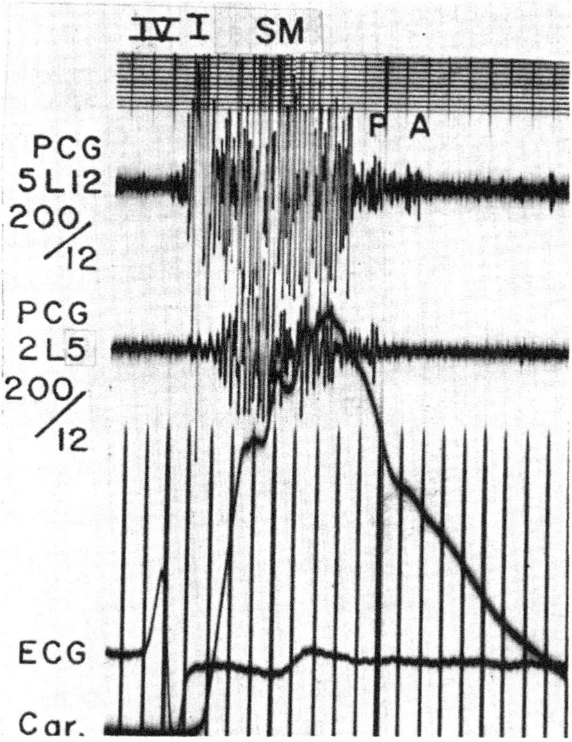

FIGURE 86. Tracings of a 66-year-old man with severe calcific aortic valvular stenosis (aortic gradient = 100 mm Hg). Large diamond-shaped murmur and carotid tracing (Car) with anacrotic depression and "carotid shudder." The aortic component of the second sound (A) is small and delayed causing reverse splitting. The first heart sound is of good quality at the apex (above), small at the base (center).

50 mm Hg), the aortic component is markedly delayed as a result of hemodynamic factors (Figs. 86-88). Then it is a small group of vibrations that occurs 20 to 60 msec. after the pulmonary component. This phenomenon, called *reverse splitting of the second sound,* is also revealed by the fact that the splitting decreases and may even disappear in inspiration (Leatham, 1954). A special study of the second sound in aortic stenosis was made by Kumar and Luisada (1971A). It was demonstrated that the delay of the aortic component in this condition is largely caused by a hemodynamic phenomenon (Fig. 61). However, a further delay is related to a delayed aortic rebound over the

closed valve, which is caused by slower speed of the pulse in the aorta. Smaller volume and slower rise of the pressure pulse cause necessarily a less deep and more rounded incisura, hence a smaller sound.

In *subaortic muscular stenosis,* the behavior of the second sound is similar to that found in valvular stenosis. Thus, close splitting, single sound, or reverse splitting can be found according to the severity of the process and the magnitude of the pressure gradient across the narrowed area (Braunwald *et al.*, 1964).

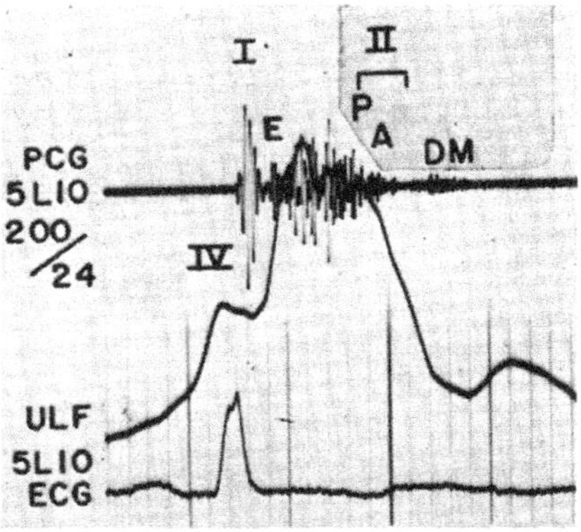

FIGURE 87. Tracings in a 62-year-old woman with calcific aortic valvular stenosis. Small ejection sound (E), diamond-shaped systolic murmur, reverse splitting of the second sound. The apex cardiogram (ULF) shows a large presystolic wave. There is also a small diastolic murmur.

Systolic Murmur. The systolic murmur is a nearly constant finding in aortic stenosis, the exception being represented by rare cases of extremely severe valvular narrowing with low output. Even in these cases, the murmur, which had not been noticed on auscultation because it was confused with other murmurs or was in an unusual location, is generally recorded. The murmur is often loudest and best recorded in the third left interspace but may be loudest over the fourth or fifth left interspace, especially in cases of subaortic stenosis. It is heard and recorded well over the area of the ascending aorta (second right interspace) and at the suprasternal notch (Luisada, 1937), especially in cases of valvular stenosis, and it may have the maximum amplitude over this area if post-stenotic dilatation is severe.

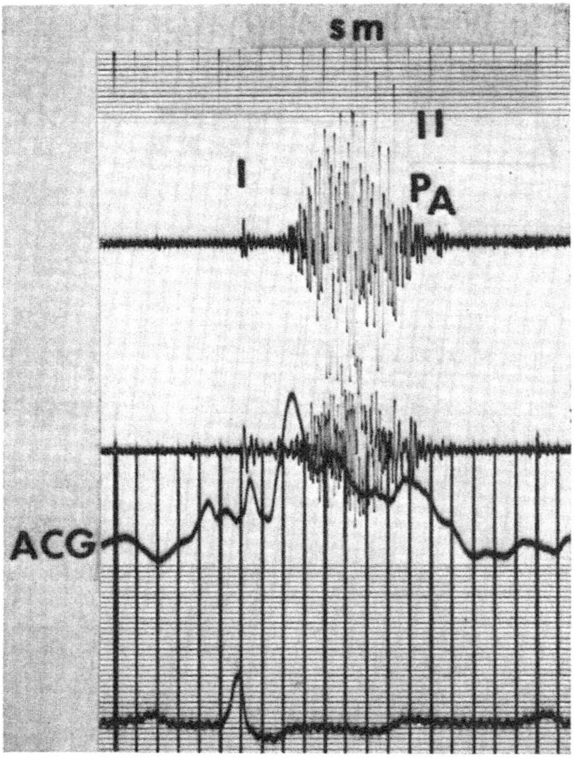

FIGURE 88. Tracings of a 73-year-old man with calcific valvular aortic stenosis. Diamond-shaped systolic murmur with late peak. Reverse splitting of the second heart sound. The apex cardiogram (ACG) has a complex presystolic wave.

The murmur is usually a grade 3 to 5, harsh murmur; is well separated from the first sound; and has a typical crescendo-decrescendo pattern (Figs. 85-88). In cases of alternation or arrhythmia, the amplitude of the murmur is related to the magnitude of flow through the aortic valve. Thus it wil alternate in case of pulsus alternans, and will be smaller in ectopic beats or in beats preceded by a short diastole if there is atrial fibrilation. In valvular stenosis or subvalvular stenosis caused by a membranous septum, a longer diastole is followed by a larger pulse and a larger murmur; on the contrary, in hypertrophic subvalvular stenosis, a longer diastole is often followed by a smaller pulse and a larger murmur (Table I). Thus, the amplitude of

Table I

Differences Between Subaortic Muscular Stenosis and Other Types of Aortic Stenosis

	Isoproterenol	**Postextrasystolic Beat**
Muscular subaortic stenosis	Larger pressure gradient	Smaller pulse
	Larger murmur	Larger murmur
Other forms of aortic stenosis	No increase of gradient	Larger pulse
	Possible increase of murmur	Larger murmur

From Brock, Goodwin *et al.*, 1960; Morrow *et al.*; Braunwald *et al.*

the murmur in arrhythmias does not allow differentiation between the two forms, whereas that of the pulse usually does.

The systolic murmur of *valvular stenosis* has a diamond shape. Its peak occurs after the anacrotic notch of the carotid pulse and is often more delayed when the stenosis is severe (Aravanis and Luisada). However, the later occurrence of the peak, often helpful in the differential diagnosis betwen organic and relative stenosis, is less useful in the evaluation of the severity of narrowing.

In *hypertrophic subaortic stenosis,* the murmur usually starts later than in valvular stenosis. It may even begin at midsystole (when narrowing becomes greater), and may have a crescendo-type or, if diamond-shaped, a late occurrence of the peak. The murmur is usually made of vibrations of various frequencies. However, some patients with a musical murmur have vibrations that predominate in one band and have a characteristic regularity. Occasionally, the murmur is musical at the apex and harsh at the base, a fact that still lacks a satisfactory explanation.

Amyl nitrite occasionally increases the magnitude of the systolic murmur in valvular stenosis by causing a more rapid flow through the aortic valve; however, often no change is noted. In muscular stenosis, an increase of the murmur has been reported (Fishleder *et al.*, 1962), and this is explained by Braunwald's observation that any decrease of aortic pressure increases the pressure gradient in such cases.

Isoproterenol typically increases the magnitude of the systolic murmur in muscular stenosis (Fig. 89) by causing a greater functional narrowing of the outflow tract. On the contrary, *methoxamine* decreases the murmur by causing hypertension

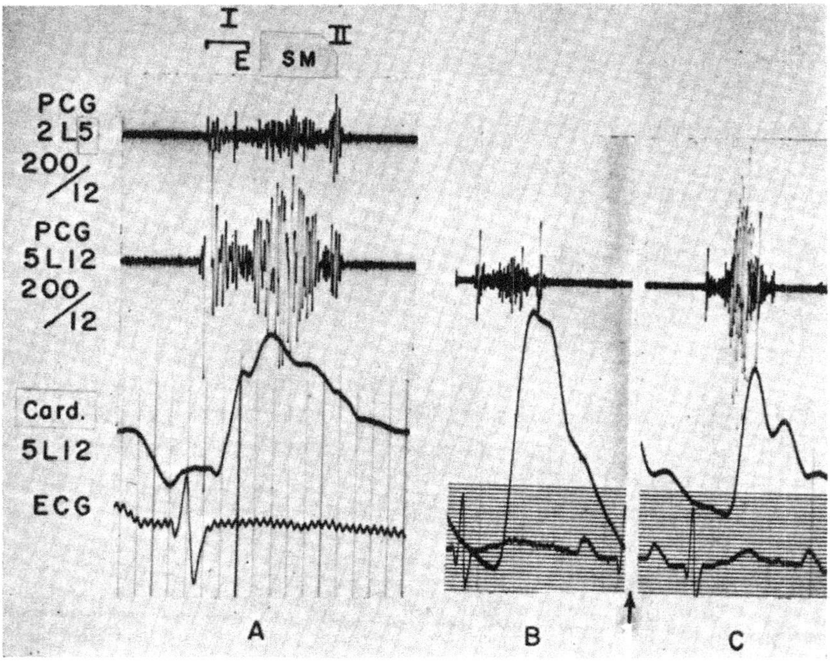

FIGURE 89. Tracings of a 41-year-old man with muscular subaortic stenosis. (A) There is a systolic murmur with late peak, preceded by an ejection sound. The apex cardiogram (Card.) has a sharp peak. (B,C) Sublingual administration of isoproterenol (arrow) causes a sharp increase of the murmur. The pulse in B and C is a digital pulse.

and dilating the outflow tract. Isoproterenol increases the murmur also in valvular stenosis, though to a lesser extent. Methoxamine either leaves the murmur unchanged or causes a minor decrease in valvular stenosis.

Squatting decreases the murmur in subaortic muscular stenosis, not in valvular stenosis.

In some patients with valvular stenosis, aortic insufficiency is also present. Then a diastolic murmur with a typical configuration follows the aortic component of the second sound.

Third Sound. A third sound is usually not recorded. When present, it is evidence of dilatation of the left ventricle with increase of the residual blood as a result of heart failure. It is more common in subaortic muscular stenosis, where it is evidence of elevated left atrial pressure (Shaw *et al.*, 1968).

Fourth Sound. A *fourth sound (atrial gallop) is often found in aortic stenosis* (Calo, 1938, 1950). A large presystolic wave of the apex cardiogram is even more typical (Figs. 87 and 88), and may be observed even in the absence of a fourth sound. According to Goldblatt *et al.*, a large fourth sound indicates that the obstruction is severe (the gradient exceeds 70 mm Hg and LV pressure is above 160 mm Hg), left ventricular end-diastolic pressure is elevated (12 mm Hg or higher), and the A wave of the left atrial pressure tracing is tall (14 mm Hg or more).

According to Perloff, the significance of a fourth sound is greater in young patients.

INTRACARDIAC PHONOCARDIOGRAM

The tracings recorded in the ascending aorta and left ventricle are extremely significant. They show that the systolic murmur is present in both chambers but is larger in the aorta (Luisada and Liu, 1958; Luisada *et al.*, 1964). They reveal the ejection sound best in the aorta. They often show well the aortic component of the second sound, thus permitting easy identification of this group of vibrations in the external tracing (Luisada *et al.*, 1964).

APEX CARDIOGRAM

The typical pattern of the apex cardiogram in aortic stenosis was recognized long ago (Weitz, Luisada, 1953; Aravanis and Luisada). It shows a high presystolic wave and a sustained systolic wave. In severe cases, it shows a slow rise, sometimes a plateau, and a series of notches during the rise and the plateau. These data have been confirmed by Goodwin *et al.*, and by Benchimol *et al.*, (1963).

In typical cases of subaortic muscular stenosis, the early part of the apical pulsation is normal and rapid, and there may be a drop at midsystole followed by a second positive wave in late systole (*double-hump or camel back profile*). The late systolic bulge is caused by the late obstruction to flow (Benchimol *et al.*, 1963).

CAROTID TRACING

The pulse tracing reveals a typical pattern in valvular stenosis, such as a slow rise, *deep anacrotic notch,* and a series of systolic vibrations (*carotid shudder*), which are often inscribed on a flat plateau and are the equivalent of the systolic thrill (Evans and Lewes, Feil and Gilder, Luisada, 1953; Donoso *et al.*) (Fig. 86). These data were confirmed through direct arterial catheterization by Brofman and Feil and by Brachfeld and Gorlin. A delayed upstroke time and a small dicrotic notch have been emphasized.

In subaortic muscular stenosis, the first part of the pulse is typically normal, whereas there is a flat section or a second bulge in late systole, which is accompanied in some cases by a late systolic shudder.

The different behavior of the pulse in regard to premature beats (Morrow and Braunwald; Braunwald *et al.*, 1962; Goodwin *et al.*, 1960) has been already mentioned.

The carotid tracing has great value for the identification of the aortic component of the second sound. The tracing should be recorded close to the right clavicle or at the suprasternal notch, together with the most significant phonocardiogram. The aortic component falls at the time of the incisura of the carotid pulse.

RELATIVE AORTIC STENOSIS (AORTIC FLOW MURMUR)

Relative aortic stenosis occurs when both the outflow tract of the left ventricle and the aortic valve are normal while the ascending aorta is dilated. It is found in patients with *severe systemic hypertension* (Fig. 90), *aortitis, atherosclerosis of the aorta,* or *Marfan syndrome,* as well as in patients with *severe aortic insufficiency*. There may be an ejection sound. The systolic murmur is diamond-shaped but has an early peak (Aravanis and Luisada) (Fig. 90). The second sound has a normal or large aortic component, and splitting is consistent with the age of the patient (single sound, more common in elderly patients or patients of any age having systemic hypertension, and

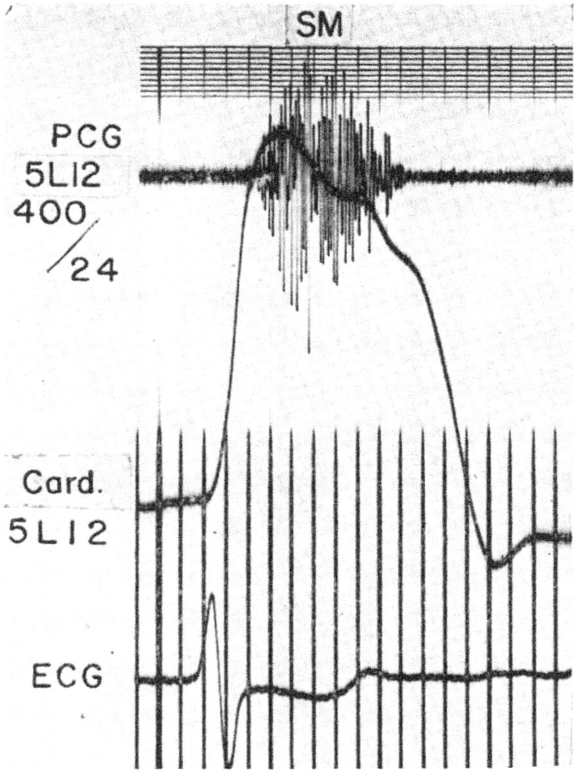

FIGURE 90. Tracings of a 58-year-old man with hypertensive heart disease and an aortic flow murmur. Diamond-shaped murmur with early peak. Sustained and lifting apex beat (Card.).

a normal type of splitting in young subjects).

The *apex cardiogram* shows a normal configuration of the apex beat, usually a tall, conical wave (Fig. 90).

The *carotid tracing* has a normal configuration (no anacrotic notch) (Aravanis and Luisada).

PULMONARY STENOSIS

Pulmonary stenosis is usually congenital. It may be supravalvular, valvular, or infundibular. Coarctation (or stenosis) of the distal pulmonary arteries is another possibility.

Several data, already reported in regard to aortic stenosis, are also present in pulmonary stenosis. However, the behavior of the second sound and other characteristics usually permit differentiation between the two conditions.

First Sound. The first sound is normal or increased in loudness in mild pulmonary stenosis but is usually small in severe stenosis. The possibility that, in severe cases, the hypertrophied right ventricle contributes to the vibrations of the first component has been considered but has not been substantiated.

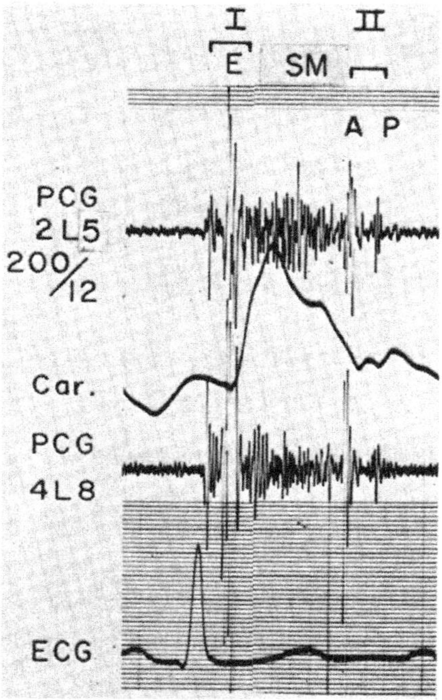

FIGURE 91. Tracings of a 16-year-old girl with moderate pulmonary stenosis. There is an ejection sound (E), a diamond shaped murmur, and a widely split second heart sound. The pulmonary component is normal. The carotid tracing is normal.

Ejection Sound. An *ejection sound* is frequently heard and recorded, though it is sometimes mistaken on auscultation for an accentuated first sound.

The ejection sound was described by Lian and Welti (1937), and Orias and Braun Menendez, and was further studied by Leatham and Weitzman. Certain authors (Facci *et al.*) distinguish between an opening click of the pulmonary valve and an ejection sound, but their documents are not convincing.*

This ejection sound is a new component because it has no correspondence among the groups of vibrations of the normal heart. It is especially found in patients with poststenotic dilatation of the pulmonary artery; it may be absent when stenosis is severe because the early systolic jet is too small to induce a marked vibration of the pulmonary arterial wall (Fig. 92). It is less common in cases of infundibular stenosis, possibly because of lesser or absent dilatation of the pulmonary artery.

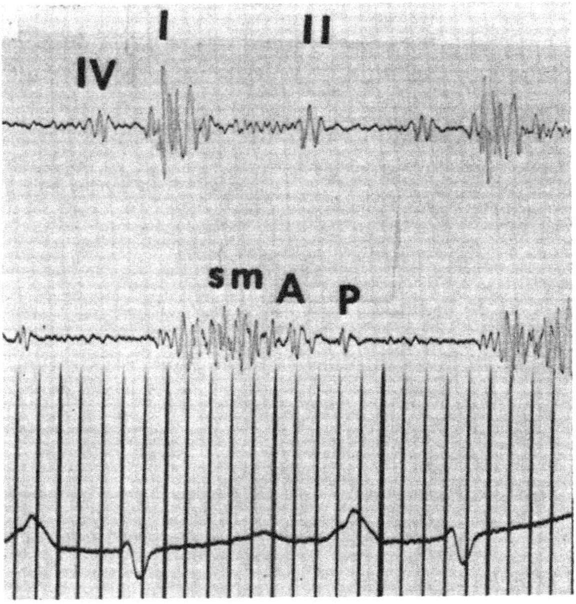

FIGURE 92. Tracings in a 26-year-old woman with severe pulmonary stenosis (RV pressure = 134/6 mm Hg; PA not entered).
(A) Tracings recorded with high pass filter at 50 Hz. Above = apex; center = 2nd left i.c.s. The apical tracing shows a large IV sound. The tracing at the base shows a wide splitting of the II sound.

Systolic Murmur. The murmur is diamond-shaped. In severe cases, the peak of the murmur occurs *after* the middle of systole, so that one has the impression of a crescendo murmur (Fig. 93). The vibrations are of various frequencies, and there may be a

*The opening click of their tracings seems to be the normal first component of the first sound.

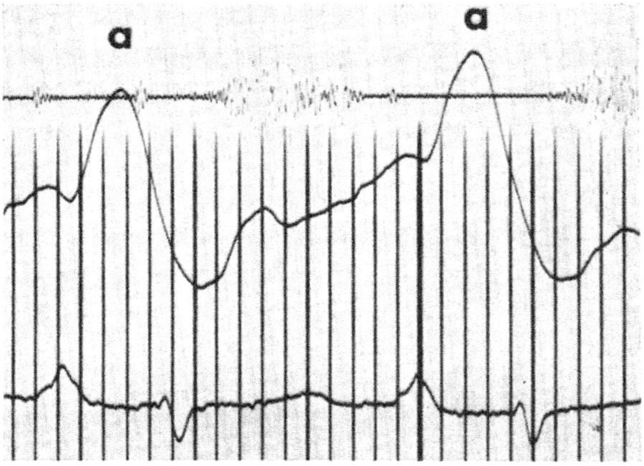

(B) Tracing recorded with high pass filter at 100 Hz over 2nd left i.c.s. Jugular tracing: very tall A wave.

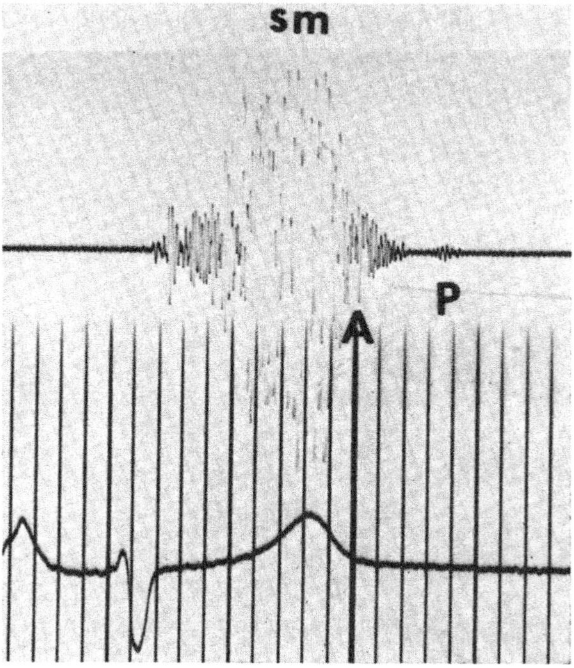

(C) Tracing recorded with high pass filter at 200 Hz over 2nd left i.c.s. Diamond-shaped murmur with late peak. The murmur continues *after* the aortic component of the II sound. Diminutive and delayed pulmonary component of the II sound.

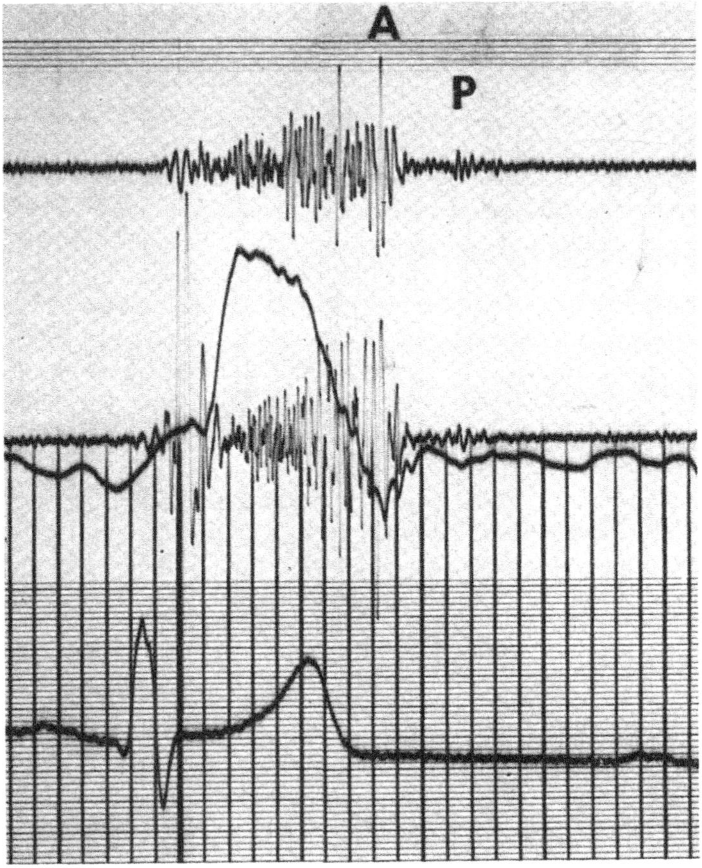

FIGURE 93. Tracings recorded in a 19-year-old boy with severe pulmonary stenosis (RV = 137/5 mm Hg; PA = 11/4 mm Hg). High pass filters at 100 Hz.
Above = 2nd left i.c.s.
Center = apex
Below = carotid tracing
Crescendo murmur. Delayed and small pulmonary component of II sound.

predominance of high-frequency components (200 to 400 Hz or higher). The murmur is recorded best in the second and third left interspaces but may have a wide diffusion, especially in children. It often radiates upward to the left clavicle and, if loud, can be heard and recorded at the two sides of the spine (posterior pulmonary area). The vibrations of the murmur may extend to the pulmonary component of the second sound (Fig. 92C). In some cases, the peak of the murmur occurs earlier over the midprecordium than over the 2nd left i.c.s. This is most likely due to some degree of associated muscular subpulmonic stenosis.

Second Sound. The second sound is typically split in all phases of respiration but splitting is not fixed. *The pulmonary component is small and delayed,* occurring from 60 to 120 msec. after the aortic component (Figs. 92 and 93). However, decrease of output in severe cases may cause such a decrease in magnitude of both components of the second sound that this sound seems to disappear. The same hemodynamic and peripheral factors that cause delay of the aortic component in aortic stenosis result in a delay of the pulmonary component in pulmonary stenosis.

Third Sound. A third sound (ventricular gallop) is not commonly recorded. When present, it is a right-sided sound and is related to right ventricular dilatation and failure. It is then recorded best over the fourth left interspace close to the sternum or over the xiphoid.

Fourth Sound. A fourth sound (atrial gallop) is common (Giroux, Toja, Vogelpoel and Schrire, 1955; Dimond and Benchimol, 1960). It is a right-sided fourth sound and is related to a mechanism similar to that already described in aortic stenosis (Fig. 92A).

Functional Tests

Inspiration, by increasing venous return, often increases right ventricular output and the magnitude of the systolic murmur of pulmonary stenosis. This may be prevented by either too severe narrowing of the valve or right ventricular failure; in such conditions the right ventricle is unable to increase the flow. The same is true of *amyl nitrite.* This drug often causes an early increase of the magniture of the murmur but may not do so if there is severe narrowing or right ventricular failure.

Subpulmonic Stenosis

Subpulmonic stenosis may be caused by eithr a congenital fibrous restriction or hypertrophic muscular narrowing of the infundibulum. This may be associated to either valvular stenosis or chronic pulmonary hypertension.

It has been stated that the murmur of subpulmonic stenosis occurs earlier than that of valvular stenosis but this is not always true.

Catheterization and angiocardiography will be needed for a complete evaluation of doubtful cases.

Coarctation (or Stenosis) of the Pulmonary Arteries

Unilateral or bilateral narrowing of the pulmonary arteries can be distinguished from pulmonary stenosis because the murmur has a later onset and may ride over the entire second sound

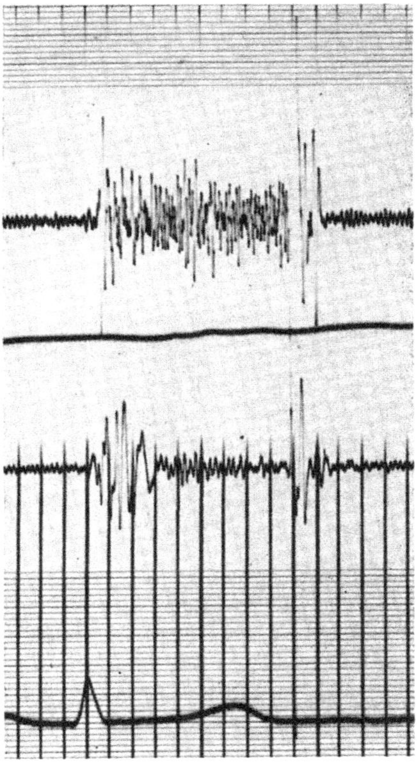

FIGURE 94. Tracings recorded in a 17-year-old boy over the 2nd left i.c.s. (upper) and at the apex (center).
There is a pansystolic murmur at the base, which is increased by amyl nitrite and is a *pulmonary flow murmur*. A diastolic-presystolic murmur (functional) appears after the inhalation.
At left = control
At right = after amyl nitrite inhalation.

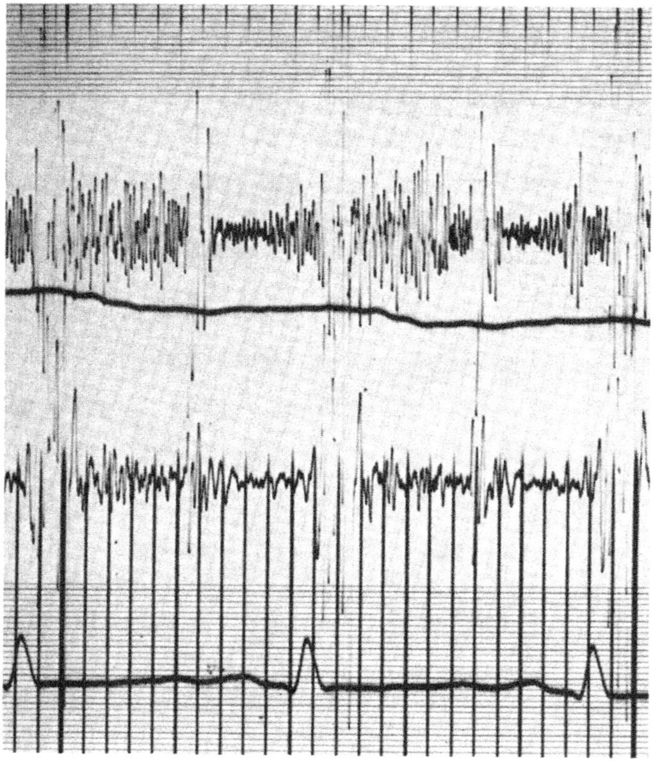

(late systolic-early diastolic murmur); it may be heard and recorded better on one side of the chest (*unilateral coarctation*) or particularly well on both sides of the chest (*bilateral coarctation*), as well as in the back.

The pulmonary component of the second sound is small, like in pulmonary stenosis. It should be remembered that, in coarctation, while the systolic pressure is high in the trunk of the pulmonary artery, the diastolic pressure is very low in this vessel.

Catheterization and angiocardiography are necessary for a complete evaluation.

Relative Stenosis

This term has been applied to cases in which there is a *pulmonary flow murmur* caused by either increased volume or increased rapidity of flow in the main pulmonary artery, or increased pulmonary artery pressure. Both usually cause some degree of dilatation of the main pulmonary artery and its stems. Typical cases are left-to-right shunts (especially atrial septal defects), idiopathic dilatation of the pulmonary artery, and pulmonary heart disease.

The murmur is usually diamond-shaped and has an early peak. Splitting of the second sound is normal for the age and, unless there is a single second sound, one can see that the pulmonary component has a normal magnitude. An ejection sound can be found in cases with increased flow.

Amyl nitrite causes an increase of the murmur (Fig. 94).

These cases will be further discussed in the various chapters dealing with the above conditions, as well as in the chapter on "innocent murmurs" (Chap. 26).

AORTIC INSUFFICIENCY

Aortic insufficiency may be of rheumatic, rheumatoid, syphilitic, atherosclerotic, traumatic, bacterial, or congenital* etiology. In addition, cystic medionecrosis of the aorta may cause this process.

The aortic diastolic murmur has been recognized since the early part of the last century; important clinical studies were published by Hope, Hodgkin, Corrigan, Wills and Leatham.

The typical murmur of this condition is described as a soft, blowing, decrescendo murmur. It may be musical (sea-gull cry, dove-coo murmur, Fig. 95). It is heard best along the left sternal

*One form of congenital aortic insufficiency is isolated (Frahm *et al.*), another is associated with a *ventricular septal defect* (Nadas *et al.*), and a third is found in the *common trunk*. In the last, the term "aortic" insufficiency is improper because the incompetent valve is the valve of the trunk, which is the equivalent of both the aortic and pulmonary valves. A *bicuspid aortic valve may be incompetent*, but this is most often the result of eithr superimposed bacterial endocarditis or the hypertension of coarctation (Runco and Booth). The *Marfan syndrome* frequently involves the ascending aorta and occasionally the aortic cusps.

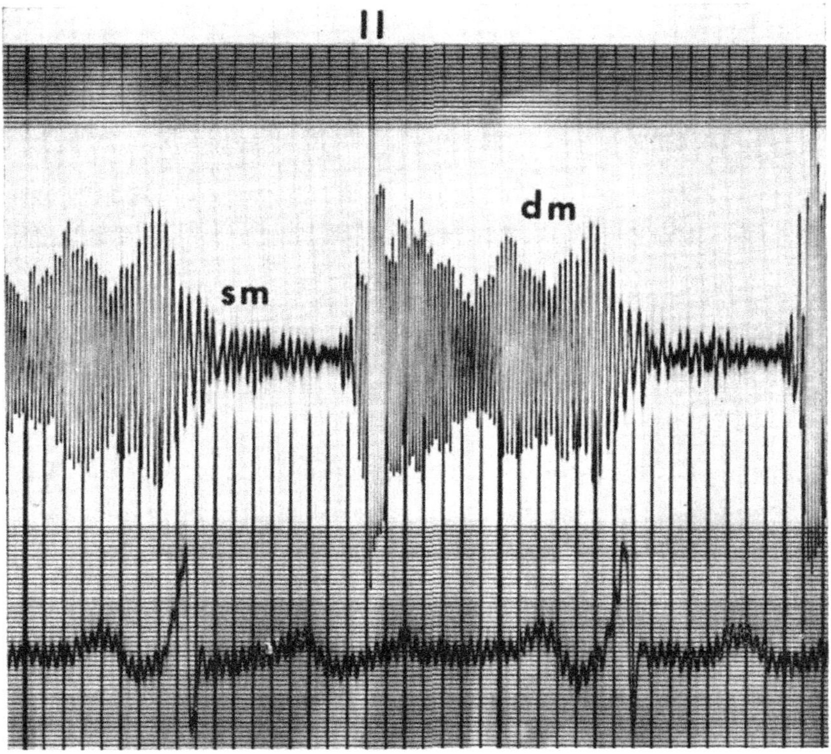

FIGURE 95. Aortic insufficiency caused by rupture of an aortic leaflet in a 54-year-old man (surgery). Musical pandiastolic murmur of medium frequency.

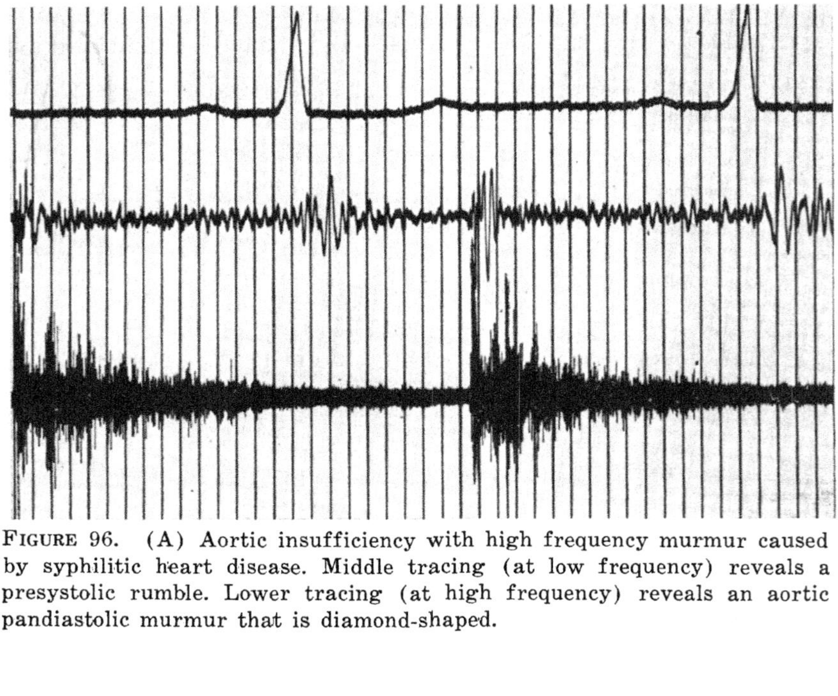

FIGURE 96. (A) Aortic insufficiency with high frequency murmur caused by syphilitic heart disease. Middle tracing (at low frequency) reveals a presystolic rumble. Lower tracing (at high frequency) reveals an aortic pandiastolic murmur that is diamond-shaped.

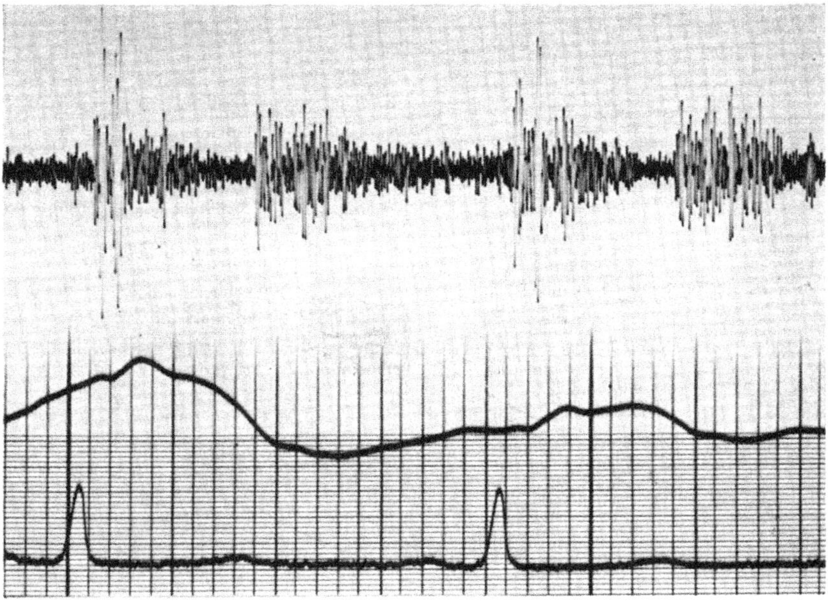

(B) Aortic insufficiency (rheumatic) with diamond-shaped systolic murmur (preceded by an ejection sound) and a diamond-shaped pandiastolic murmur.

border in cases of rheumatic etiology and has its maximal intensity at the third left intercostal space. In syphilitic aortitis, atherosclerosis, and Marfan syndrome, the enlarged ascending aorta often transmits the murmur best along the right costal border (Harvey *et al.*, 1963).

First Heart Sound. The first heart sound is usually normal but may be decreased in amplitude.* Incomplete closure of the aortic valve may cause a less rapid rise of left ventricular pressure, and thus a smaller first sound. Left ventricular enlargement might also explain this decrease, which was actually noted in a few of our cases.

*The low amplitude of the first sound, noticed by others in clinical tracings, can be also the result of inadequate recording technique. Meadows *et al.* attributed it to premature mitral valvular closure, an interpretation that is not consistent with a more modern theory of the mechanism of the first sound. In cases of aortic insufficiency, it is common to notice a small first sound when recording at the base, and a normal first sound over the left ventricular area.

Second Heart Sound. The second heart sound is often of normal amplitude. It was found reduced in amplitude in only 4.5% of our cases. It is frequently single but may show splitting. It is recorded best with medium-low filtration. Its poor clinical audibility is caused by the fact that, in the best audible range, the high frequency vibrations of the murmur seem to obscure the sound. The second sound is often of increased amplitude in cases of syphilitic origin.

The second sound may have a normal type of splitting, occasionally wide, with normal respiratory variations; it may be single; or it may be reversely split (Fig. 98). Reverse splitting was observed in 11 of our cases (10%); of these cases, 3 had LBBB and 8 had hypertensive heart disease. Thus, the latter

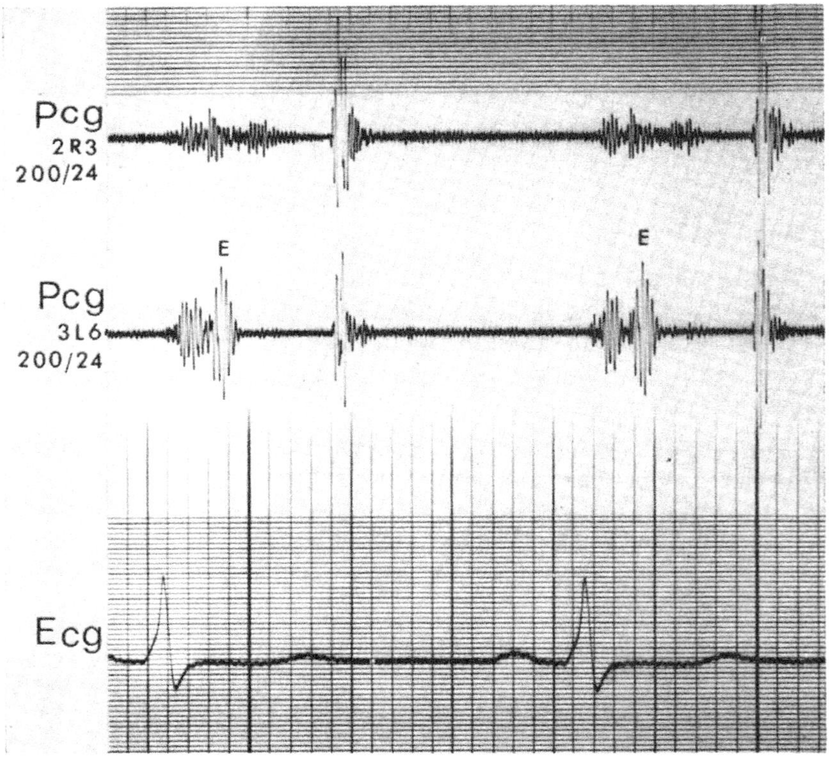

FIGURE 97. Large ejection sound in a 22-year-old woman with rheumatic aortic insufficiency.

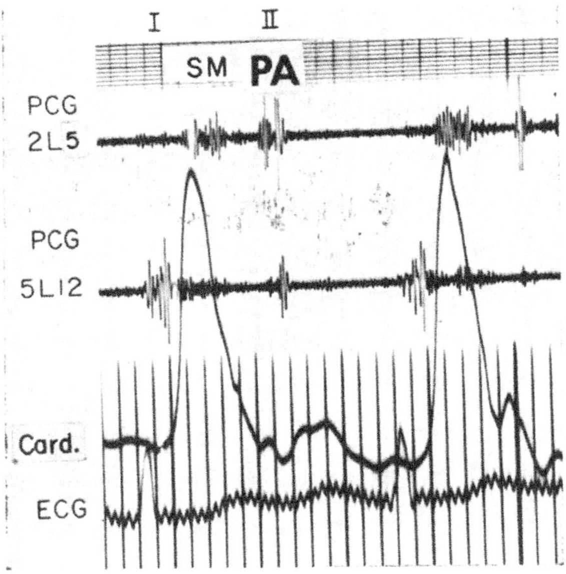

FIGURE 98. Tall and peaked apex cardiogram in a 65-year-old man with aortic insufficiency (the murmur is not revealed by this tracing). Reverse splitting of second sound.

seems to greatly favor a late occurrence of the aortic component (Argano and Luisada, 1971).

Ejection Sound. This sound was found in over 25% of our cases (Fig. 96B). However, this sound occurred in 50% of the cases of syphilitic origin (Argano and Luisada, 1971). A lesion of the ascending aorta would explain this frequency by causing more severe dilatation of the vessel.

Third and Fourth Heart Sound. The diastolic sounds are rarely present. They may occur if there is a concomitant hypertensive heart disease or severe left ventricular failure (Fig. 99).

Systolic Murmur. The systolic murmur is practically noted in all cases. It is usually an *ejection murmur* with a diamond shape and an early peak. It is an *aortic flow murmur,* related both to increased flow and dynamic dilatation of the aorta (Fig. 100). Of course, cases with even mild aortic stenosis would have both a structural and a functional cause for the systolic murmur. Occasionally there is a diamond-shaped murmur over the 2nd right i.c.s. and a decrescendo murmur at the midprecordium and apex. It is likely that the latter is caused by associated

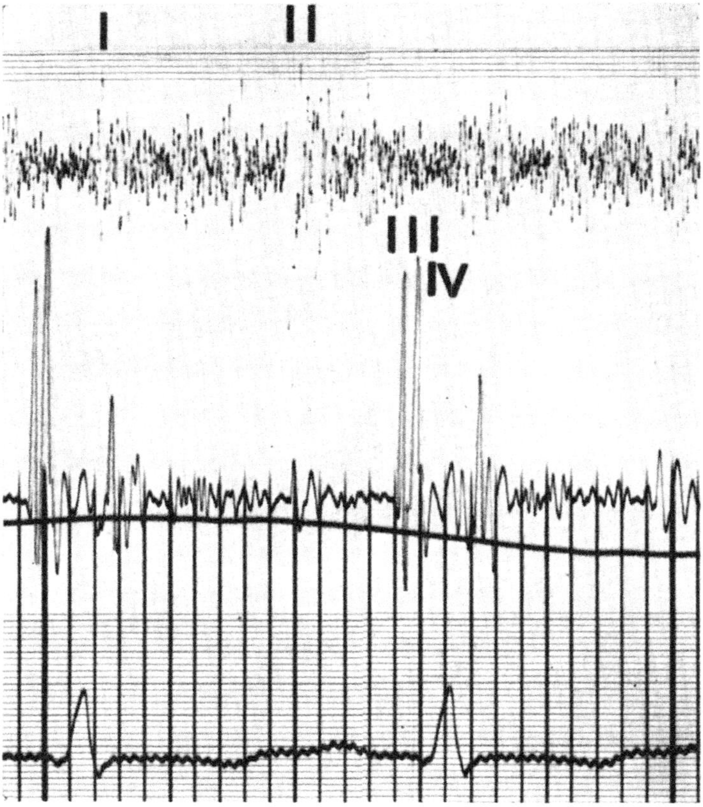

FIGURE 99. Aortic insufficiency in a 32-year-old woman. There is a gigantic summation gallop (III-IV) in the low frequency range (high pass filter = 50 Hz) and both a systolic and a diastolic murmur in the high frequency (high pass filter = 200 Hz).
Above = filter 200.
Center = filter 50.
At left = control.

mitral insufficiency of either organic or functional nature. Stretching of the papillary muscles caused by left ventricular dilatation actually often causes "relative" mitral insufficiency.

Diastolic Murmurs. Both the typical decrescendo, high-pitched murmur of aortic insufficiency and the possible diastolic rumble (*Austin Flint murmur*) may assume a variety of patterns and configurations (Figs. 96 and 99).

The most typical is a *pandiastolic murmur*, that may have

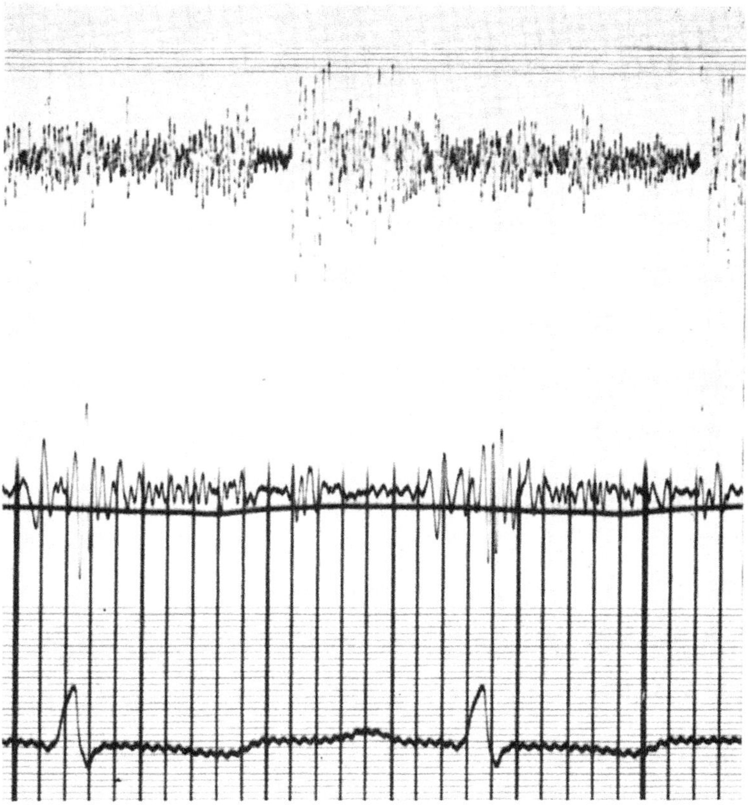

At right = after inhalation with amyl nitrite.
The drug causes disappearance of the low frequency murmur but does not change the high frequency murmurs.

either a *decrescendo* or a *diamond-shaped* appearance (31% and 21% in our series, respectively).

Tracings recorded in the low frequency bands may already reveal the existence of the murmur (Chapter 17). However, the best results are obtained by recording high frequency tracings (200 to 400 Hz). In certain cases, very high frequency tracings (1000-1500 Hz) may still reveal the murmur.

The best area for recording the "blowing murmur" of aortic insufficiency is along the left sternal border (3rd or 4th i.c.s. = aortic area). Exceptions are represented by patients with markedly dilated ascending aorta (mostly related to syphilitic but occasionally caused by atherosclerotic lesions), in which the

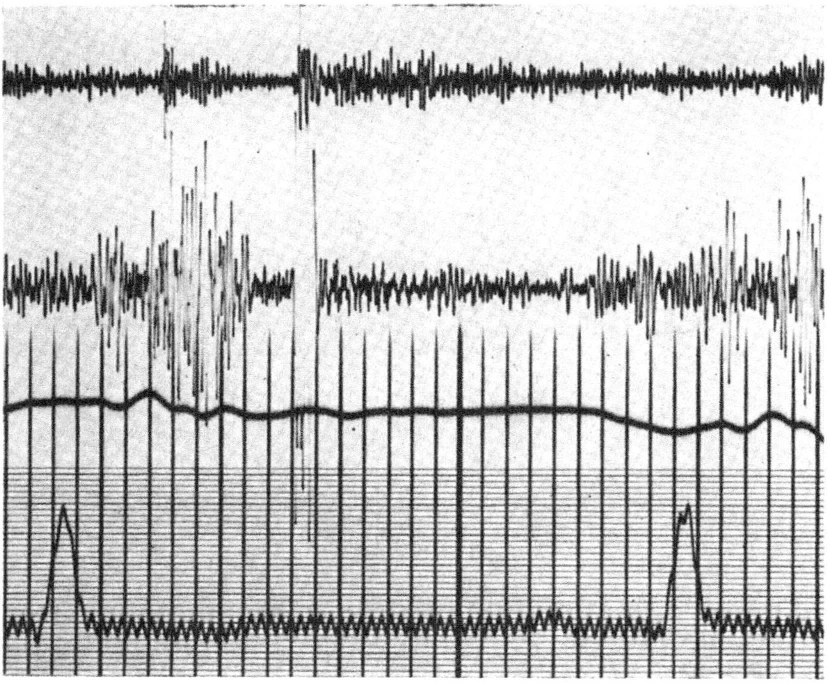

FIGURE 100. Large systolic aortic flow murmur and smaller aortic diastolic murmur.
Above = tracing over 3rd left i.c.s.
Center = tracing at the apex.
At left = control.

best record is obtained over the 2nd or 3rd right i.c.s. (extension of the aortic area).

When the murmur is diamond-shaped, the peak of the murmur coincides with both the peak of the dicrotic wave and the lowest point of the left ventricular presure curve, and also the O point of the apex cardiogram (Luisada, 1953). It is obvious that the maximum gradient of diastolic pressure across the aortic valve would occur at this time (Fig. 64).

It should be mentioned that occasionally the "blowing" murmur starts slightly before the aortic component of the second sound (McKusick, Watanabe and Sakamoto) in a phase in which the aortic valve has already closed but the aortic component of the second sound has not yet occurred.

A low-pitched rumble can be recorded over the apex or mid-

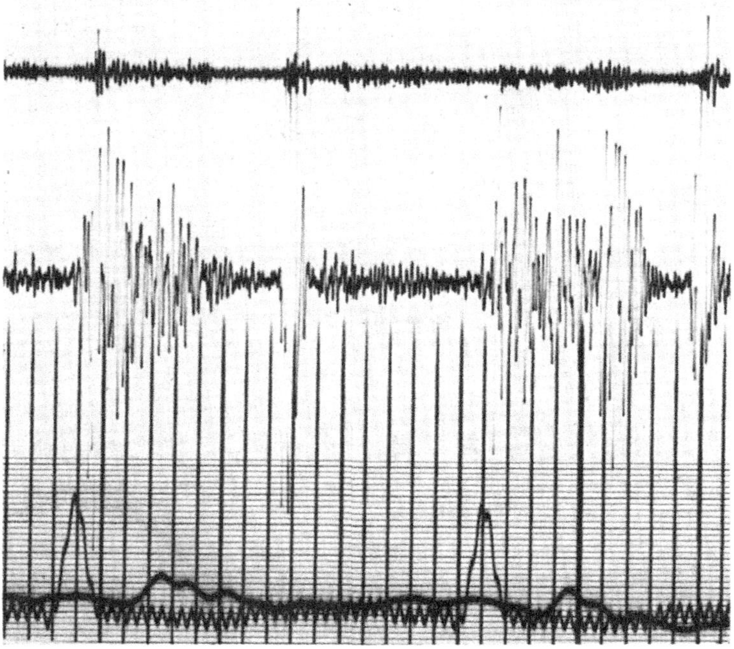

At right = after inhalation with amyl nitrite.
The drug causes an appreciable decrease of the basal murmurs but does not change those at the apex.

precordium (left ventricular area) in cases with markedly enlarged left ventricle (*Austin Flint murmur*) (Figs. 96A and 99). It may be either middiastolic or middiastolic and presystolic. It is not preceded by an opening snap. This rumble is usually well recorded between 20 and 60 Hz so that, when simultaneous records are taken at the base and apex (or over the midprecordium with two different filters), one can see early-diastolic vibrations predominating in the high frequency tracing, and middiastolic or presystolic vibrations predominating in the low frequency tracing. While the presystolic, "functional" murmur is usually low-pitched, occasionally one can find high-frequency components in this phase. It could be postulated that the vibrations of the aortic murmur are "reinforced" by ventricular vibrations, caused by atrial systole.

The possible occurrence of a functional diastolic (or diastolic-presystolic) rumble over the left ventricular area has been repeatedly mentioned since Austin Flint and was the object of several publications by our group (Luisada and Perez Montes; Luisada *et al.*, 1955, 1959, 1971). Although a good case can be made for the similarity between the functional diastolic rumble of patients with myocardial, hypertensive, or coronary heart disease with dilatation of the left ventricle and that of aortic insufficiency (White, 1926; Bland *et al.*; Coombs; Weinstein and Lev; Ravin and Darley; Luisada and Perez Montes), there is no doubt that, in the latter, a different factor is present as a result of the meeting of two streams of blood. It is open to question whether or not this is the most important element or whether approximation of the mitral leaflets (as postulated by others) causes a "functional" mitral stenosis.

Patients with a ruptured, perforated, or everted leaflet of the aortic valve usually have a *musical murmur*. In man, this is mostly observed in cases of traumatic or bacterial origin (Fig. 95). On the other hand, musical diastolic murmurs are not uncommon in animals with calcification of the aortic valve. The vibrations of the murmur then are regular and start immediately after the aortic component of the second sound.

The cooing quality of musical murmurs seems to be caused by changes in the frequency of the fundamental harmonics of the murmur (Feruglio, 1962). In comparing the type of murmur with the hemodynamic and radiological evidence of severity of disease, we have found the following: a large pandiastolic murmur is usually found either in severe insufficiency or in cases having a medium insufficiency plus left heart failure.

Intracardiac Phonocardiogram

The intracardiac phonocardiogram records best the regurgitant murmur within the left ventricle while the tracing of the ascending aorta shows a murmur of similar or lesser amplitude (Lusada and Liu, 1958, Luisada *et al.*, 1964) (Fig. 101). According to Feruglio (1962), the intracardiac phonocardiogram reveals best the changes in frequency of the murmur whereas the external tracing usually shows only the fundamental harmonic of the murmur, which is conducted to the chest wall.

Apex Cardiogram

The apex cardiogram shows, in severe regurgitation, a typically high and conical pulsation (*choc en dome*) (Fig. 98), as first

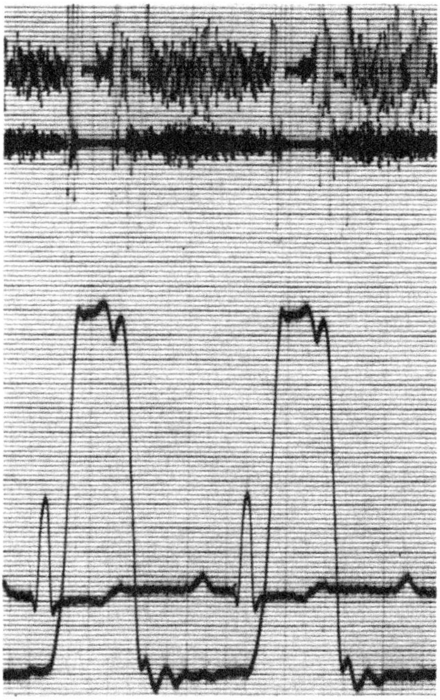

FIGURE 101. Experimental aortic insufficiency in a dog.
Above = intracardiac phonocardiogram (LV) with an FCT microphone (high pass-filter at 400 Hz; slope 24 db/octave).
Center = intracardiac phonocardiogram with an A.E.L. microphone (no filter).
Below = ECG, left ventricular pressure.

shown by Weitz. In severe cases, the phase of rapid filling is accompanied by a high wave in early diastole because the blood regurgitating from the aorta increases the rapidity of filling.

AORTOGRAMS

Tracings recorded over the second right interspace or the suprasternal notch reeval a high systolic wave. The *carotid tracing* also shows a typical pattern characterized by a rapid rise, a rapid drop, and a low and small dicrotic wave (Corrigan pulse, *collapsing pulse*). If aortic insufficiency is associated with stenosis, a double-peak pulse is often recorded (*pulsus bisferiens*).

Dilatation of the aorta is revealed by typical changes of the phonocardiogram in the low frequency bands (Chapter 17).

DRUG TESTS

Methoxamine increases the amplitude of the aortic diastolic murmur.

Amyl nitrite decreases the diastolic murmur but may leave unchanged the systolic murmur. In some cases, the high frequency murmurs are unchanged by this test (Fig. 99).

RELATIVE AORTIC INSUFFICIENCY

Relative aortic insufficiency is likely to occur in patients with severe, systemic hypertension. These patients present a small and short blowing diastolic murmur that often disappears when blood pressure is lowered by means of hypotensive agents. The murmur seems caused by dilatation of the aortic ring. It is recorded as a short series of vibrations in decrescendo that start soon after a large, snapping aortic component of the second sound.

A typical murmur of aortic insufficiency appears sometimes in children during the acute stage of rheumatic fever and disappears later with the recovery of the patient. It is not known whether this is caused by dilatation of the aortic ring or edema of the aortic leaflets, both of which are reversible processes.

A minimal murmur of aortic insufficiency is occasionally found in normal subjects. One may speculate whether this is caused by *congenital fenestrations of the aortic valve.*

PULMONARY INSUFFICIENCY

Pulmonary insufficiency may be the result of a rare congenital lesion* or the stretching of the valvular ring by long-lasting pulmonary hypertension. Bacterial or rheumatic endocarditis may cause it in extremely rare cases.

The murmur of pulmonary insufficiency is similar to that of aortic insufficiency, except that it starts barely before, or soon after, the pulmonary component of the second sound instead of having the same relationship with the aortic component (Fig.

*Pulmonary insufficiency is occasionally observed as a congenital, isolated lesion. Such a case was observed by us. It may be associated with a high ventricular septal defect or with the dilatation of the pulmonary artery that occurs in the Eisenmenger syndrome. In the last case, the insufficiency may be caused by fibrosis of the valve cusps but is usually caused by dilatation of the valvular ring, thus falling in the category of "relative" pulmonary insufficiency.

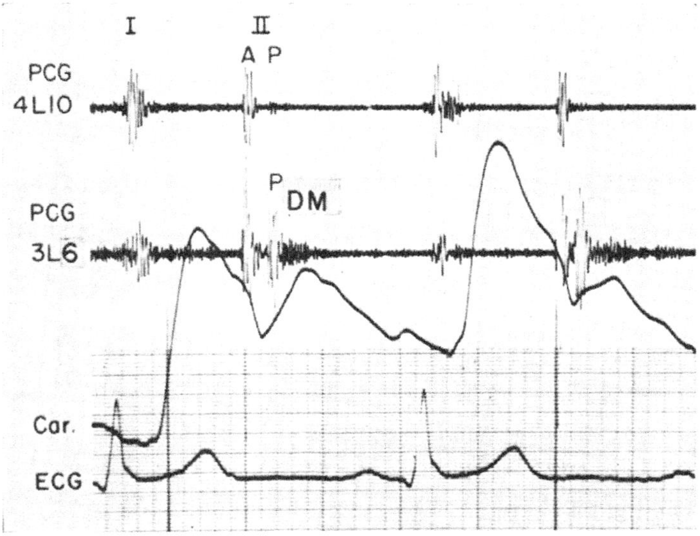

FIGURE 102. Tracings of a 22-year-old man with congenital pulmonary valve insufficiency (PA end-diastolic pressure was only 4 mm higher than RV end-diastolic pressure). Early diastolic murmur starting after the pulmonary component of the second sound.

102).** This fact is revealed by high-speed phonocardiography*** but a tracing at 200 or even 100 mm per second may be adequate. The murmur is usually in decrescendo, and only occasionally does it show the short-crescendo-long-decrescendo pattern that is common in aortic insufficiency. The vibrations of this murmur may be of a lower frequency (rumbling murmur) than in aortic insufficiency (Luisada and Wolff) but this can-

**In recent experimental observations on pulmonary insufficiency caused by high pressure in the pulmonary artery, we have noted that the murmur started soon after the aortic component of the II sound. This strange fact seems related to a prolongation of the ejection phase by a short phase of "injection" due to the extreme tension existing in the pulmonary artery and causing early onset of regurgitation as soon as right ventricular pressure drops but before the pulmonary component of the second sound. This fact unfortunately renders more difficult the differentiation from the murmur of aortic insufficiency.

***The first author to suggest this procedure was Sterz. Curiously enough, and making reference to a special, unnecessary technical procedure (recording on tape and playing it back with a slower speed of tape), he called it *slow-motion phonocardiography*. It is unfortunate that occasionally the aortic component is prolonged and the murmur seems to start immediately after it.

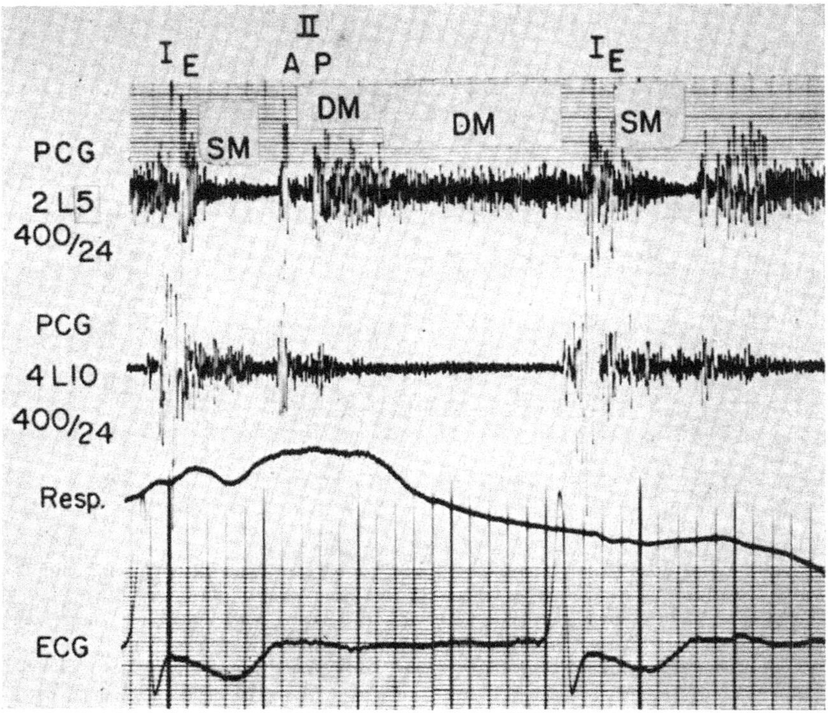

FIGURE 103. Tracings of a 45-year-old woman with mitral stenosis and "relative" pulmonary insufficiency *(G. Steell murmur).*
The murmur recorded over the 2nd left i.c.s. starts with the pulmonic component and is increased by inspiration.

not be considered a rule. In the cases associated with pulmonary hypertension, the pulmonary component of the second sound is large in spite of the regurgitation. The most typical murmur is found in cases of mitral stenosis with pulmonary arteriosclerosis having a pulmonary ssytolic presure similar to or higher thon the aortic pressure (*Graham Steell murmur*) (Fig. 103). In such cases, the pulmonary insufficiency is caused by dilatation of the ring of the pulmonary valve, a fact that may not be apparent at autopsy. A relative pulmonary insufficiency is then the cause of the murmur.

Differential Diagnosis Between Aortic and Pulmonary Diastolic Murmurs

Several differential criteria are based on data of blood pressure, roentgenography, fluoroscopy, catheterization, and angiocardiography. Other differential data are based on the changes of the murmur caused by respiration and the methoxamine and amyl nitrite tests.

Respiration. The pulmonary murmur is louder in inspiratory apnea and vice versa. The aortic murmur behaves in the opposite fashion or fails to show respiratory changes.

Methoxamine Test. This drug increases the magnitude of the aortic regurgitant murmur but does not change the pulmonary murmur.

Amyl nitirite Test. This drug increases the magnitude of the pulmonary regurgitant murmur in an early phase while it decreases that of the aortic murmur in such phase.

Other Early Diastolic Murmurs Simulating Aortic Insufficiency

A few conditions may cause sound vibrations that fall (or seem to fall) in early diastole, thus simulating on auscultation aortic insufficiency.

Severe Mitral Insufficiency. Severe mitral insufficiency is accompanied by a large flow from the left atrium to the left ventricle. This flow has he most rapid phase in early diastole, where an early diastolic gradient is typically found (Nixon and Wooler). In some of these cases, it is possible to observe that a few early-diastolic vibrations of high frequency occur immediately after the aortic component of the second sound as a prolongation of the pansystolic murmur. Although this fact can be recognized easily in the phonocardiogram because of the continuation of the systolic murmur into the early diastolic murmur, occasionally there may be a diagnostic problem.

Severe Aortic Stenosis. If there is a reverse splitting of the second sound, the vibrations of the systolic murmur continue beyond the large and loud pulmonary component of the second sound. As the latter is the sound that is heard best (or is the only one heard), part of the murmur will seem to occur in diastole. The graphic diagnosis is based on the location of the aortic component of the second sound and by observing its relationship to the incisura of the carotid pulse.

Severe Pulmonary Stenosis. The pulmonary component of the second sound is faint and markedly delayed. The murmur in rare cases rides over the large aortic component and simulates an early-diastolic murmur.

Patent Ductus Arteriosus with pulmonary hypertension. Although in some cases only a systolic murmur is present, in other cases the murmur may continue slightly after the second sound. The graphic diagnosis is based on the recognition that there is only one murmur that is largest in late systole and continues after the second sound.

Chapter 9

Left-to-Right Shunts: The Continuous Murmur and Coarctation of the Aorta

THE AUSCULTATORY AND PHONOCARDIOGRAPHIC aspects of the non-cyanotic forms of congenital heart disease have been clarified in the last twenty-five years so that now they contribute a great deal to the clinical diagnosis.

ATRIAL SEPTAL DEFECT (ASD)

The auscultatory and phonocardiographic findings of ASD have been studied by Wood (1956), Barber *et al.*, Leatham and Gray, Eisenberg and Hultgren, and Dimond and Benchimol (1959). It should be kept in mind that *partial anomalous venous return* is characterized by similar data.

The main cause of the cardiovascular changes resides in a left-to-right shunt at the atrial level. On the one hand, this causes a right ventricular and pulmonary flow that is much greater than the left ventricular and aortic flow. On the other, it results in less sensitivity to respiratory changes because the two atria communicate and represent a kind of joint and balancing "reservoir." Right ventricular hypertrophy is constant, and right bundle branch block may be occasionally present.

It was considered in the past that increased right ventricular flow would result in a longer duration of right ventricular systole. On the contrary, recent studies of Kumar and Luisada

(1971) have shown that right and left ventricular systoles end with a similar relationship to that occurring in normal people and that the observed changes result only from the greater flow of the pulmonary artery.

First Heart Sound. The first heart sound in ASD was considered to be typically "split" because its normal first component (so-caled "mitral") was followed by another (so-called "tricuspid") (Leatham and Grey, 1956). This fact was considered as the result of the right ventricular overload.

However, Zakrzewski *et al.* have studied in our laboratory the dynamics of cases of ASD and shown that, whenever present,

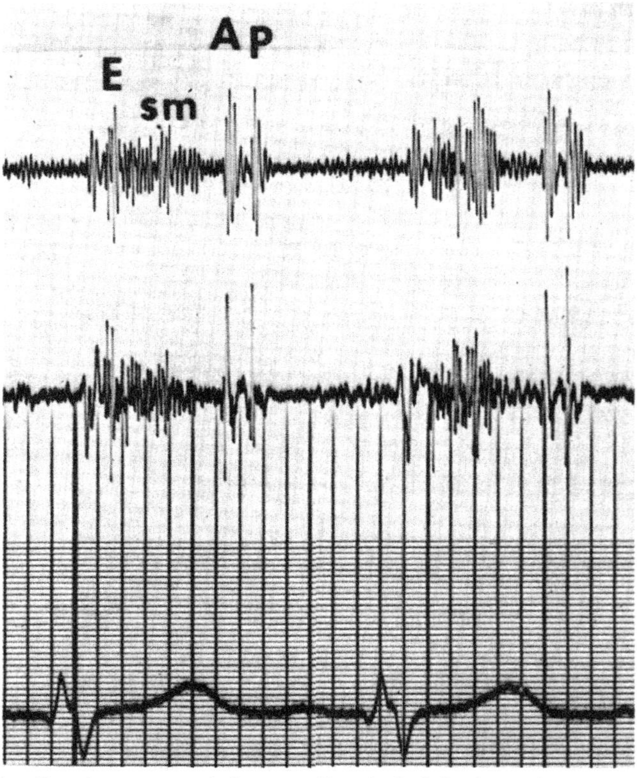

FIGURE 104. Tracings recorded over the 2nd left i.c.s. in a 3-year-old child with atrial septal defect. The upper tracing is a velocity PCG; the middle tracing is an acceleration PCG. Both are filtered in the 50-400 Hz band. Ejection sound, diamond-shaped systolic murmur, widely split second sound. Small vibrations in presystole.

the second group of vibrations of the first heart sound coincided with or followed the opening of the pulmonic valve. This was confirmed by Plass *et al.* Thus, it could only be considered as a *pulmonary ejection sound,* and not a sound related to the right ventricle or the tricuspid valve (Figs. 104-106).

Second Heart Sound. Wide and "fixed" splitting of the second sound is typical of this condition (Figs. 104 and 105). This splitting is of the normal type (A,P) and is characterized by the fact that, clinically, it is audible both in inspiration and expiration. However, small respiratory variations in the degree of splitting can be often observed in the phonocardiogram, occasionally approaching normal limits. The dynamic reasons for the wide splitting have been studied by Kumar and Luisada (1971). They ascertained that the increased distance between

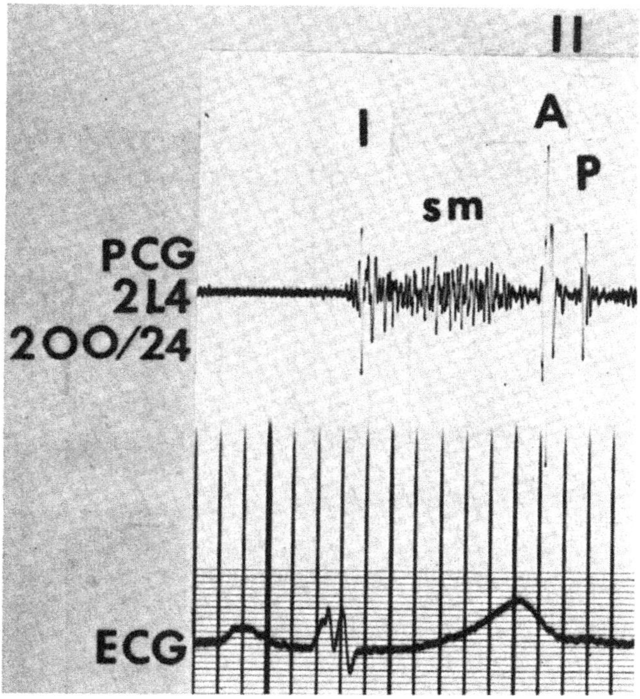

FIGURE 105. Ten-year-old child with atrial septal defect (secundum type). Velocity tracing recorded over the 2nd left i.c.s. Diamond-shaped systolic murmur, wide splitting of the second heart sound.

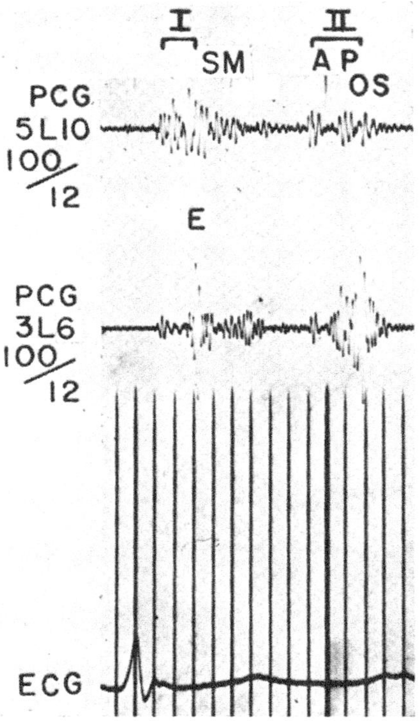

FIGURE 106. Tracings recorded in a 39-year-old woman with ASD. (Catheterization; normal pressures.)
Systolic murmur in decrescendo, wide splitting of the second sound, and opening snap (from the right heart?) are noted in the upper tracing, recorded at the apex.
An ejection sound is recorded at the base (middle tracing).

the aortic and pulmonic components is caused, not by increased duration of right ventricular systole and delayed closure of the pulmonic valve (as maintained by previous authors), but by a peripheral factor: the distended pulmonary artery has a greater compliance and delays the rebound over the previously closed pulmonary valve. This was demonstrated for both cases of "secundum" and of "primum" type. The absence of respiratory variations is of course due to the "common atrial reservoir"; increased inspiratory flow to the right atrium decreases the flow coming from the left atrium and vice-versa, so that changes in flow related to respiration are minimized.

In certain cases, right bundle branch block contributes to the wide and fixed splitting of the second sound.* When this block is present, right ventricular systole actually ends after the left. Following repair of an ASD without RBBB, the splitting may disappear but it usually persists. This fact can be caused by two mechanisms; (1) dilatation of the pulmonary artery persists; (2) surgically-induced RBBB has developed. In the first instance, splitting may disappear some months or years after surgical repair. In the second, it persists.

Opening Sound. This was described by Aravanis and was often found by us (Fig. 106). It may be related to the increased dynamics of the right heart. On the other hand, instead of being a "tricuspid" opening snap, it may be a "mitral" opening snap, i.e., an extra-sound related to left atrial volume and dynamics (see Chap. 7). Actual demonstration of the side of origin is still needed.

Systolic Murmur. In ASD of the *secundum* type, a systolic murmur of moderate or medium intensity is nearly always found. This murmur is crescendo-decrescendo in type, and is maximal over the pulmonary area (Figs. 104 and 106). It is obviously a *flow murmur,* related to the greater volume of blood that is ejected in the pulmonary artery and to the dilatation of this artery. The phonocardiogram shows a diamond-shaped murmur of moderate amplitude composed of various frequencies. However, medium low-frequency vibrations are fewer than in the murmur of pulmonary stenosis. This murmur may start after an *ejection sound* that is probably related to the dilatation of the pulmonary artery.

In ASD of the *primum* type, the systolic murmur is larger, is often pansystolic, and is recorded best over the third and fourth left interspaces, as well as over the apex. Actually, there are two different murmurs. One is caused by either mitral regurgitation (cleft mitral valve) or a shunt between left ventricle and right atrium or ventricle (Nadas, 1959); the other is a flow murmur of the pulmonary artery.

In cases with extremely severe shunts, a gradient develops across the pulmonary valve. In these cases, the flow murmur is probably increased by "relative pulmonic stenosis."

*In the past, it was assumed that bundle branch block was typically present in ASD. In fact, many of the electrocardiograms have the pattern of right ventricular hypertrophy (diastolic overload type). However, some of the cases actually have right BBB.

Presystolic Murmur. This murmur was observed by Luisada *et al.* (1959) in several patients with large shunts. It is recorded over the right atrial area, is made of medium low-pitched vibrations, is well separated from the first sound, and is *not in crescendo*. It is difficult to state whether this murmur (Figs. 104 and 107) is caused by acceleration of blood through the defect or by "relative tricuspid stenosis." However, intracardiac phonocardiography seems to favor the former (see later).

Diastolic Rumble. This rumble was first observed by Nadas and Alimurung and was then confirmed by others. It is a low-pitched rumble, recorded over the third and fourth left interspaces close to the sternum (right ventricular area), and is obviously an A-V flow murmur (so-called murmur of "relative tricuspid stenosis"). It starts relatively late after the second sound and may be initiated by a large third sound of right ventricular origin.

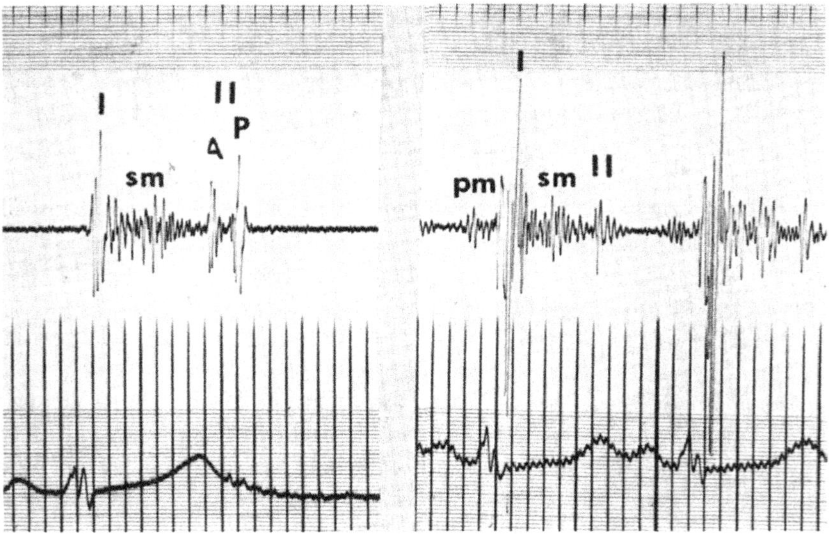

FIGURE 107. Amyl nitrite test in a child with ASD. After inhalation (at right), one can see a presystolic murmur and a somewhat decreased systolic murmur. The pulmonary component of the second sound has disappeared.

Soft, Blowing, Early Diastolic Murmur. Occasionally observed, this soft, blowing, early diastolic murmur is considered evidence of pulmonary insufficiency caused by stretch of the pulmonary ring (Barber *et al,*; Nadas, 1959; Dimond and Benchimol, Mouquin and Py).

Functional Tests

The systolic murmur of ASD is often increased by *amyl nitrite* and by inspiration. However, the amount of blood passing through the shunt may be so large that neither the drug nor inspiration may be able to further increase it. In certain cases, the murmur even decreases (Fig. 107). The pulmonary component of the second sound may disappear after amyl nitrite. A presystolic, non-crescendo murmur may appear (Fig. 107).

Intracardiac Phonocardiography

Intracardiac phonocardiography reveals that, in patients with left-to-right shunt, the systolic murmur is recorded best in the pulmonary artery; however, it is also recorded in the right ventricle and atrium (Liu and Jacono, Plass *et al.*). A *fourth sound* or a *presystolic murmur* is often recorded in the chambers of the right heart (Liu and Jacono; Luisada *et al.*, 1964), and especially in the right atrium at the level of the shunt (Guenther *et al.*, 1971).

VENTRICULAR SEPTAL DEFECT (VSD)

In ventricular septal defect, the shunt is between the ventricles. If the communication is at a high level, the blood shunts practically into the pulmonary artery. The amount of shunted blood is usually less than in ASD in contrast with the murmur that is louder.

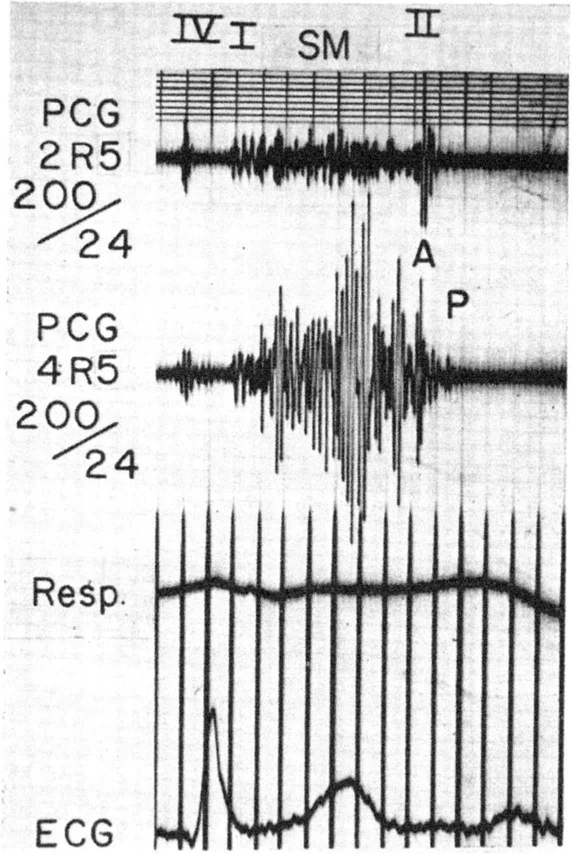

FIGURE 108. Tracings of a 39-year-old woman with VSD. High-frequency IV sound. Large, diamond-shaped systolic murmur. The vibrations of the latter end slightly after the aortic component of the second sound.

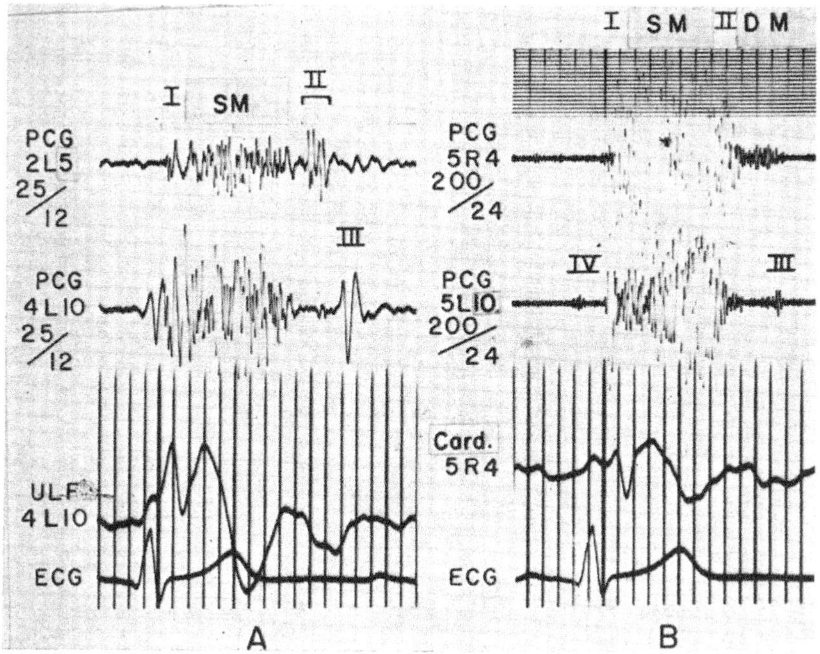

FIGURE 109. Tracings of a 37-year-old man with VSD. Large III sound (A and B); closely split II sound (A); large systolic murmur, which is crescendo at apex (B), pansystolic at the right of the sternum (B, upper tracing). Early-diastolic murmur at the right of the sternum (B, upper tracing).

Aortic pressure = 107/79, pulmonary pressure = 38/19.

A large, high VSD was surgically repaired a few weeks later.

Heart Sounds

The *first sound* in VSD may be of poor intensity but is often normal. The *second sound* is fairly widely split but shows respiratory variations (Figs. 109 and 112). One of the reasons why this fact is not usually recognized is that the aortic component may be confused with the end of the systolic murmur both on auscultation and in graphic tracings (Fig. 112 C). The pulmonary component is often larger than normal (Nadas, 1959). However, if there is severe pulmonary hypertension, a single, large second sound is often observed. A *third heart sound* may be present (Fig. 109).

Murmurs

Pansystolic Murmur. The typical murmur of VSD is a pansystolic murmur, heard over the midprecordium, well transmitted toward the right (right ventricular area), and less well transmitted to the apex and the left axilla (lef ventricular and left atrial areas). This murmur has been well known since the description of Roger. The murmur starts with the first sound, contains medium-low, medium-, and high-frequency vibrations,

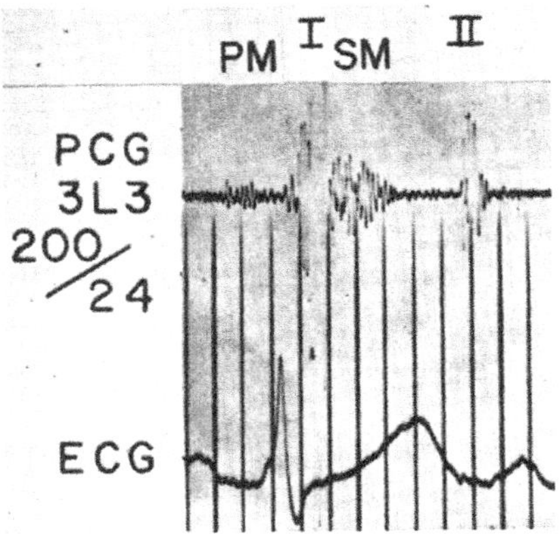

FIGURE 110. Tracings of a 4-year-old child with VSD, probably in the muscular part of the septum. The systolic murmur ends at mid-systole. Functional presystolic murmur (catheterization).

lasts until the aortic component of the second sound (Figs. 109 and 111), or may continue slightly beyond it (Fig. 109 B). The predominant frequency seems to be at or above 200 Hz, both in the *muscular* and in the *membranous* septal defetcs. The mur-

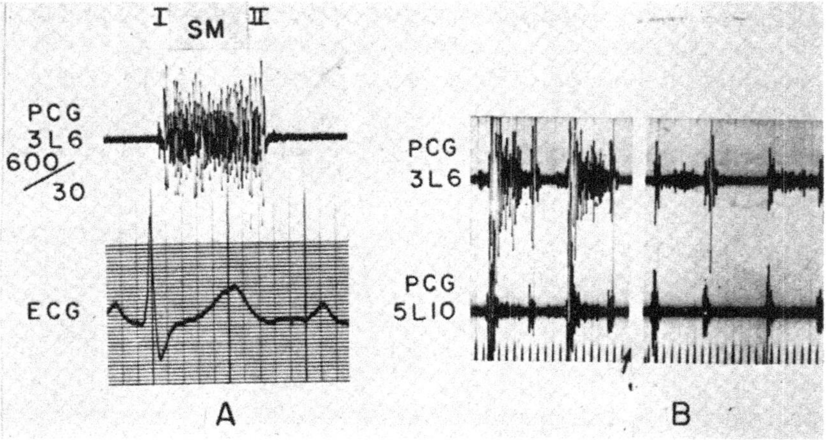

FIGURE 111. Tracings of a 16-year-old girl with a small VSD. Typical pansystolic murmur (A) that decreases with amyl nitrite (B).

mur may be shorter (early- and mid-systolic) if either the opening is small or is in the muscular part of the septum, and thus is liable to close with systolic contraction (Fig. 112). Some differentiation from the pansystolic murmur of mitral insufficiency can be made (apart from the area of maximal intensity and the radiation of the murmur) because the murmur of VSD has a greater number of medium-frequency vibrations, i.e., is larger in the 30 to 100 Hz band. High-frequency tracings reveal more often a diamond-shaped configuration in contrast with medium-frequency tracings. Small muscular septal defects have often a shorter murmur terminating in the central third of systole (Figs. 110 and 112A,B). In contrast, in membranous septal defects, the murmur is longer and may extend, beyond the aortic component, to the pulmonary component of the second sound. The murmur has a similar amplitude throughout systole (*plateau murmur*) but may have high and low groups of vibrations (*concertina murmur*).

The factors that govern the intensity of the murmur are the diameter of the defect and the pressure gradient between the

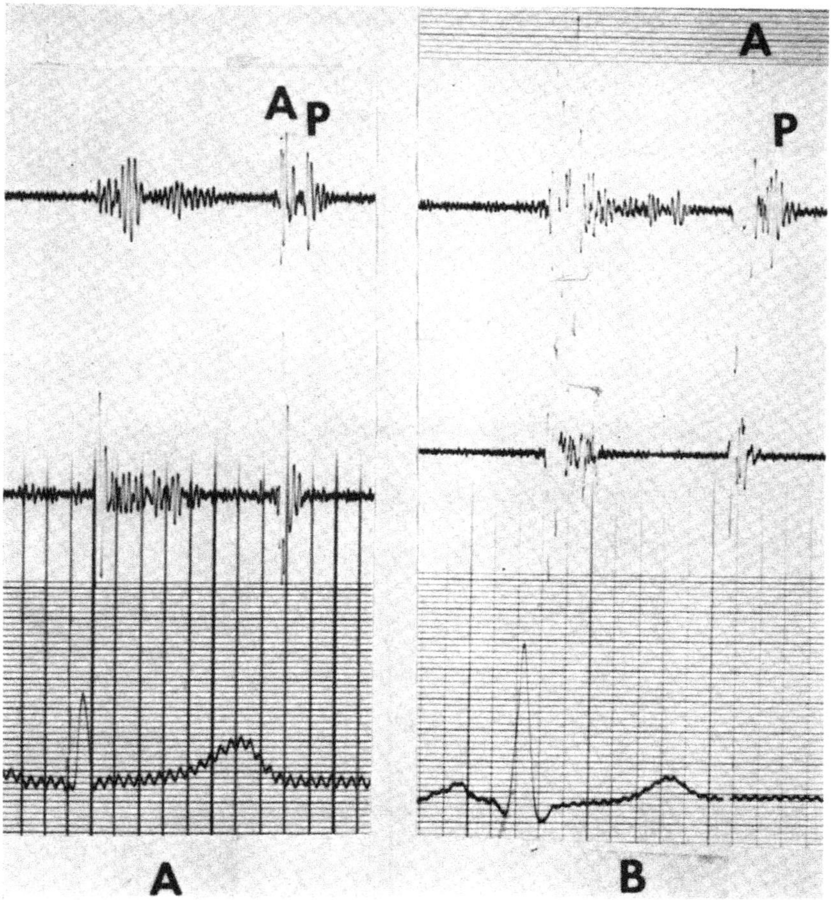

FIGURE 112. (A) Small ventricular septal defect in a 10-year-old boy. Wide splitting of second sound. Upper tracing = 2nd left i.c.s. Short systolic murmur.
(B) Ventricular septal defect with medium-severe shunt in a 20-year-old man. Upper tracing = 2nd left i.c.s. Wide splitting of second sound. Short systolic murmur.

two ventricles, the latter being related to both systemic and pulmonary resistances. It is obvious that, the higher the pressure in the right ventricle, the lesser is the magnitude of the systolic murmur.

In patients with pulmonary hypertension, the murmur is fainter (as a result of a smaller flow through the shunt) and diamond-shaped, and the second sound is single and loud (Mannheimer *et al.*; Py; Bleifer *et al.*, 1960). Some patients with large shunts, and others having a small shunt and mild pulmonary hypertension, may also have a diamond-shaped murmur (Hollman *et al.*, Craige). In some cases, these characteristics may be

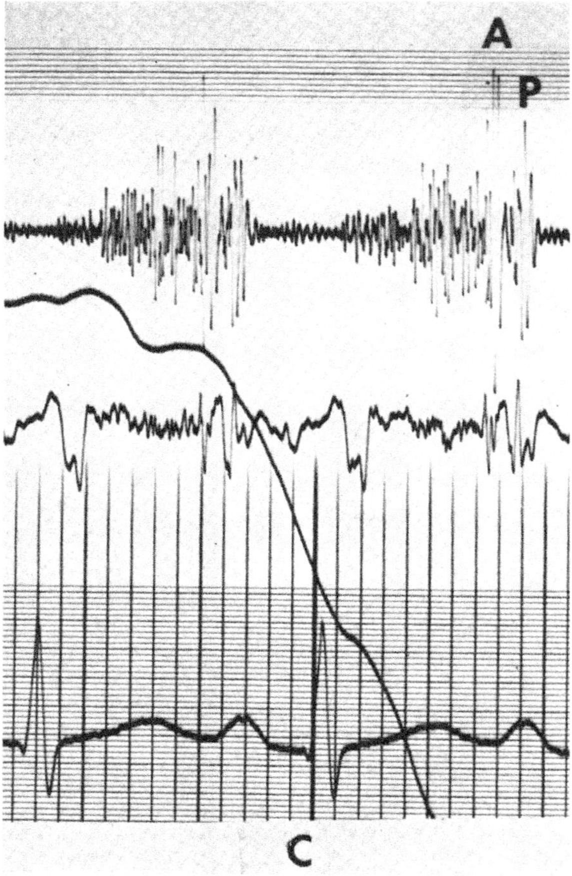

(C) Ventricular septal defect with large shunt in a 9-year-old girl. Tracings at the 2nd left i.c.s. Upper tracing = acceleration 80-600 Hz. Center tracing = velocity, unfiltered. Large systolic murmur terminating with the aortic component of the second sound. Wide splitting of second sound.

due to the superimposition of a pulmonary, diamond-shaped, systolic murmur on the murmur of the defect (Leatham and

Segal). However, the shape of the murmur can be explained also by hemodynamic factors (Hollman *et al.*) (Fig. 65).

Mid-diastolic or Presystolic Rumble. A mid-diastolic or presystolic rumble can be recorded near the apex. It is probably a mitral flow murmur, related to the diastolic overload of the left ventricle (Nadas, 1959; Holman *et al.*).

Soft, Blowing, Early-diastolic Murmur. A soft, blowing, early diastolic murmur of pulmonary regurgitation was described by Lessof in 40 per cent of his patients with pulmonary pressure higher than 30 mm Hg. A similar murmur may be caused by

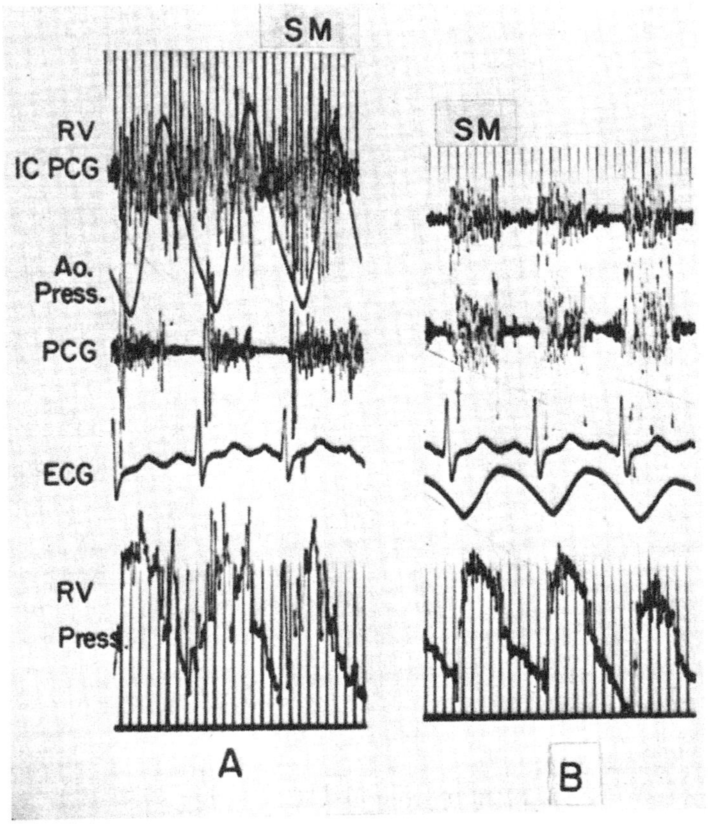

FIGURE 113. Intracardiac and external phonocardiograms in a 2-year-old child with ventricular septal defect. The intracardiac murmur (upper tracing) is much larger in the right ventricle (A) than in the pulmonary artery (B).

associated aortic insufficiency (Laubry and Pezzi, 1921; Claypool *et al.*; Wood *et al.*; Hollman *et al.*). The former is usually the result of a stretch of the pulmonary ring caused by pulmonary hypertension, whereas the latter is caused by an associated deformity of the aortic valve (bicuspid aortic valve, herniation of an aortic leaflet into the VSD, or aneurysm of a sinus of Valsalva) (Fig. 109 B),

FUNCTIONAL TESTS

The murmur of VSD is typically increased by *pressor amines* and decreased by *amyl nitrite* (Fig. 111). This fact is due to the increased amount of blood passing through the shunt whenever left ventricular pressure is raised, and vice versa (Cumming; Bousvaros; Schrire *et al.*, 1961).

INTRACARDIAC PHONOCARDIOGRAPHY

Intracardiac phonocardiography reveals that the typical systolic murmur is larger in the right ventricle than in the left ventricle or pulmonary artery (Fig. 113) (Feruglio and Gunton; Sears *et al.*; Luisada *et al.*, 1964). This technique may permit localization of the murmur in patients with small shunts, in which both catheterization and selective angiocardiography fail to demonstrate the lesion.

PATENT DUCTUS ARTERIOSUS

MURMUR

The typical finding of patent ductus arteriosus is present in the uncomplicated form (no associated lesions, no pulmonary hypertension), in both adults and children (Haring *et al.*, 1954). It consists of a loud machinery murmur, best head and recorded in the first and second left interspaces but also transmitted to the left supraclavicular space (in children), the suprasternal notch, and the second right interspace. The murmur is *systolic-diastolic* but seems to be louder in systole on auscultation. In phonocardiograms, it is apparent that it has the greatest magnitude in late-systole and early-diastole because it "rides over" the second sound without interruption (Figs. 114 and 115). It is important to note that *there is no gap between the systolic and diastolic phases of the murmur.*

The characteristics of this murmur were described by Routier and by Hubbard, and were then confirmed by many others,

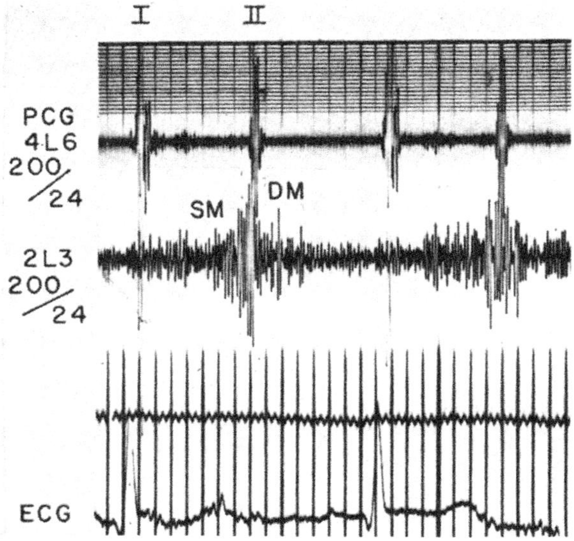

FIGURE 114. Tracings of a 6-year-old child with patent ductus. The middle tracing is recorded at the 2nd left i.c.s. and shows a continuous murmur with late systolic-early diastolic accentuation.

including Luisada (1953), Haring *et al.* (1954), Nadas (1959), Kjellberg *et al.*, Neill and Mounsey, Fishleder and Friedland, and De Micheli *et al.* This murmur is caused by the blood flowing from the aorta to the pulmonary artery through the ductus. The amount of flow is proportional to the pressure gradient between the aorta and the pulmonary artery, and so is the magnitude of the murmur.

In patients with pulmonary hypertension, the pressure gradient between the aorta and pulmonary artery decreases. Therefore, with increasing pulmonary pressures, first the diastolic phase of the murmur disappears; then the systolic phase of the murmur decreases; and finally even this phase may disappear (systemic and pulmonary pressures are equal) (Nadas, 1959). When only a systolic murmur is present, the diagnosis may be made by observing a late-systolic murmur in crescendo or a murmur overriding the second sound (Fig. 116).

Intravascular phonocardiography recorded during catheterization typically reveals the vibrations of the murmur (Feruglio, 1960), which are maximal in the ductus and less large in the aorta and pulmonary artery (Fig. 117).

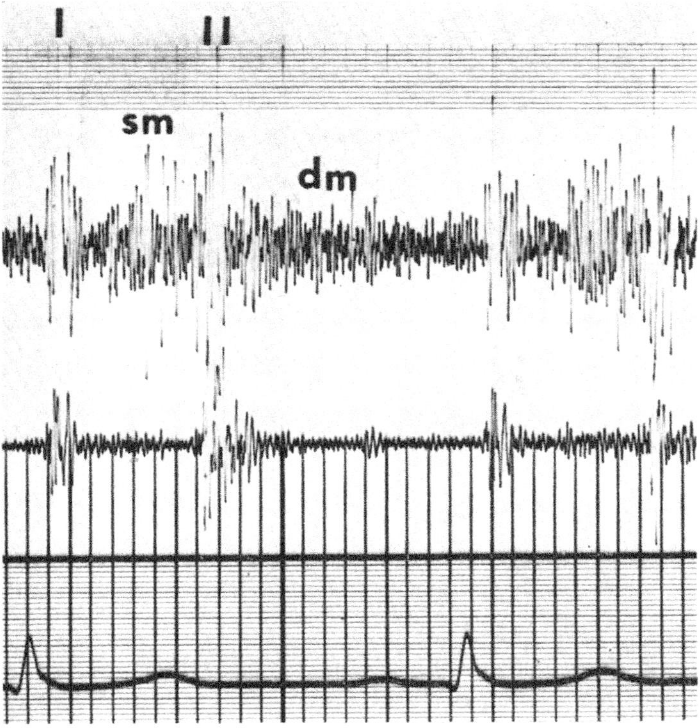

FIGURE 115. Tracings of a 23-year-old man with patent ductus. The upper tracing, recorded at the 1st left i.c.s. shows a typical continuous murmur. The middle tracing, recorded at the apex, shows a minimal systolic and a small early-diastolic murmur. Tracings with high pass filter at 100 Hz with a 24 db/oct. slope.

OTHER RECORDED PHENOMENA *in* PATENT DUCTUS ARTERIOSUS

Several other phenomena are ofter recorded in patients with patent ductus arteriosus.

Pulmonary Ejection Sound. A pulmonary ejection sound (Schmidt-Voigt; McKusick, 1958; De Micheli *et al.*) can be observed, and is best recorded over the second left interspace. This usually occurs about 9 msec. after the beginning of the first sound and it can be debated whether it is an aortic or a pulmonic sound.

Midsystolic Click. A midsystolic click may be recorded over the third left interspace close to the sternum. This occurs about 0.14 to 0.16 seconds before the second sound (McKusick, 1958;

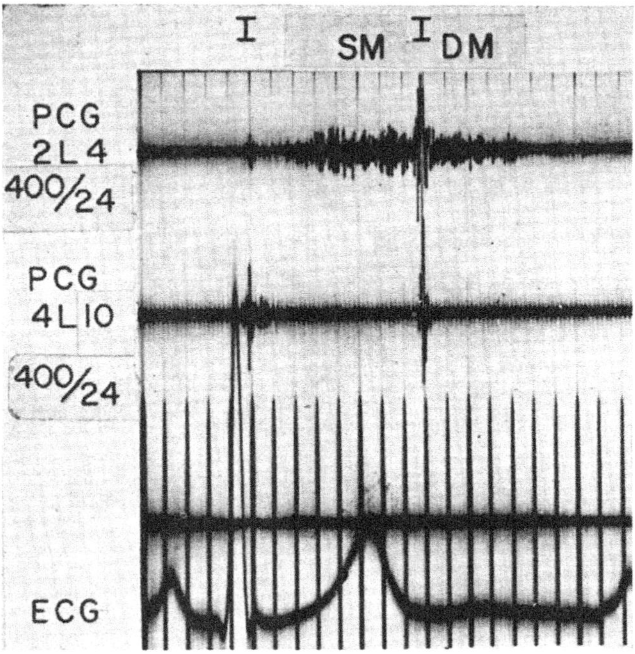

FIGURE 116. Tracings of a 26-year-old man with patent ductus. The murmur is not continuous but "rides over" the second sound.

De Micheli *et al.*). It is a high-frequency, large vibration for which an explanation is still lacking. It seems related to the degre of dilatation of the wall of the pulmonary artery and may be a late vibration of this wall.

Large Third Sound. A large third sound, often followed by a diastolic or presystolic rumble at the apex, can be observed (Ravin and Darley, Fishleder and Medrano, Nadas and Alimurung, Rosa and Kunos). The sound is obviously related to the diastolic overload of the left ventricle. The murmur is a flow murmur of the left ventricle (so-called *relative mitral stenosis*).

Opening Sound. A vibration may be recorded after the main components of the second sound; it is probably an opening sound of the mitral valve (Luisada, 1953; Rosa and Kunos). This vibration is better seen in the medium-frequency bands.

Diastolic Murmur. A large and rough diastolic murmur can often be recorded in the second left interspace (Fishleder and Friedland). As this murmur is heard and recorded only in

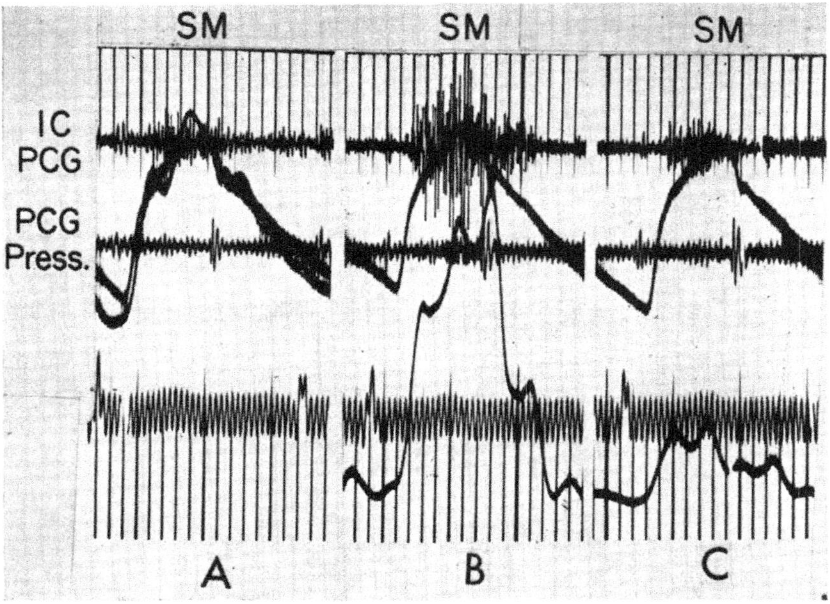

FIGURE 117. Intracardiac phonocardiograms (upper tracing) in a 6-year-old child with patent ductus during a pullback maneuver upon catheterization.
(A) Small murmur in the aorta.
(B) Large murmur in the ductus.
(C) Small murmur in the pulmonary artery.

patients with severely elevated pulmonary hypertension, it was interpreted as caused by relative pulmonary insufficiency. However, this is not as yet proven.

Patients with an *aortopulmonary window* have auscultatory and graphic signs that are similar to those of patients with patent ductus arteriosus. However, the murmur is usually loudest in the third and fourth interspaces.

THE CONTINUOUS MURMUR

A continuous murmur is a murmur that is present in the entire cardiac cycle, though with acentuation in a certain phase. It is typically caused by blood flowing in an artery (occasionally a vein) or through a communication between two arteries, or between an artery and a vein. The most commonly observed and best known continuous murmur is that of patent ductus arteriosus. However, a similar type of murmur has been described in the following conditions.

(1) Bronchopulmonary arterial collaterals (n patients with pulmonary stenosis or atresia or coarctation of the branches of the pulmonary artery or tetralogy).
(2) Pulmonary or systemic arteriovenous fistula (including a coronary AV fistula, rupture of an aortic aneurysm into the pulmonary artery, and the surgical fistulas of Potts or Blalock).
(3) Aneurysm of a sinus of Valsalva (intact or ruptured).
(4) Aortopulmonary window.
(5) Ventricular septal defect with aortic or pulmonary insufficiency.
(6) Flow through the coarcted area in coarctation of the aorta or in coarctation of the pulmonary arteries.
(7) Partial block of a subclavian artery.
(8) So-called venous hum.
(9) The mammary souffle of pregnancy or lactation (Tabatznik *et al.*).
(10) Small atrial septal defect with left atrial hypertension (Ross *et al.*, 1963) including that following transseptal left heart cateterization in mitral stenosis.
(11) Murmur of common arterial trunk.

The phonocardiogram is of moderate help in differentiating these conditions, except for the fact that the murmur is recorded best, or exclusively, over certain areas of the neck, chest, abdomen, or limbs. Therefore, correlation between history, physical findings, electrocardiogram, and phonocardiogram is necessary, and, in this context, the sound tracing may be of use. For example, the fiding of a continuous murmur over one side of the chest may confirm the diagnosis of *a-v fistula of the lung or coarctation of one pulmonary artery.*

The *venous hum* is typically found in children and is usually limited to the right side of the neck or, at most, extends to the second and third right interspaces. In our experience it is a soft, high-pitched murmur, best revealed by high-frequency tracings. The changes of this murmur, and its possible disappearance, following rotation of the head are too well known to need detailed explanation.

Further details will be given in Chapters 20 and 21.

Intracardiac phonocardiography is of help in differentiating (by intracardiac localization) patent ductus arteriosus from aortopulmonary window, aortic insufficiency with VSD, coarctation of a pulmonary artery, or rupture of a sinus of Valsalva (Feruglio, 1960).

COARCTATION OF THE AORTA

In coarctation, the most common finding is a diamond-shaped systolic murmur of medium or low intensity over the second left interspace. An *ejection sound* and a *loud aortic component of the second sound* are usually recorded (Figs. 118 and 119). Tracings recorded over the back between the left scapula and the spine (area of descending aorta) often reveals a *continuous murmur* with late systolic-early diastolic accentuation (Wells *et al.*, 1949) (Fig. 196). It is questionable whether this murmur is caused by flow through the narrow section or through large collateral vessels. For example, this murmur was recorded in

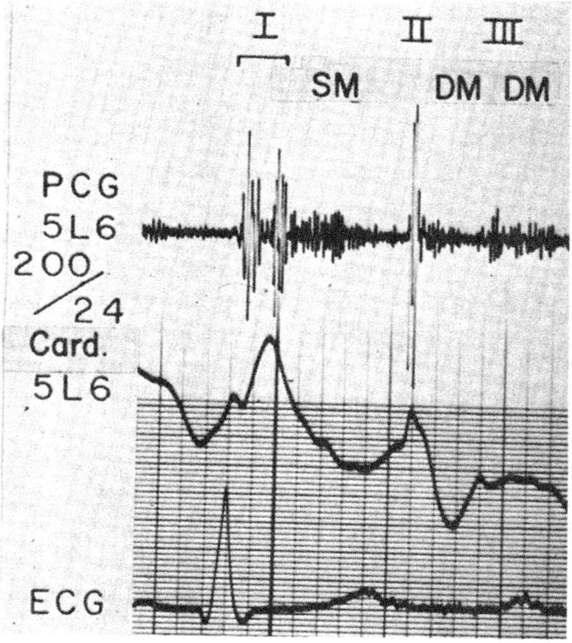

FIGURE 118. Tracings of a 6-year-old boy with coarctation of the aorta. Ejection sound, diamond-shaped systolic murmur, large single second sound, early-diastolic vibrations.

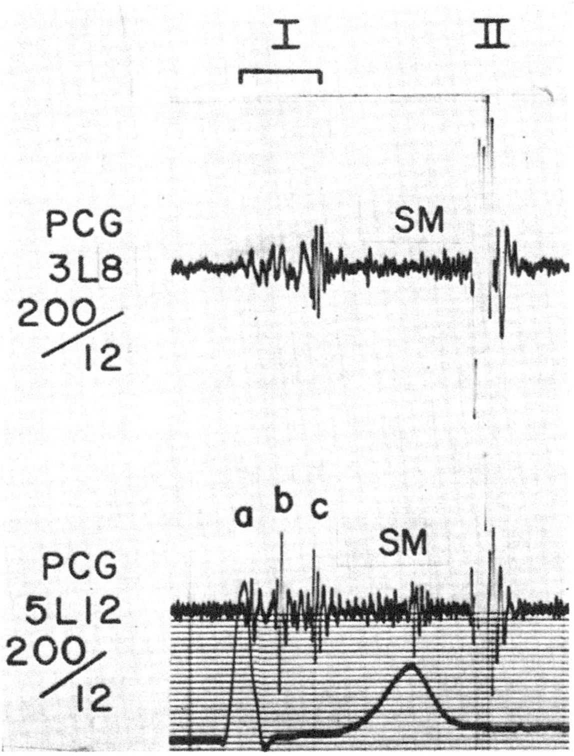

FIGURE 119. Tracings of a 34-year-old woman with coarctation of the aorta. Three components of the first sound are visible. The third (c) is accentuated and represents an ejection sound. There is a crescendo systolic murmur over the 3rd left i.c.s. (upper tracing), and a large, single second sound.

one of our patients where subsequent surgical intervention revealed a complete occlusion of the coarcted section of the aorta, and was recorded in another patient after successful graft and repair.

A murmur of aortic insufficiency may be observed as a result of the frequent association with a *bicuspid aortic valve.** A diastolic rumble has been recorded in several cases. It is a functional rumble, probably caused by left ventricular dilatation.

*A bicuspid valve may be incompetent either because of its dynamics or following a superimposed endocarditis.

Chapter 10

The Complex Congenital Malformations

THE COMPLEX CONGENITAL MALFORMATIONS include several abnormalities, often based on the combination of one or more shunts with narrowing of one of the arterial valves, and other structural changes.

The modifications of the heart sounds and the various murmurs can be understood only if they are correlated with the dynamic changes of the heart and vessels. Numerous studies have led to the recognition of certain sound abnormalities that may have diagnostic importance.

TETRALOGY OF FALLOT

This syndrome is the most common after adolescence and, even if not corrected, may be found in mature people. The most constant combination is: pulmonary stenosis (usually infundibular), ventricular septal defect (usually high), dextroposition of the aorta (overriding aorta), and, of course, right ventricular hypertrophy. Additional lesions may be a single ventricle and a persistent right aortic arch. The right-to-left shunt causes a typical cyanosis. The two ventricles have about the same systolic pressure. However, if the pulmonary stenosis is moderate, there may be a left-to-right shunt at the ventricular level and the patient will not be cyanotic (*acyanotic tetralogy*).

The tetralogy of Fallot has typical acoustic and graphic findings, which have been particularly clarified through the work of Vogelpoel and Schrire (1960).

First Sound. The first sound may be followed by a large ejection sound. This is an *aortic ejection sound.* as shown by its distance from the first component of the first sound (80 to 100 msec.), its wide transmission, its lack of change with respiration, and the fact that it is found even in patients with the

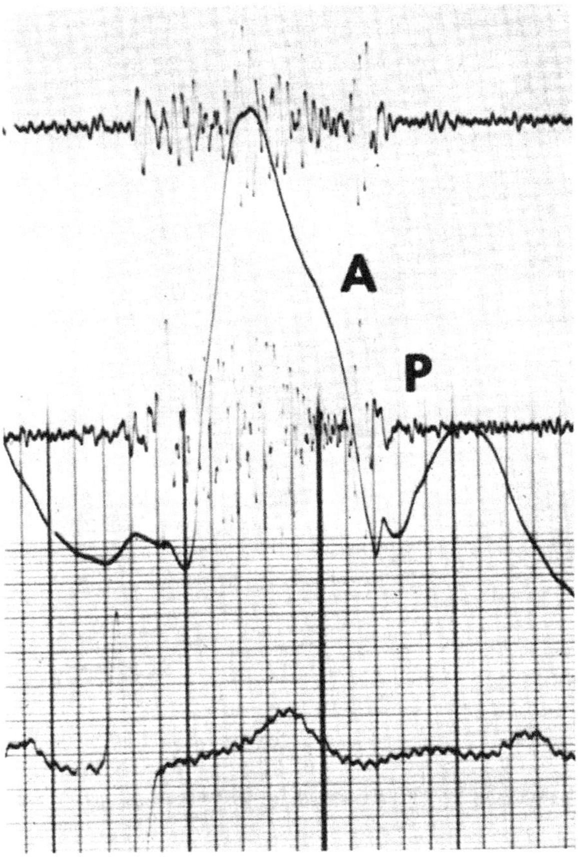

FIGURE 120. Tracings obtained in a 24-year-old man with tetralogy of Fallot with high pass filters at 100 Hz. Upper tracing recorded over the 2nd right i.c.s. There is a larger group of vibrations coinciding with the peak of the carotid pulse. The medium tracing is recorded at the apex. There is a large diamond-shaped murmur decreasing in late-systole. A tiny vibration in early-diastole may be a pulmonary component of the second sound.

most severe pulmonary stenosis. This sound is usually well audible and may give the auditory impression of a split first sound.

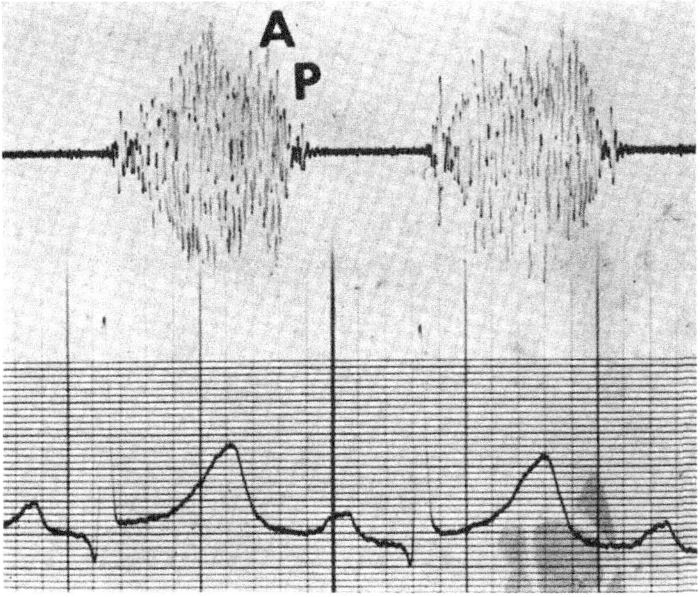

FIGURE 121. Tracings recorded over the 4th left i.c.s. in a 1-year-old child with tetralogy of Fallot. High pass filter at 200 Hz. There is a small pulmonary component of the second sound. The systolic murmur is diamond-shaped and has a late peak.

Second Sound. It has been stated that the second sound is single and consists only of the aortic component; the pulmonary component would be absent as a result of the low pressure of the pulmonary artery. However, frequent exceptions are reported in the literature (Vogelpoel and Schrire, 1960; Parizel; Macieira-Coelho and Coelho; Tofler) and have been observed in some of our cases. The pulmonary component is recorded from the fourth to the second left i.c.s.; it is a small, high-frequency vibration ocurring from 80 to 120 msec. after the aortic component (Figs. 120-122). This vibration is inaudible, not only because it is small, but also because it is preceded by the much larger aortic component. Therefore, differentiation between pure pulmonary stenosis and the tetralogy of Fallot, based on the

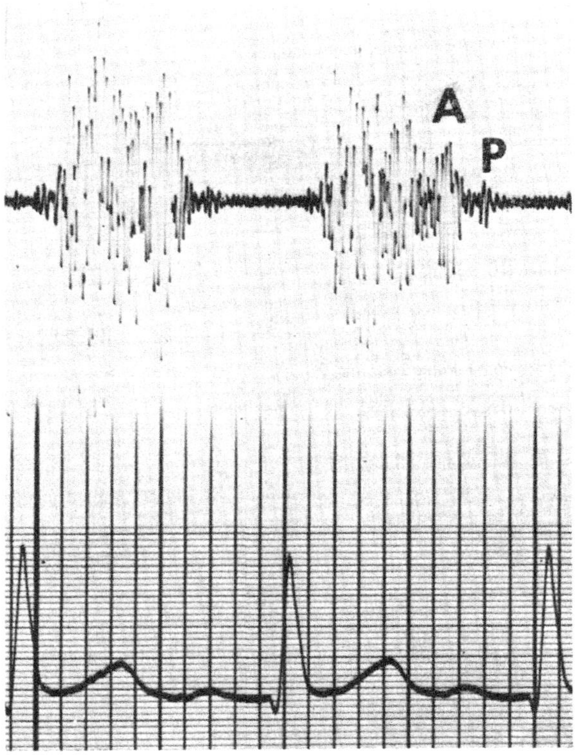

FIGURE 122. Tracings of a 1-year-old child with tetralogy of Fallot. Velocity tracing (80-600 Hz) recorded over the 2nd left i.c.s. Distinct, late pulmonary component of the second sound.

presence or absence of the pulmonary component (as tried by Grishman *et al.*), is not conclusive.

Systolic Murmur. The systolic murmur is heard and recorded in the second, third, and fourth left interspaces near the sternal border. It starts with the first sound, is usually diamond-shaped, and is mainly caused by the pulmonary stenosis (Figs. 120-122). The fact that it is often heard best in the third left interspace has been explained by the infundibular type of stenosis that is frequently present (Nadas, 1959). It is a murmur of medium and high frequency, which seldom continues beyond the aortic component of the second sound. It is far less loud on auscultation (and smaller in graphic tracings) in severe stenosis than it is in moderate stenosis. The murmur may have an early peak in contrast with cases of isolated pulmonary stenosis or *Fallot's trilogy* (Kjellberg *et al.*; Parizel). It may be in crescendo. However, it is possible to observe that the murmur has a late peak at the 2nd left i.c.s. and an early peak in the 3rd or 4th left i.c.s. This fact was observed by us also in pure pulmonary stenosis. It is likely that the lower murmur is related to the right ventricular dynamics (infundibular stenosis) while the higher murmur is caused by valvular stenosis.

Even though patients with left-to-right shunt at the ventricular level (*acyanotic tetralogy*) could have a second murmur caused by the septal defect, both these and those with right-to-left shunt seem to have only a pulmonary stenotic murmur. The murmur is diamond-shaped and usually has a late peak, like in pure pulmonary stenosis.

The intensity of the murmur, which is in the inverse proportion of the severity of the disease, is determined by the severity of pulmonary obstruction and by the pressure gradient across the ventricular septum. The latter can be modified by drugs acting on the systemic circulation. Thus, *amyl nitrite,* by decreasing the peripheral resistance and increasing the right-to-left shunt, will decrease pulmonary flow and decrease the magnitude of the murmur. This again is in contrast with the increase of the murmur in pure pulmonary stenosis but is similar to the decrease observed in pure VSD.

Older children and adults with maximal pulmonary obstruction often have a continuous murmur over the back and under the left clavicle, probably related to *bronchopulmonary vascular anastomoses* (Nadas, 1959).

PULMONARY STENOSIS WITH ATRIAL SEPTAL DEFECT

The *systolic murmur* of pulmonary stenosis with ASD is similar to that of pure pulmonary stenosis. Splitting of the second sound is wide, but the *delayed pulmonary component is small.* A large *fourth sound* can be recorded.

EISENMENGER SYNDROME

In the Eisenmenger syndrome, the two ventricles have the same systolic pressure, equivalent to the systemic systolic pressure. The pulmonary resistance is elevated, similar to the systemic. Therefore, the rebounds of the pulses in the aorta and pulmonary artery will occur at about the same time, thus explaining the reason for a *single second sound.*

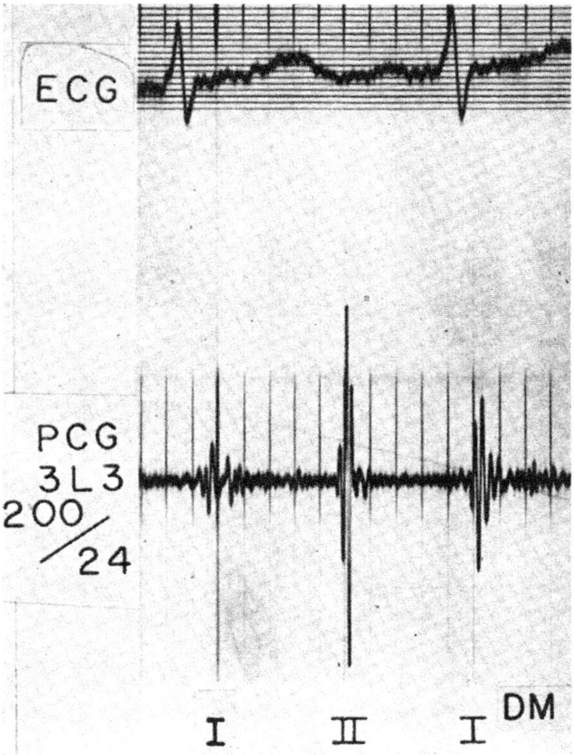

FIGURE 123. Tracings of a 21-month-old girl with Eisenmenger complex. Minimal systolic murmur are present. There is a large, single second over pulmonic area (catheterization, angiocardiogram). Pulmonary artery pressure is equal to aortic pressure. (Courtesy of Drs. Miller and Arevalo, Cook County Hospital, Chicago, Ill.)

The main auscultatory and phonocardiographic data are the following:

(1) The first component of the first sound is followed by a large *ejection sound arising in the dilated pulmonary artery.*

(2) There is a large, *single second sound* due to the fact that the incisurae of the aortic and pulmonary pulses are simultaneous (Fig. 123).

(3) *Diastolic sounds* (either the third or the fourth) can be recorded.

(4) A *crescendo-decrescendo* systolic murmur of moderate intensity is often present over the pulmonary area; it is a flow murmur related to the pulmonary hypertension and to the dilatation of the pulmonary artery.

(5) There may be a soft, blowing *early-diastolic murmur* over the pulmonary area. This is caused by moderate *relative pulmonary insufficiency* and is revealed by a few vibrations in decrescendo.

(6) In the advanced stage, a *functional diastolic rumble* can be recorded over the midprecordium and apex (right ventricular area) as a result of right ventricular dilatation.

LUTEMBACHER'S SYNDROME

This syndrome, thought initially to be of purely ccngenital nature, includes mitral stenosis and atrial septal defect. However, subsequent descriptions have revelaed cases in which a rheumatic mitral stenosis was associated with a congenital atrial defect. More recently, cases of rheumatic mitral stenosis that developed atrial septal defect following trans-septal atrial catheterization were also described.

In general the radiologic and acoustic signs of mitral stenosis are less obvious than in cases without the septal defect. On the other hand, the signs of the congenital malformation are also present.

The acoustic and phonocardiographic data are the following.

(1) The *first heart sound* is accentuated.

(2) The *second heart sound* is widely split and "fixed." The pulmonary component is often large.

(3) There is often a *mitral opening snap.*

(4) There is a *diastolic-presystolic rumble.*

(5) There is a diamond-shaped basal murmur with early peak (*pulmonary flow murmur*).

(6) There often is a blowing, early-diastolic murmur of *pulmonic insufficiency.*

(7) A *continuous murmur* is recorded in certain cases (Asano *et al.*). This occurs if the mitral stenosis is severe and the atrial defect is small, so that a high gradient of pressure occurs across the atrial septum.

This syndrome was considered fairly common in the past. However, cardiac catheterization has shown that many cases of atrial septal defect having an opening snap and a diastolic rumble have *no* mitral stenosis, the rumble and the snap being of functional nature.

TRANSPOSITION OF GREAT VESSELS

The heart sounds are of good quality. There is no ejection sound. The *second sound is normal or accentuated* and may show either a variable or a fixed splitting in patients without septal

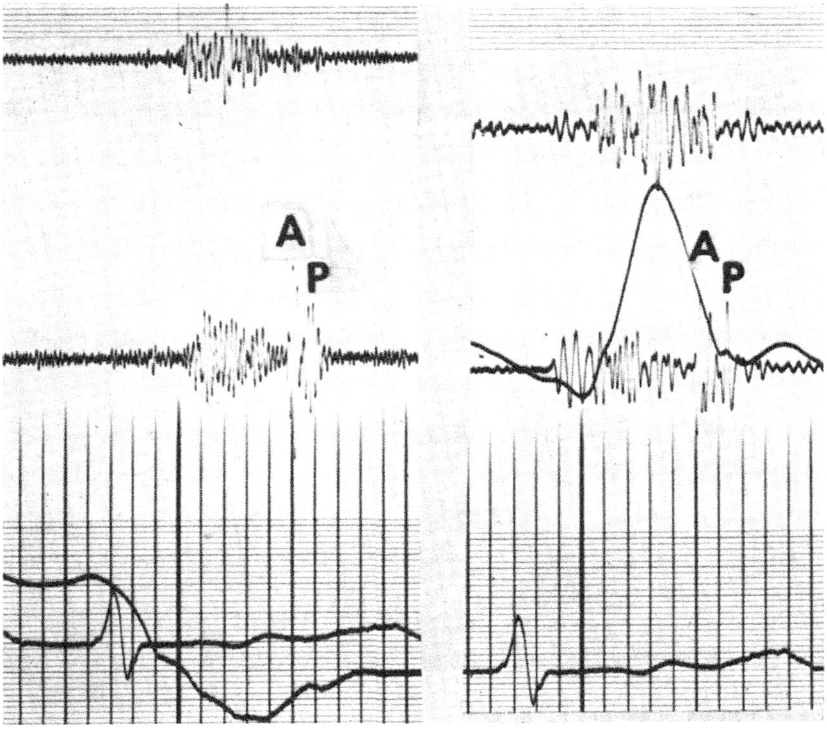

FIGURE 124. Tracings recorded in a 16-year-old male with transposition of the great vessels.
Upper tracing = 2nd right i.c.s.
Medium tracing = 2nd left i.c.s.
At left = filter at 100 Hz.
At right = filter at 40 Hz.
There is a systolic murmur and a visible splitting of the second sound in the medium tracing. The upper tracing shows a late vibration after A that is probably an opening sound of the AV valves.

defect or pulmonary stenosis (Keith *et al.*; Miller *et al.*; Wells, 1963) (Fig. 124). It was thought in the past that splitting would be better recorded at the right of the sternum, in contrast with all other clinical cases, but this was not confirmed (Fig. 124).

There often is a *systolic murmur,* which is either in decrescendo or diamond shaped but has an early peak and is well separated from the second sound (Wells, 1963). It is frequently a murmur of small magnitude.

A *diastolic* or *presystolic murmur* is frequently found at the apex (Wells, 1963).

If there is an associated *ventricular septal defect* (Fig. 12), the systolic murmur is larger and an ejection sound is occasionallly found. There is no basic difference, however, between these patients and those without such defect.

If there are both a *ventricular septal defect* and a *pulmonary stenosis,* there may be (1) no murmur and a single second sound; (2) a longer diamond-shaped murmur of low intensity and a single second sound; or (3) a longer and larger systolic murmur extending beyond the aortic component of the second sound, and followed by a small pulmonary component (Wells, 1963).

If there is right bundle branch block, *reverse splitting of the second sound* is observed because the right ventricle ejects its blood in the aorta.

EBSTEIN'S SYNDROME

Several auscultatory and phonocardiographic phenomena have been described in Ebstein's syndrome.

First Sound. The first sound has a typical crescendo quality being made of a first component of low frequency and a second component of high frequency (Meyer and Schaede). The duration of the first sound is greater than normal.

Second Sound. The second sound occurs early (short I-II interval), a fact that is revealed by comparison with the T wave of the electrocardiogram. It is usually split, but the pulmonary component is small and delayed (Fishleder, 1962).

Early Diastolic Sound. An early diastolic sound of high frequency is often heard and recorded over the xiphoid. This sound occurs about 0.12 seconds after the aortic component of the second sound and has been interpreted as a *tricuspid opening snap* (Vacca *et al.*; Medd *et al.*; Aron *et al.*; Fishleder, 1962).

Third and Fourth Sounds. A third sound and a fourth sound are often recorded. The latter is more often noted in patients with prolonged AV conduction (Aron *et al.*; Fishleder, 1962). The fourth sound is frequently complex in nature, a fact that explains its "scratchy" character on auscultation (Fig. 125).

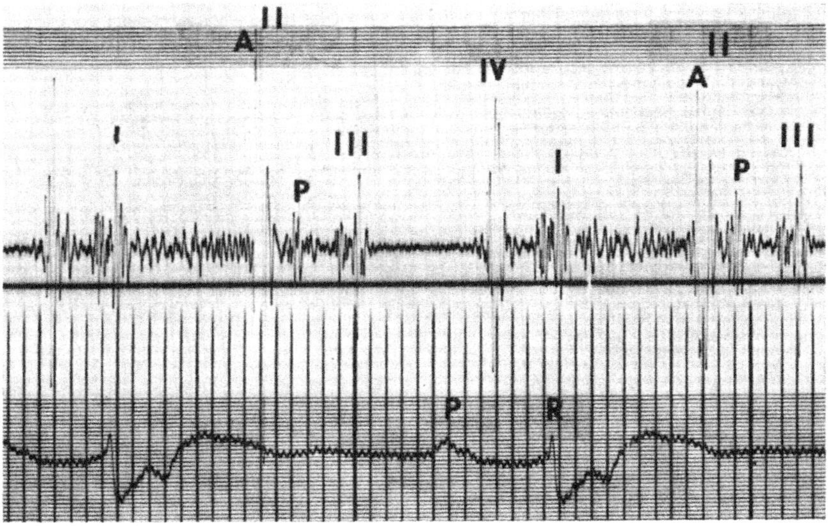

FIGURE 125. Tracings recorded over the mid-precordium in a 24-year-old man with Ebstein's syndrome. High pass filter at 100 Hz (slope 12 db/octave). Large and complex fourth sound. Further group of vibrations in late presystole just prior to the first sound. Systolic murmur. Widely split second sound (RBBB). Mid-diastolic complex of vibrations made of two components.

Murmurs. A *systolic murmur* or a *diastolic-presystolic rumble* (Aron *et al.*; Fishleder, 1962; Meyer and Schaede; Nadas, 1959) are often recorded. The former is usually early-systolic and is *not* increased by inspiration* (Aron *et al.*), even though it is probably caused by the tricuspid insufficiency. However, in many cases, the wide opening of the tricuspid valve in systole seems to cause no murmur, and the recorded murmur is a *pulmonary flow murmur.*

*This fact is probably related to the existence of an atrial shunt, which tends to prevent a marked change of filling during inspiration.

The rumble is a *tricuspid flow murmur* of functional nature (with the exception of the few cases in which the tricuspid valve is stenotic).

COMMON TRUNK

These subjects, usually infants, have a common trunk from which arise both the aortic arch and either the pulmonary artery or its two stems. There is a single semilunar valve with several cusps, often four.

The phonocardiographic findings are the following (Victorica *et al.*):

(1) The *first sound* is followed by a large *ejection sound* arising in the trunk.

(2) The *second sound is split* in the majority of the cases. This fact is due to non-simultaneous rebounds in the aorta and pulmonary arteries due to different compliance of the two vascular systems on account of different resistance. If the pulmonary resistance is high, then the second sound tends to become single.

(3) A systolic *crescendo-decrescendo* murmur of ejection type.

(4) A low frequency *functional diastolic murmur* often initiated by a summation "gallop" (III plus IV sound) or a fourth sound.

(5) In occasional cases, the valve of the trunk is incompetent and an *early-diastolic decrescendo* murmur is recorded.

Chapter 11

Disturbances of the Heart Rate, Rhythm, and Conduction (Tachycardias, Bradycardias, Arrhythmias). Paced Beats

ECTOPIC BEATS

THE ODDITIES AND ERRATIC behavior of heart sounds and heart pulsations in ectopic beats have been known for a long time. In order to find some explanation for the changes of the heart sounds and pulses, we shall separate the supraventricular ectopic beats from the ventricular, and the right from the left ventricular ectopic beats.

SUPRAVENTRICULAR ECTOPIC BEATS

In this type of ectopic beats, the atrial and ventricular contractions are either simultaneous or very close to each other. In some of the cases, they have such a relationship that the ventricular filling caused by the atrial contraction is either interrupted or prevented by the onset of the ventricular contraction. Mitral valve closure in these cases is directly caused by the ventricular contraction in contrast with the sequence of events of normal sinus rhythm. The first sound occurs at a time in which bulging of the mitral valve in the left atrium is decreased

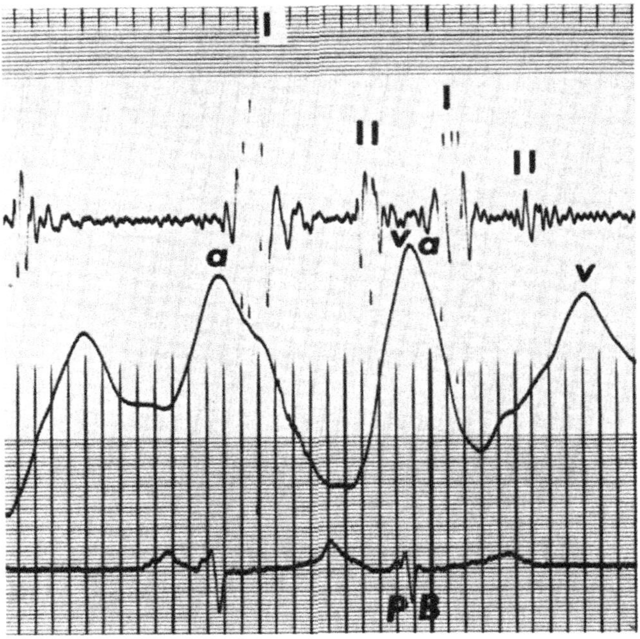

FIGURE 126. Supraventricular ectopic beat. Man of 52 with coronary heart disease. Medium-low frequency velocity. Tracing at apex recorded with filter at 20-100 Hz over midprecordium. Jugular tracing. The first sound of the ectopic beat is of nearly normal amplitude while the second is much smaller. There is a giant VA wave caused by the ectopic beat.

by the higher pressure of this chamber. In such cases, theoretically the first sound should be of greater magnitude than the normal first sound.* However, other factors may be involved; in particular, early appearance of the contraction will cause the left ventricle to be smaller on account of less complete filling.

In clinical cases, the *first heart sound* of a supraventricular ectopic beat is often of nearly normal amplitude, though less so for high frequency vibrations (Figs. 126 and 127). The second sound is smaller. There is no theoretical reason for abnormal splitting of the second sound of a supraventricular ectopic beat. However, occasionally one can see a wider splitting than in a normal beat (Fig. 127).

The *jugular tracing* shows often a giant wave that is either an

*A similar sequence of events occurs in certain cycles during complete AV block. It will be discussed later in this Chapter.

a + *c* wave or a *v* + *a* wave (Fig. 126). The *carotid pulse* of the ectopic beat is usually smaller (Fig. 127) but may be of nearly normal amplitude if the ectopic beat does not occur too early.

VENTRICULAR ECTOPIC BEATS. PACED BEATS.

First Heart Sound. Splitting of te first heart sound has been described in ectopic, as well as in paced beats, similarly to what was considered typical of bundle branch block (Haber and Leatham). However, a new study of the first heart sound in bundle branch block made in our laboratory has shown that this sound has a normal relation to the onset of the QRS complex in *right* bundle branch block, whereas it is delayed over the QRS complex in *left* bundle branch block (Oravetz *et al.*) This

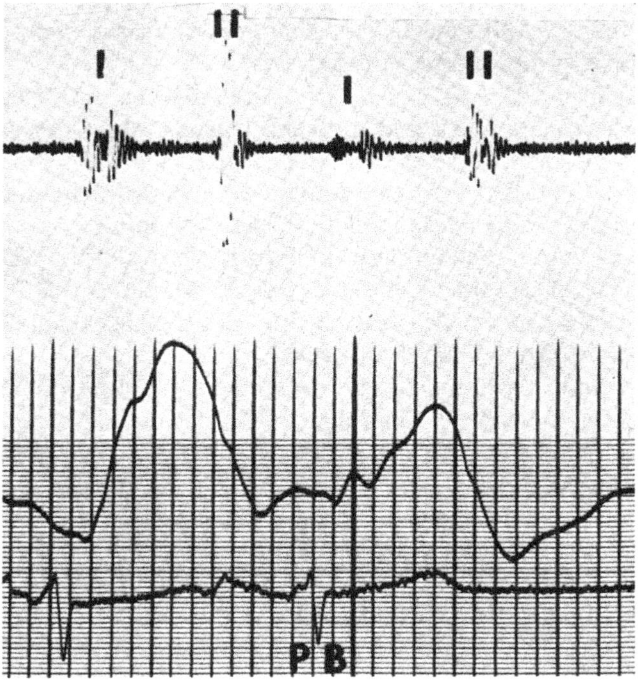

FIGURE 127. Supraventricular ectopic beat. Same patient as in the previous illustration. Higher-frequency velocity tracing over midprecordium (recorded with filter at 100-200 Hz). Carotid tracing.
The first sound of the ectopic beat is much smaller than the normal. The second is also smaller and split. There is a smaller carotid pulsation.

has led to a new study of the first sound and the Q-I interval in ectopic beats (Luisada *et al.*, 1970).

Laubry and Bouvrain noted prolongation but no splitting of the first heart sound of ventricular ectopic beats. Cossio *et al.* found that the first sound was delayed in ventricular ectopic beats of 14 patients; it was louder than normal in 9 cases, less loud in 3, and equal or slightly decreased in 2. The sound was occasionally split (4 of 14 cases) if the ectopic beat occurred after the end of the T wave of the previous normal beat (late ectopic beats) instead of during the T wave (early ectopic beats). Louder first sounds were usually found in late ectopic beats.

Little and co-workers found that the first sound was delayed in ectopic beats of dogs. Schaefer and Little confirmed this in both dogs and man, and stated that the first sound presented greater delay in ectopic beats arising at the base than in those arising at the apex of the heart.

Di Perri *et al.* stated that, in ventricular ectopic beats, both the electro-mechanical and tension periods show an increase, whatever the point of origin of the ectopic beat. Therefore, there is a constnt delay of the first sound. According to them, splitting may be due to separate audition of the first and last component, the latter being louder than in normal beats. Calo found the first sound of ectopic beats delayed as a rule (QRS-I interval was 80 to 120 msec); in his tracings, it was occasionally small or absent, often louder, and occasionally split.

Haber and Leatham found that the ectopic beats from the *right* ventricle behaved like cycles with left bundle branch block, and *usually had a low intensity first sound.* They found that the ectopic beats from the *left* ventricle behaved like cycles with right bundle branch block, and *usually had a loud first sound* that was occasionally split. *Artificial pacing* of the heart gave the same results, but stimulation from the epicardium caused a wider splitting than stimulation from the endocardium. The delay of the first sound was explained by the longer time required for the stimulus to spread through the ventricular walls, and splitting was explained by asynchronous closure of the atrioventricular valves.

In a new study (Luisada *et al.*, 1970), the electrocardiograms and phonocardiograms of 27 patients presenting ventricular

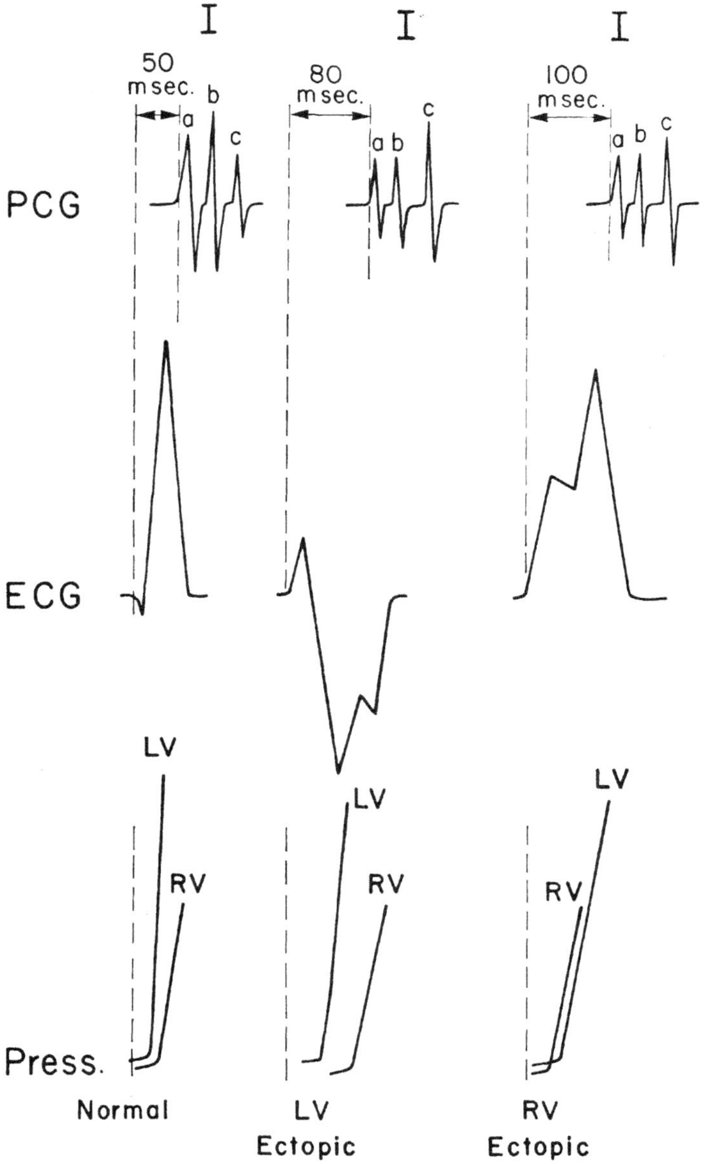

FIGURE 128. Schematic drawing of the relation of the various components of the first heart sound to the pressure events of the two ventricles and the electrocardiogram in ventricular ectopic beats.

ECG = electrocardiogram, lead 1. PCG = phonocardiogram.
1 = first heart sound. press. = pressure curve.
LV= left ventricular pressure. RV = right ventricular pressure.
(From *Am J Cardiol*, *25*:529, 1970.)

ectopic beats were recorded; the electrocardiograms, intracardiac phonocardiograms, and pressure tracings from both ventricles were also compared in 11 patients presenting ventricular ectopic beats during catheterization. The electrocardiograms and clinical phonocardiograms plus the carotid tracings of 5 patients with artificially paced hearts were also studied.

Our study confirmed that *the first sound is delayed* in ectopic beats arising from either the right or the left ventricle. However, *this delay is greater when the origin of the beats is in the right ventricle than when it is in the left* (Fig. 128). This delay was found to be 45.3 msec. (origin in the right ventricle) compared to 26.0 msec. (origin in the left ventricle) in clinical external phonocardiograms; and 65.7 msec. (right ventricle) compared to 49.1 msec. (left ventricle) in clinical (intracardiac) phonocardiograms. The difference betwen the two sets of figures can be explained by the different method of recording.

Separate analysis of the *intervals* between the onset of the QRS complex and the various components of the first heart sound (these have been called with different names and were attributed by others to the dynamic events of the two ventricles) was made in order to confirm or exclude the possibility of vibrations arising in the right ventricle. If this were the case, the first two components would have presented opposite variations of their interval in ectopic beats arising in either the right or the left ventricle.

If we accept that the first sound is only of left ventricular origin, as proven by our studies, the delay of the first sound in left ventricular ectopic beats can be explained only by the longer time required for the stimulus to activate the entire left ventricular wall; on the other hand, the even greater delay for the right ventricular ectopic beats is explained by the same fact plus the interval necessary for the stimulus to cross the ventricular septum and reach the left ventricle (16.6 to 19.3 additional msec. according to the method being used).

The longer delay for the first sound of *basal* ectopic beats in comparison with *apical* beats, described by Schafer and Little, was not confirmed. However, the point of origin of the ectopic beat cannot be determined as accurately in clinical observations as in open chest animal experiments; therefore, their statement must be considered more reliable.

In regard to *splitting of the first sound,* reference should be made to our present concept: the first sound of normal beats has three components (*a*, *b*, and *c*) ; the first two are of left ventricular origin, and the third is of aortic origin. The ectopic beats of our study showed no consistent change in the interval between the *a* and *b* components but presented a greater interval between the *a* and *c* components in 12 of 14 cases in which the ectopic beat arose in the right ventricle. As this interval corresponds grossly to the isovolumic or tension period of the left ventricle, the observations of Di Perri *et al.* seem to be confirmed.

Paced hearts behaved similarly to hearts with ectopic beats. *Endocardial* pacing from the *right* ventricle is followed by a

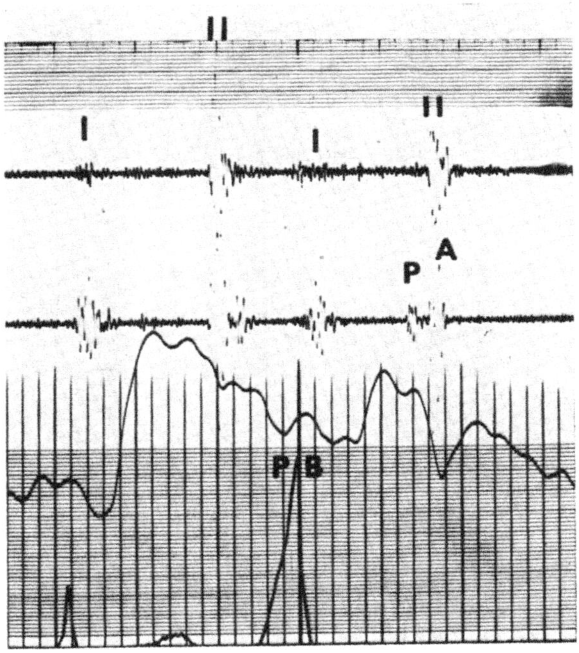

FIGURE 129. Right ventricular ectopic beat. Thirty-year-old man with a pulmonary flow murmur. Phonocardiograms recorded over 2nd right i.c.s. (above) and midprecordium with high pass filters at 200 Hz.
Carotid tracing.
The ectopic beat causes a much smaller and quite delayed first sound, a second sound of good amplitude, and reverse splitting. The carotid pulsation of the ectopic beat is much smaller.

delay of the first sound similar to but greater than that of left bundle branch block. *Epicardial* pacing from the *left* ventricle (1 case) is followed by a lesser delay, similar to that of right bundle branch block.

The relationship of the first heart sound with right and left ventricular pressure rises in normal subjects and ectopic beats is illustrated in Figure 128.

In conclusion, the clinically measured QRS-Ia interval has a significant prolongation in four conditions, and such prolongation follows a definite progression. The average QRS-Ia interval, measured from the chest wall, was as follows: (a) left ventricular ectopic beats = 83 msec.; (b) left bundle branch block = 90 msec.; (c) right ventricular ectopic beats = 103 msec.; and (d) right ventricular pacing = 115 msec. The delay of (a) is due to the time required for the spreading of the impulse to the left ventricular wall; that of (b), to that required for crossing the septum; and that of (c) and (d), to the both factors.

In a recent study (1971), we have observed that, in experimental animals, the following can be observed. *Right ventricular*

A

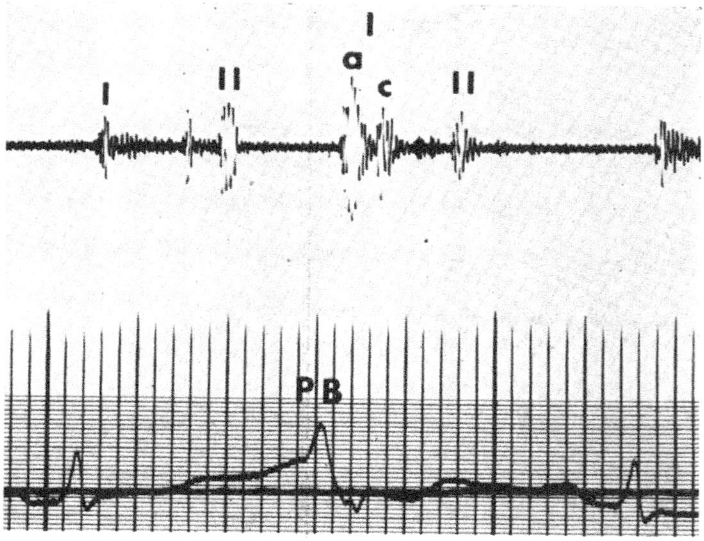

FIGURE 130. Left ventricular ectopic beats of rather late occurrence in a 32-year-old pregnant woman without heart disease. The ectopic beat has a larger first sound than the normal beats both in the medium (A) and in the high frequency range (B). An ejection sound (c) occurs with

B

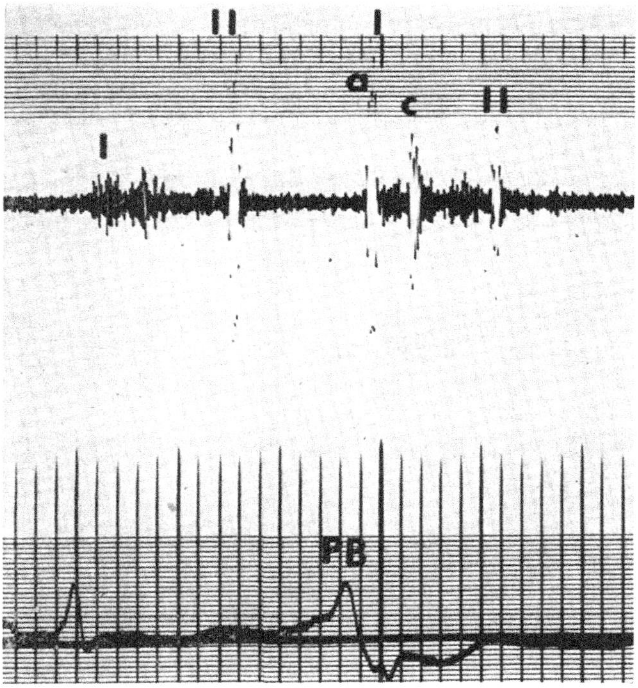

the ectopic beat while a late-systolic click is present in the normal beats. No visible splitting of the second sound. Tracing (A) is a velocity tracing at apex with a band pass filter at 100-600 Hz; tracing (B) is a similar tracing at 200-600 Hz.

ectopic beats always have a *smaller* first heart sound. The dp/dt of the left ventricle is also smaller for the ectopic than for the normal beats. *Left ventricular ectopic beats* may have a first sound which is *larger* than that of the normal beats. However, if the ectopic beat occurs very early, then the first sound is smaller than in normal beats. The later the left ventricular ectopic, and the more chance of a larger first sound. In contrast with this, the dp/dt of the left ventricle usually shows a *smaller* early-systolic rise than in normal beats.

After an ectopic beat, there is a potentiation, so that the following 2-3 beats, either normal or ectopic, have larger first sound and larger systolic wave of the dp/dt. An exception is represented by patients with hypertrophic muscular stenosis where the post extra-systolic beat has a smaller first sound.

Clinical tracings confirm the above data. The first sound is *smaller* in RV ectopic beats (Fig. 129); it may be *larger* in LV ectopic beats (Figs. 130, 131, 133, 134 and 135), and the larger component may be either the *a* or the *b* component, the latter

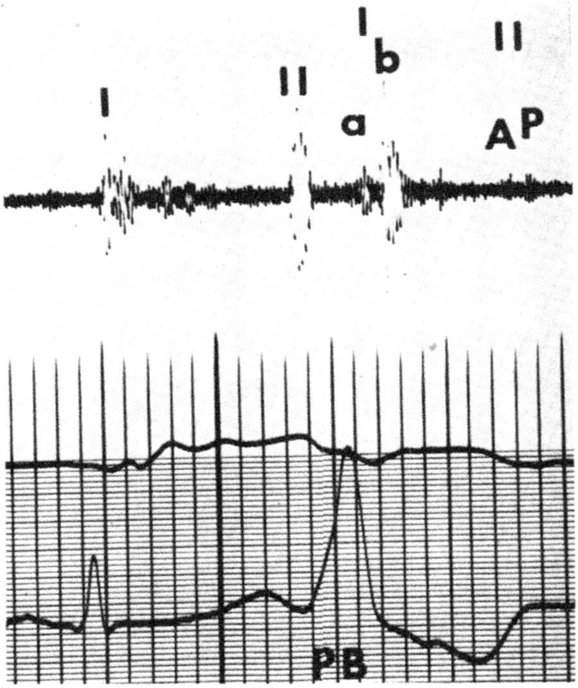

FIGURE 131. Left ventricular ectopic beat of early occurrence recorded at the apex.
This is a displacement tracing at 150-800 Hz. The first sound of the ectopic beat has a small *a* component and a *b* component much larger than the normal. The second sound is small and clearly split (A,P).

being the most common; it is often *smaller* in very early LV ectopic beats (Fig. 132).

The *arterial pulse* is smaller than normal (Fig. 135) but has a greater amplitude if the ectopic beat occurs late (Fig. 134). The pulse of the postextrasystolic beat is usually larger than normal (exceptions occur). However, in hypertrophic muscular stenosis, the pulse is smaller than normal.

Apex Cardiogram. The ectopic beat is usually manifested by a smaller pulsation, which often has a bizarre configuration (Fig. 12).

The *jugular tracing* may show a *cannon wave* whenever an atrial contraction (sinus controlled) occurs during the ventricular contraction (of ectopic origin).

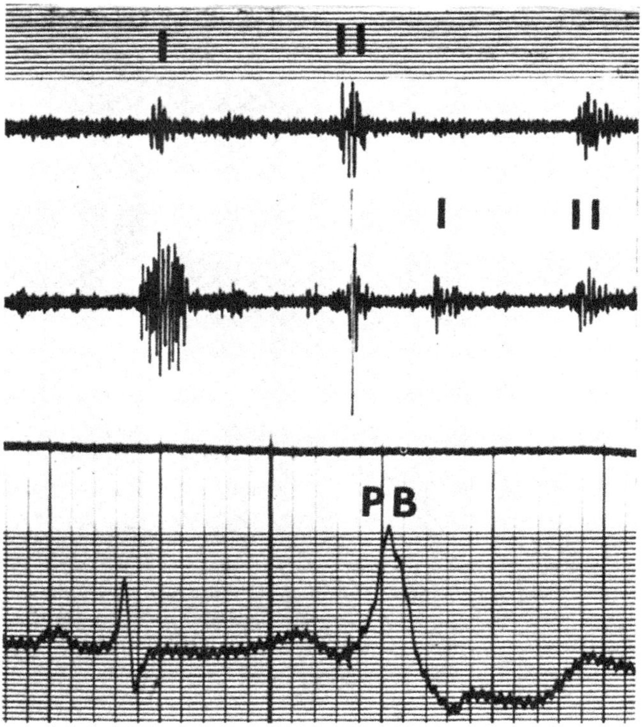

FIGURE 132. Left ventricular ectopic beat of early occurrence recorded over the 2nd left and 5th left i.c.s. The first sound of the ectopic beat is smaller than the normal. The second sound is also smaller and split.

In the interpretaion of these facts, one should consider the following factors. A *right* ventricular ectopic beat would induce contraction of the septum prior to contraction of the free left ventricular wall; this may by itself cause a less rapid rise of pressure in the left ventricle. On the other hand, a *left* ventricular contraction would induce a more harmonic contraction of the entire left ventricle including the septum. In both right and left ventricular contractions, the left ventricle contains a smaller

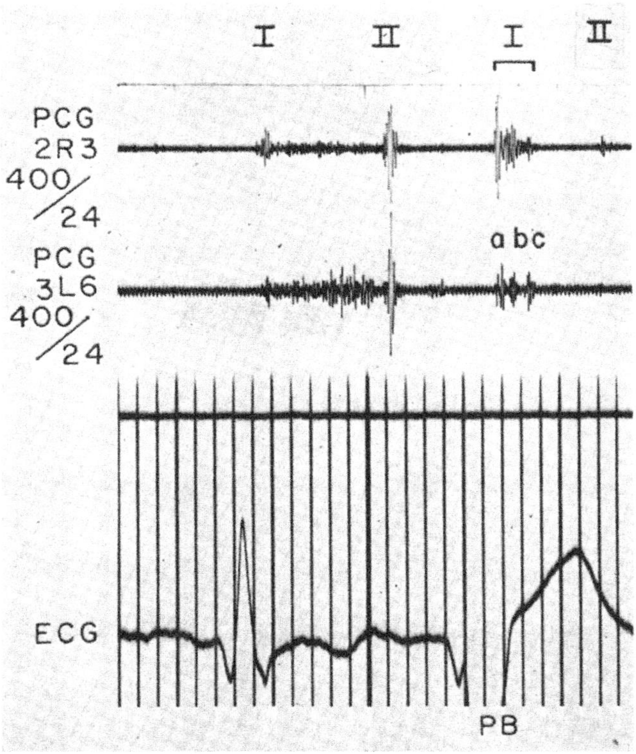

FIGURE 133. Left ventricular ectopic beat (PB) of late occurrence. Three components are recorded; the first (a) is much larger at the 2nd right i.c.s. (upper tracing) than that of the normal beats, less so at the 3rd left i.c.s. (middle tracing).

amount of blood. Lesser mass requires less wall tension to create the same rise of pressure, and thus causes a larger sound for the same amplitude of the systolic wave of the dp/dt of the ventricular pressure.

In spite of these considerations, further studies are required for a complete evaluation of the factors responsible for the changes of the first heart sounds in ectopic beats.

Second Heart Sound. The second heart sound of ventricular ectopic beats is generally small and may be absent. This fact is obviously related to the small amount of blood ejected by the ectopic beat. When the second sound is visible, it is often clearly split. In particular, *right* ventricular origin of the beat is associated with *reverse splitting* (Fig. 129) because the left ventricle is activated late and the aortic ejection ends after the pulmonic (like in LBBB). On the contrary, *left* ventricular origin of the beat is often associated with *wide splitting* of the normal type because the right ventricle is activated late and the pulmonic ejection ends after the aortic (Figs. 131 and 132).

Third and Fourth Sounds. An ectopic beat may be followed by either a third sound or a murmur, which do not occur after a normal cycle or were masked by the rapid sequence in bigeminal rhythm (Luisada, 1953).

In ventricular ectopic beats, if the normal beat is preceded by a fourth sound, this disappears in the ectopic cycle because of the lack of atrial contraction prior to the ventricular contraction.

An ectopic beat may occur so early that its first sound coincides with the second sound of the previous normal beat; this then becomes larger and longer than in previous cycles (Calo, 1959). More common is the coincidence of the first sound of the ectopic beat with the third sound of the previous beat. This would cause an unusual type of summation gallop.

Clicks. In several types of ectopic beats, the second (*b*) component of the first sound is unusually large (Fig. 131). This may be due to the higher pressure existing in the aorta at the moment of opening of the aortic valve. In some, the third (*c*) com-

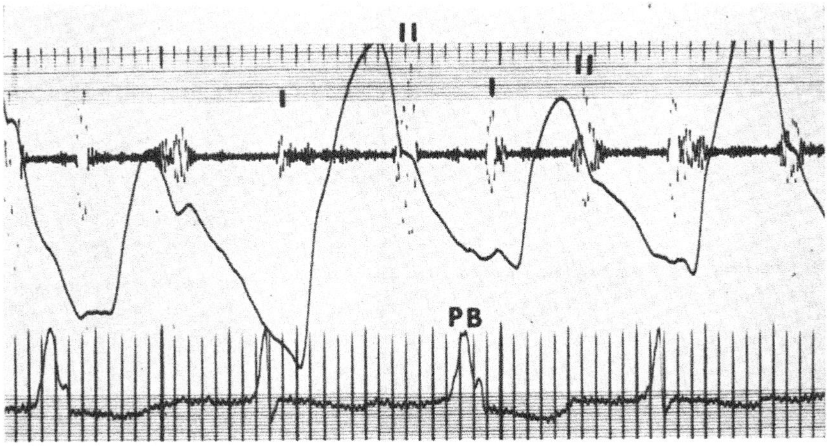

FIGURE 134. Left ventricular ectopic beat of late occurrence (probably a fusion beat). Larger first sound, good amplitude of pulse. Velocity tracing over 3rd left i.c.s. at 100-600 Hz.

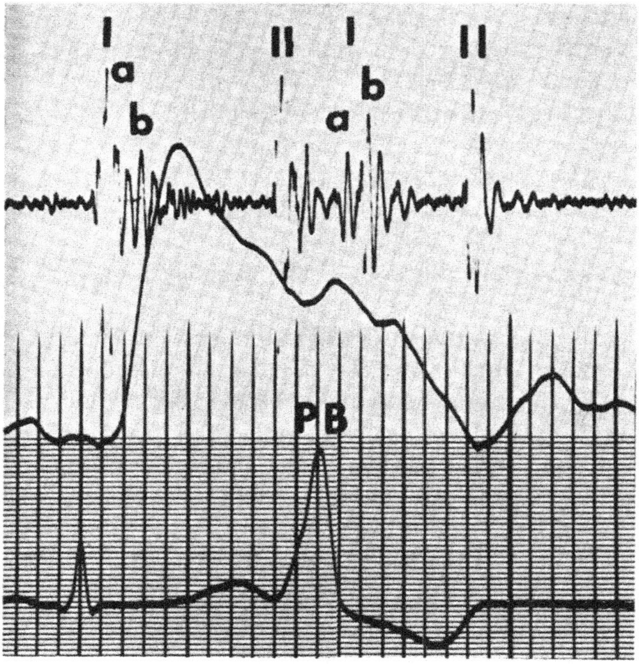

FIGURE 135. Left ventricular ectopic beat of early occurrence. The first sound of the ectopic beat has a *b* component much larger than that of the normal beats while the *a* component is much smaller. The pulse is very small.
Velocity tracing over 3rd left i.c.s. at 30-900 Hz.

ponent is large (Figs. 130 A,B), and one can call it an *ejection sound.* The reason for this is still unclear.

A *midsystolic or late-systolic click* existing in normal beats may disappear in an ectopic beat (Fig. 130A).

Murmurs. If the patient has a systolic murmur, whether originating at the AV valves level or at the semilunar valves level, it tends to be smaller in the ectopic beat than in normal beats and may be absent, though this is not always the case. This fact is related to the smaller ejection of the ectopic in comparison with the normal beat.

SUPRAVENTRICULAR TACHYCARDIA

During an attack of supraventricular tachycardia, *increased loudness* of the first sound is commonly found (Fig. 136). This is probably the result of the sympathetic stimulation causing the attack: this stimulation increases the repidity of contraction and the rate of pressure rise in the left ventricle. On the other hand, simultaneous contraction of the atria and ventricles adds a mechanical factor, like that already discussed for the supraventricular ectopic beats.

In *low junctional tachycardia with retrograde conduction,* the atrial contraction occurs during ventricular ejection, and a *fourth (atrial) sound* may follow the first sound simulating

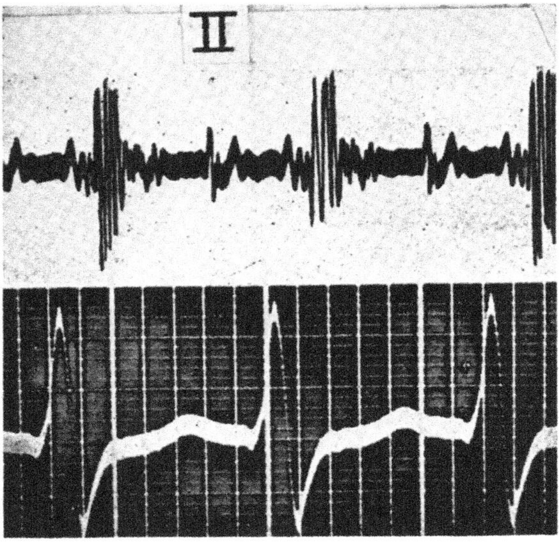

FIGURE 136. Phonocardiogram of a 24-year-old man during an episode of supraventricular tachycardia.
(From Luisada, A. A.: *The Heart Beat,* New York, P. B. Hober, 1953.)

splitting of the latter (Weyler and Dustin). Because diastole is short and tumultuous, *triple rhythm* is often observed, due to the occurrence of a *third sound.*

Murmurs may appear, either a pulmonary flow murmur or a mitral regurgitant murmur related to left ventricular dilatation. On the other hand, pre-existing murmurs may become attenuated or disappear (Calo, 1959).

VENTRICULAR TACHYCARDIA

During an attack of ventricular tachycardia, the atria and ventricles beat with separate, independent rhythms, which now

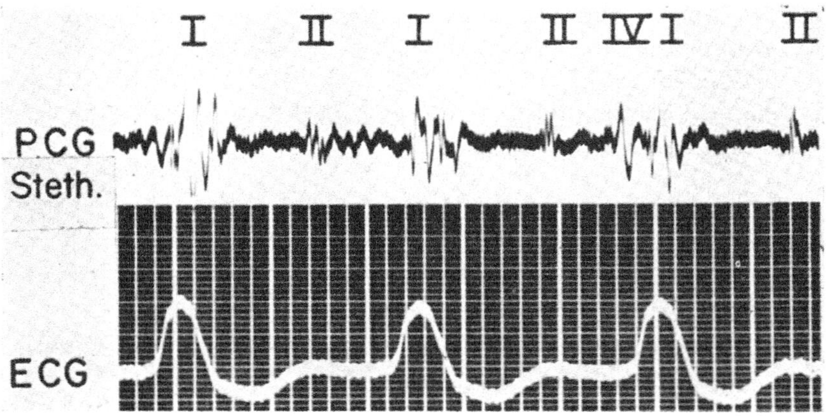

FIGURE 137. Phonocardiogram of a 60-year-old woman during an episode of ventricular tachycardia. The third beat is preceded by a large IV sound. (From Luisada, A. A.: *The Heart Beat*, New York, P. B. Hoeber, 1953.)

and then cause a casual atrioventricular sequence. When an atrial contraction takes place shortly *before* a ventricular contraction, a typical *sudden increase in loudness of the first sound* occurs (Levine and Harvey). This fact is explained by the fact that atrial contraction leads to better ventricular filling and greater output, as documented by Wilson *et al.*

The *first sound* is usually prolonged, often small and may be split; the considerations that have already been presented for the ventricular ectopic beats also apply here (Fig. 137).

One should keep in mind that ventricular tachycardia may arise on account of a focus located in either the right or the left ventricle. When the focus is in the *right* ventricle, the first sound is delayed and of lesser magnitude, like in right ventricular ectopic beats and for similar reasons. The delay is due to the time necessary for the stimulus to reach the left ventricle across the septum plus that due to abnormal apread of the impulse. The lower amplitude of the first sound is most likely due to the non-simultaneous activation of the two ventricles, so that the septum is contracting before the free left ventricular wall. When the focus is in the *left* ventricle, the amplitude of the first sound is greater (see left ventricular ectopic beats).

The *second sound* is often widely split in left ventricular tachycardia, and single or reversely split in right ventricular tachycardia. These changes are due to the asynchronism of contraction of the two ventricles; obviously, the ventricle contracting later would cause the occurrence of a delayed component of the second sound in comparison with the other ventricle. In particular, in left ventricular tachycardia, the pulmonary component is delayed while, in right ventricular tachycardia, the aortic component is delayed.

ATRIAL FLUTTER

(1) *Atrial Flutter with Regular Block.*

First Sound. If slight irregularities of the ventricular rate occur, a *changing intensity of the first sound* may be noted (Harvey and Levine). This fact is connected with slight changes of the P-I (and of the P-R) interval, the maximum intensity being observed for P-R values between 0.08 and 0.12 seconds.

Second Sound. At times the second sound may be extremely small and may even disappear because of small cardiac output.

Atrial Sounds. The atrial sounds* may be abnormally large and clinically audible (Calo, 1939; Contro; Hecht and Myers; Lian and Welti). They are often well recorded both in diastole and systole. They occur from 0.14 to 0.24 seconds after the beginning of the P waves (Calo). When the ventricular rate is rapid, a summation type of *triple rhythm* is common. Either presystolic or early-systolic occurrence of an atrial sound may simulate splitting of the first sound. There may be either a normal or an increased Q-I interval, and the first sound may be alternating (Calo, 1959).

Murmurs. While the presystolic murmur of mitral stenosis usually *increases* during atrial flutter, the systolic murmur of mitral insufficiency usually *decreases* in this condition. A functional diastolic rumble can be observed *without* mitral stenosis (Fig. 138).

(2) *Atrial Flutter with Variable Block.*

The amplitude of the *first* and *second sounds* varies from cycle to cycle like in atrial fibrillation (see later) (Fig. 139).

Atrial sounds may be observed and may be the only clue

*In this circumstance, the term "atrial sound" is more appropriate than "fourth sound."

to the diagnosis in cases with equivocal electrocardiograms (Fig. 139).

Murmurs vary from cycle to cycle.

ATRIAL FIBRILLATION

Heart Sounds. There is a complete irregularity of the two main heart sounds in atrial fibrillation. In cases of atrial fibrillation without valvular lesions, the amplitude of the heart sounds varies from cycle to cycle.

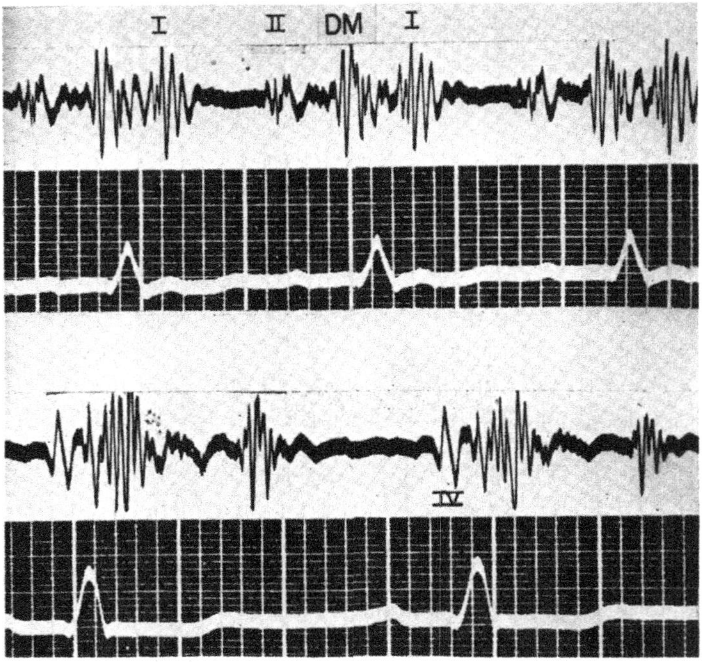

FIGURE 138. Phonocardiogram of a 36-year-old man during an attack of atrial flutter with regular block (above) and after resumption of sinus rhythm (below). During the attack, there is a short mid-diastolic rumble (DM). After the attack, there is a large IV sound and no murmur.

Since the graphic studies of Th. Lewis (1925) and Battaerd, it has been stated that the phonocardiogram reveals a certain relationship between the variable cycle length and the amplitude of the heart sounds, particularly the first sound. Both Wolferth and Margolies (1930) and Herkel and Zur stated that the first heart sound is fainter after a long diastole and louder after a

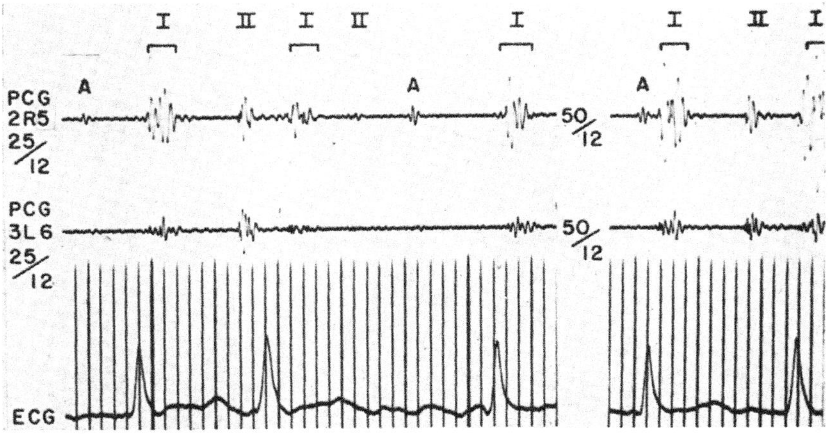

FIGURE 139. (A) Phonocardiogram of a 43-year-old man during an episode of atrial flutter with variable block. Atrial sounds (A) are occasionally observed in mid-diastole (at left) and in presystole (at right). (From Luisada, A. A.: *From Auscultation to Phonocardiography*, St. Louis, Mosby, 1965.)

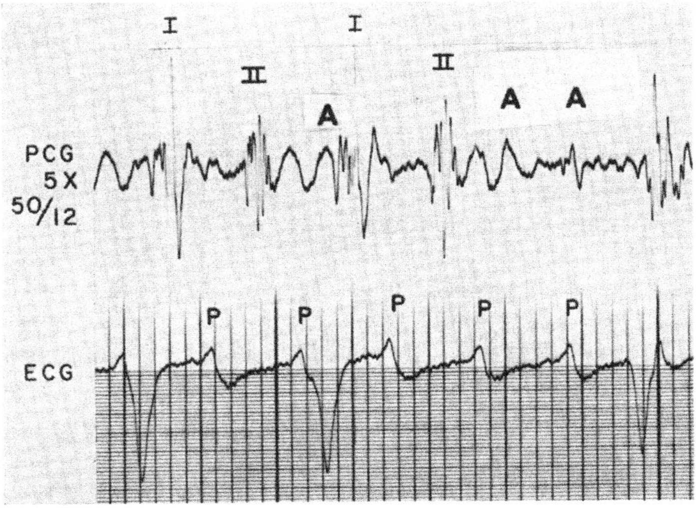

(B) Tracings of 54-year-old man with atrial flutter and variable block. The phonocardiogram is a medium-low tracing recorded over the lower part of the sternum and shows low-pitched atrial sounds (A,A). Variable amplitude of the first sound.

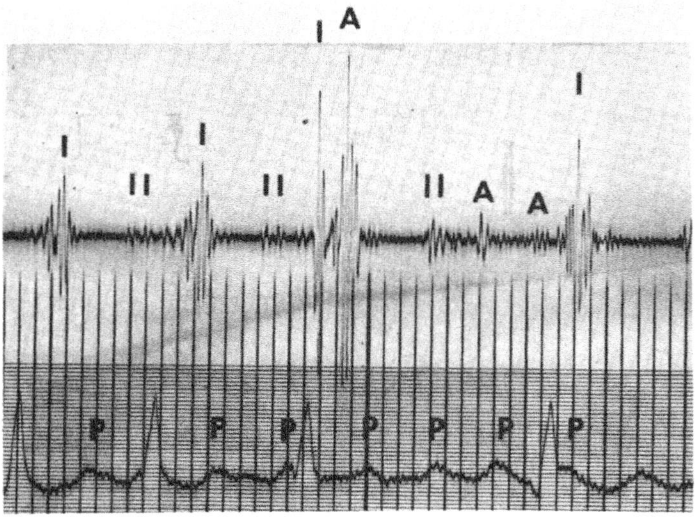

(C) Tracings of 42-year-old patient with atrial flutter and variable block. The phonocardiogram is a medium-high tracing recorded over the 3rd left i.c.s.
Marked variability in amplitude of the first sound. The atrial sounds (A,A) are usually small. However, a very large atrial sound follows the first sound after the third ECG complex.

short one. However, the former also noted faint first sounds followig extremely short diastoles; they also noted that mitral insufficiency tended to intensify the first sound that follows a short diastole and vice versa.

A good correlation between the amplitude of the first sound and that of the early-systolic wave of the first derivative of LV pressure has been found by Sakamoto *et al.* (1965A) in animal experiments* and was confirmed by clinical tracings in the absence of valvular lesions (Fig. 140). On the contrary, patients with mitral insufficiency seem to have an opposite behavior, longer diastoles being followed by smaller first and second sounds (Fig. 141). Apparently, greater filling of the left ventricle causes greater regurgitation, slower rate of pressure rise,

*The amplitude of the first sound and the length of the preceding diastolic interval tended to have a parabolic relationship. Exceptions observed during short runs of tachycardia were related to the adequacy of the previous beats. The correlation between the amplitude of the first sound and that of the early-systolic wave of the first derivative of LV pressure were similar to that in animals with sinus rhythm.

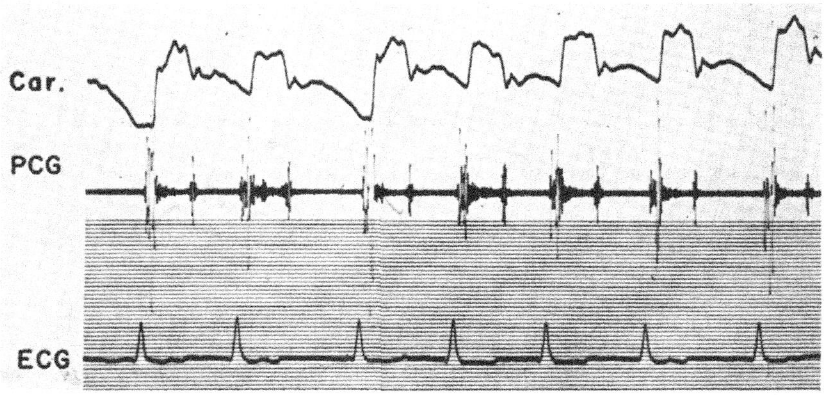

FIGURE 140. Carotid tracing and phonocardiogram of a 50-year-old man with thyrotoxicosis and atrial fibrillation.
The pulse is typically irregular and of variable amplitude. The heart sounds vary in amplitude. The first sound is larger when the previous diastole was longer.
(From Luisada, A. A.: *From Auscultation to Phonocardiography*. St. Louis, Mosby, 1965.)

and smaller sounds but larger murmur.

If there is no mitral stenosis, the Q-I interval is only slightly delayed, and shows moderate variations from cycle to cycle. If there is mitral stenosis, the Q-I interval is markedly prolonged and has a typical variability being much longer after a short diastole (Luisada, 1941)** (Fig. 142). This is also true in atrial flutter with variable AV block.

Murmurs. A short *systolic murmur* over the apex or pulmonary area is common, even in the absence of valvular lesions. It is either a pulmonary flow murmur or a murmur caused by relative mitral insufficiency.

The murmur of mitral insufficiency is usually larger following a longer diastole (Fig. 141).

Third Sound. It is frequent to observe a ventricular type of triple rhythm (large III sound). If the third sound falls close to

**According to Ferrero et al., the intensity of the first sound varies in cases with atrial fibrillation and mitral stenosis from beat to beat. If there is correlation between these changes and the length of the previous diastoles, then the pulmonary capillary pressure would be below 20 mm Hg. When such pressure is higher, the variability has no correlation with diastolic intervals.

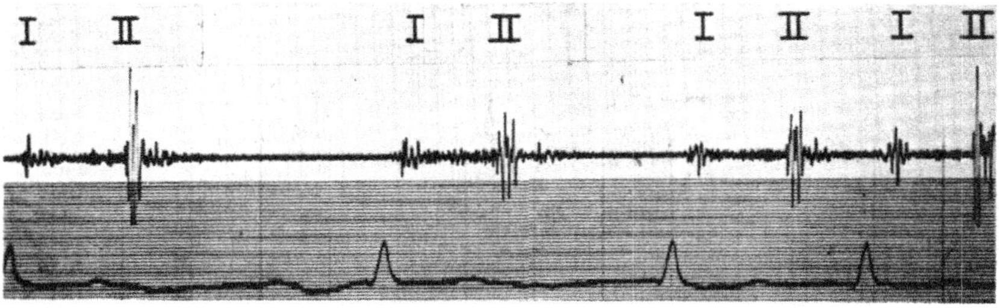

FIGURE 141. Phonocardiogram of a 37-year-old woman with pure mitral insufficiency (catheterization) and atrial fibrillation. A shorter diastole is followed by *larger* first and second sounds (see last cycle). A longer diastole causes a larger systolic murmur.

the following first sound (short diastole), it may become more prolonged and cause a low-pitched mid-diastolic rumble in the absence of mitral stenosis.

Opening Snap. In patients with mitral stenosis, the interval between the second sound and the opening snap is variable and changes so that the beat following a larger diastole also has a longer II-os interval (the opposite is true for the Q-I interval) (Fig. 143).

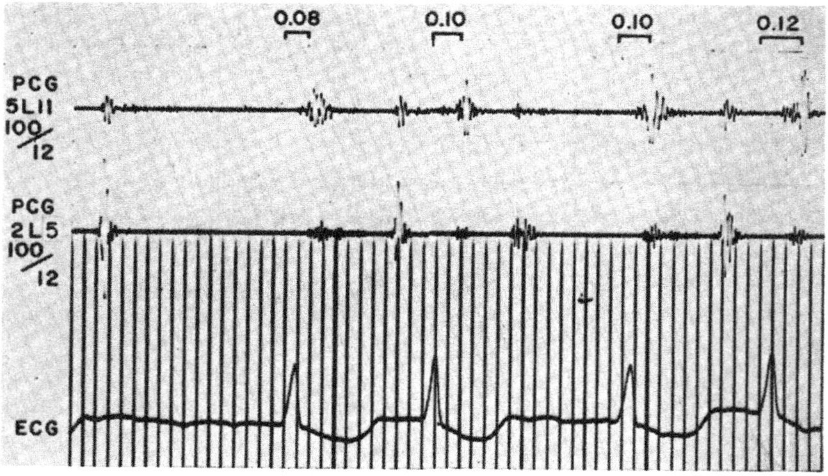

FIGURE 142. Case of mitral stenosis and atrial fibrillation with marked variability of the Q-I interval. This is particularly evident if one measures the distance between the onset of QRS and the largest vibration of the first sound.

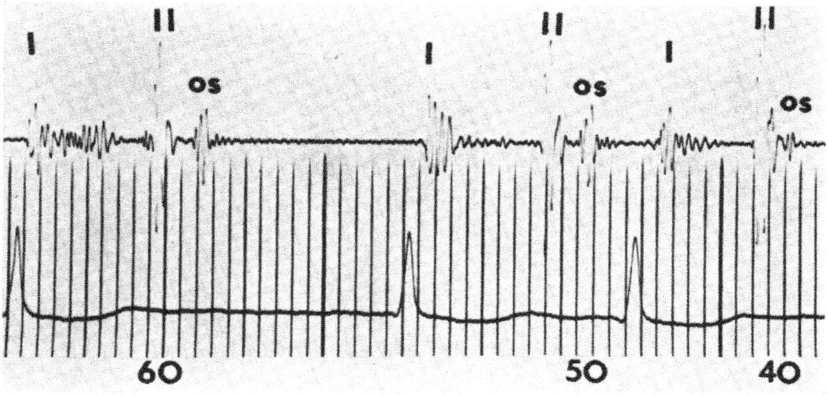

FIGURE 143. Case of pure mitral stenosis and atrial fibrillation with marked variability of the II-OS interval (from 60 msec to 40 msec).

NODAL RHYTHM

This term is usually employed in cases in which the pacemaker is at the junction between atria and ventricles. While the ventricles contract regularly and slowly, the atria may either be stimulated in a retrograde fashion (and contract simultaneously with the ventricles) or do not respond (*atrial paralysis*). In the first instance, it is likely that the *first heart sound* will be louder than normal while, in the second, it would be normal. In regard to the *second sound,* this behaves in a normal fashion in all cases.

SA BLOCK

In sinoatrial block, a complete cycle will be absent during the block. Thus, no heart sounds are recorded during the pause.

AV BLOCK

In *Grade 1 AV block (prolongation of AV conduction),* the fourth sound may be larger, better audible, and better visible than with sinus rhythm whenever the causes for an atrial type of triple rhythm are present. There is a better separation of the fourth sound from the first sound, a fact that also leads to better audibility of the former. On the other hand, the first sound may become fainter because of a mechanism that will be explained below (Levine and Harvey).

In *Grade 2 AV block (intermittent or periodic AV block),* and in *Grade 3 AV block (complete AV block,)* it is possible to hear faint sounds in coincidence with the atrial contractions. These

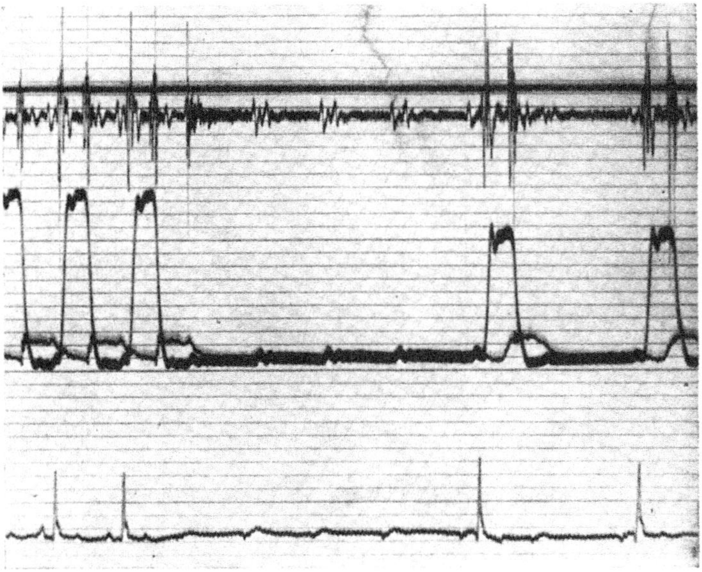

FIGURE 144. Complete AV block induced in a dog by vagal stimulation. During the ventricular pause, two atrial sounds are recorded for each atrial contraction. The last two complexes show evidence of mitral regurgitation (rise of pressure in the LA in the second part of ventricular systole).
From above = external phonocardiogram at apex
left ventricular and left atrial pressures
electrocardiogram
Film speed = 25 mm/sec.

are recorded as small, low-pitched waves in the phonocardiogram (Figs. 144-147). They are never as large as those of atrial flutter, because the slow ventricular rhythm causes better emptying of the ventricles and no atrial overload. When an atrial sound falls in early diastole, it is much larger, probably because it is *reinforced* by a third sound (Grayzel). This would be the result of a mechanism similar to that which causes a summation type of triple rhythm (*summation gallop*) in patients with a sinus rhythm (Fig. 146).

Each atrial contraction is often accompanied by 2 (seldom 3) groups of vibrations. These correspond to the various components of the fourth sound (Lewis, T., 1925; Cossio *et al.*; 1942; Gonzalez Sabathie *et al.*; Luisada and Mautner; Calo, 1959; Muiesan *et al.*) (Fig. 144). According to Calo, these

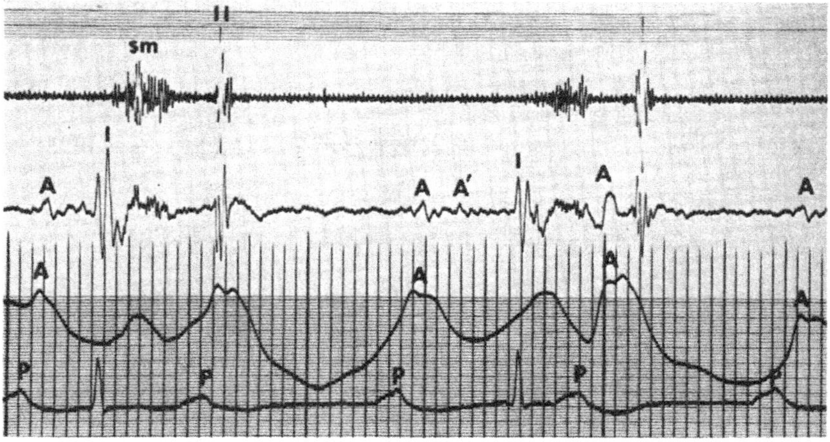

FIGURE 145. Tracings in a 34-year-old woman with complete AV block following viral myocarditis.
Upper tracing = PCG, velocity, 80-800 Hz.
Center tracing = PCG, velocity, unfiltered.
Third tracing = jugular tracing.
Lower tracing = electrocardiograms.
There is a systolic, diamond-shaped murmur. The atrial sounds are visible as low frequency vibrations, and, in the center of the illustration, two sounds are caused by an atrial contraction. The following atrial contraction gives one sound during ventricular systole.

groups occur 0.04, 0.12, and 0.27 seconds, respectively, after the beginning of the P wave. In certain cases, only a series of low-pitched vibrations is recorded.

One or two low frequency components can be recorded even during systole if the atrial contraction falls at the time of the ventricular contraction (Fig. 145). On the other hand, atrial contractions may cause a *diastolic murmur* in elderly patients (Rytand).

The *first sound* may seem to be occasionally split whenever a fourth suond occurs, either immediately before, or soon after, the first sound itself. The amplitude of the first sound varies markedly according to the relationship between atrial and ventricular contractions (Fig. 147). The following rule has been stated (Levine and Harvey): a P-R interval of 0.50 to 0.32 seconds would cause a slightly larger first sound; a P-R interval of 0.32 to 0.23 seconds would cause a smaller first sound; a P-R interval of 0.14 to 0.04 second would cause a very large first sound (*cannon sound*). A *double cannon sound* has been described in AV block with bundle branch block* as a result of ventricular asynchronism but this fact requires confirmation.

The *varying intensity of the first sound* has been related to the different positions of the AV valves at the beginning of

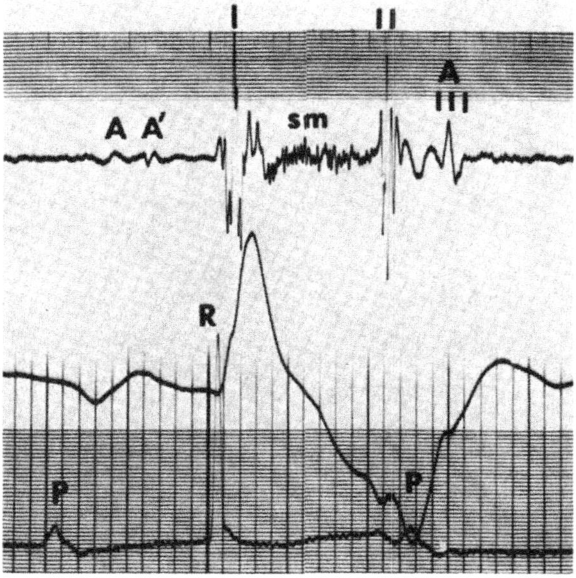

FIGURE 146. Case of congenital, complete AV block. Eight-year-old child. The upper tracing is a PCG at apex with a high pass filter at 50 Hz (6 db/oct. slope). The center tracing is an apex cardiogram. The lower tracing is an ECG. There is a double atrial sound (A,A′) with the first atrial contraction and a "reinforced" atrial sound with the second. There is a pansystolic murmur.

systole. However, different rates of pressure rise, possibly related to this valvular mechanism, provide a more logical explanation.

In experiments conducted in our laboratory, the following data were observed:

---

*In most cases there is no bundle branch block. However, as the pacemaker is in one of the ventricles (*idioventricular rhythm*), the sequence of mechanical events and the ECG pattern are similar to those of bundle branch block. It is likely that the double cannon sound was a large component *a* followed by a large component *c* (ejection sound).

(1) In *Grade I AV block* with very long P-R interval, there is an increase in the early-systolic wave of the left ventricular dp/dt and a correspondent increase in amplitude of the first heart sound.

(2) In *Grade II AV block* with a 2:1 rhythm and normal P-R once every two cycles, there is an increase of the dp/dt and an increase of the first heart sound.

(3) In *Grade III AV block,* coincidental atrio-ventricular contractions simulating junctional beats are accompanied by a normal dp/dt and a larger first heart sound. It is possible that, in such a case, as the mitral valve is, more firmly supported by the atrial contraction, there is a larger vibration of the left ventricular mass even though the pressure rise has the same rapidity.

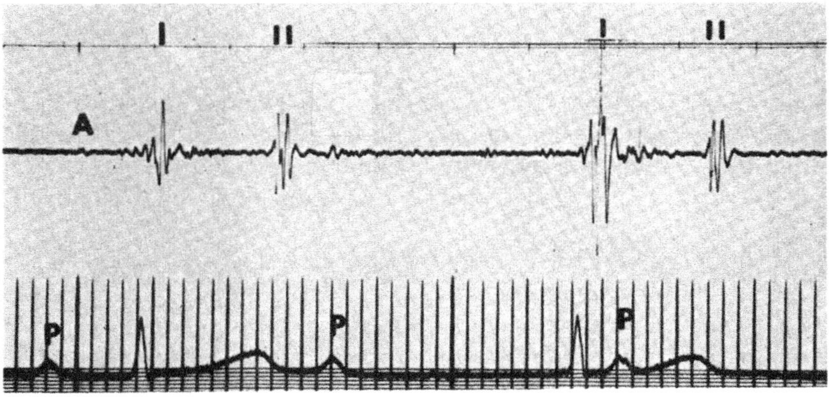

FIGURE 147. Case of congenital, complete AV block. Six-year-old child. PCG at apex with high pass filter at 50 Hz (6 db/oct. slope). Typical variable amplitude of the first heart sound.

The *second sound* may precede the end of the T wave because the prolongation of electric systole is at times greater than that of mechanical systole. If an atrial contraction falls at the end of ventricular systole, the *second sound* may be *reinforced* or *split* (Calo, 1959).

A *systolic murmur* is frequently observed in the absence of valvular lesions. This is sometimes an *aortic or pulmonic flow murmur* (Fig. 145) that is related to the large ejection and, in older patients, to aortic atherosclerosis. In other cases, there is often a *pansystolic murmur* of atrioventricular valves regurgitation (mitral or tricuspid) (Fig. 146).

In patients with mitral valve lesions, the pansystolic murmur of insufficiency may decrease when there is a short P-R interval. This has been explained as due to the fact that atrial relaxation is somewhat delayed, a fact that would tend to resist the early stage of regurgitation (Calo, 1959).

In patients with mitral stenosis, each atrial contraction may be accompanied by a distinct murmur. If more than one atrial contraction occurs in diastole, the murmurs are of decreasing amplitude due to progressive filling of the left ventricle (Calo, 1959).

Bundle Branch Block

On the basis of theoretical considerations, splitting of the first heart sound has ben considered typical of bundle branch block (BBB) for a long time in spite of inadequate evidence supporting this view.

Studies based on clinical auscultation include those of King and McEachern, Cossio and associates, Laubry and Pezzi, Lewis, Lian and associates, and Segall. Various percentages of splitting of the first sound were reported, from an occasional finding to a 56% incidence.

Graphic studies were made by Wolferth and Margolies, Battro and associates, Contro and Luisada, Kelly, Haber and Leatham, and Brachetti and associates. Three components were described within the first sound in BBB by Battro and associates and by Contro and Luisada (at that time, the division of normal first sound into three components had not been recognized). Haber and Leatham described frequent splitting of the first sound in right bundle-branch block (RBBB) in contrast with left bundle-branch block (LBBB), where there was no splitting.

Studies based on auscultation or on phonocardiography gave a clear-cut description of the common abnormalities of the second sound. These were (1) wide, fixed splitting of the second sound with delay of the pulmonary component in RBBB, and (2) single second sound or reverse splitting with delay of the aortic component in LBBB.

First Sound, Carotid Pulse. A new study was conducted by Oravetz *et al.* in our laboratory in 1967 including 20 normal subjects and 40 patients with bundle branch block.

The graphic studies on the first sound from our laboratory had already demonstrated the existence of three components or groups of vibrations within the normal first heart sound. They had shown that the first two (*Ia, Ib*) were of ventricular origin while the third (*Ic*) was of aortic origin. These components originate only in the left heart while the contribution of the right heart to the first sound is minimal and unrecognizable in the normal phonocardiogram.

This being the case, several statements of the past should be revised on the basis of our study.

(1) *The first sound is split on auscultation in bundle branch block.* This statement is meaningless because the normal first sound may be split (both on auscultation and in phonocardiograms) due to separate audition of the first (*Ia*) and second (*Ib*) components. This fact has been widely recognized since the study of Leatham on the first heart sound, even though Leatham himself did not draw a correct conclusion based on this fact.

(2) *The first sound is divided into two or three components in the phonocardiogram of bundle branch block.* This statement is meaningless because either two or three components are found in the normal first sound. Moreover, auscultation may give the impression of a splitting of the first sound due to two other possibilities:

  (a) There is a fourth sound; this plus the first component of the first sound would simulate splitting.

  (b) The third component (*Ic*) is louder than normal on account of alterations of the aorta (so-called aortic ejection sound); this, following a larger component *Ia* (and a smaller, inaudible component *Ib*), again would simulate splitting on auscultation.

Both possibilities were considered, and their occurrence was demonstrated by van Bogaert and associates in their experimental and clinical study on bundle branch block.

The theoretical possibility of a splitting of the first sound as a result of BBB was based on two assumptions: the existence of left and right ventricular (or mitral and tricuspid) components within the first sound, and the delay of activation and contraction of one ventricle in BBB.

The first assumption has been challenged by our studies,

which showed that only the left ventricle and aorta contributed to the generation of the first heart sound.

The second assumption has been challenged in LBBB (not in RBBB) by Braunwald and associates (1956, 1957) following a study of the timing of electric and mechanical events in the normal heart. These authors pointed out that lesions of the left ventricle can produce the pattern of complete LBBB by causing a conduction block in some of the left ventricular branches without delay of initiation of left ventricular contraction. For this reason, we studied only patients with a typical electrocardiographic pattern of BBB and excluded patients with congenital or rheumatic heart disease, as well as those with systemic or pulmonary hypertension. This selection tried to avoid a partial "conduction block" with an ECG pattern of left BBB. We also tried to avoid the occurrence of a one-sided overload causing prolongation of either the tension or the ejection phase of one ventricle. As experimental interruption of the entire left bundle branch causes a delayed contraction of the left ventricle (Braun Menendez and Solari), we have assumed that this may also occur in man.

Our study of the *arterial pulse* showed a normal (114 msec.) Q-carotid rise interval* in RBBB (Figs. 148, 149) and a prolonged interval (167 msec.) in LBBB (Fig. 151). One might postulate that this fact merely revealed a delay in left ventricular ejection and not a delay in left ventricular contraction. However, study of the first sound proved to be revealing for an understanding of the facts.

The beginning of the first sound was not delayed over the Q wave in *RBB* (55 msec. versus 59 msec. for normal subjects). On the contrary, this beginning was delayed over Q in *LBBB* (91 msec. versus 59 msec. for normal subjects).

In all cases, three components were observed within the first sound, and no basic difference was found in such respect between right and left bundle branch blocks, execept for the fact that the first sound is usually of better quality in RBBB (Fig. 149). The poor quality of the first sound in LBBB can be explained on two grounds, the extensive LV wall damage that is common in these cases and the fact that, when the LV is acti-

*This interval is identical with that mentioned by Braunwald and associates as representing the Q-to onset of ventricular ejection in normal man.

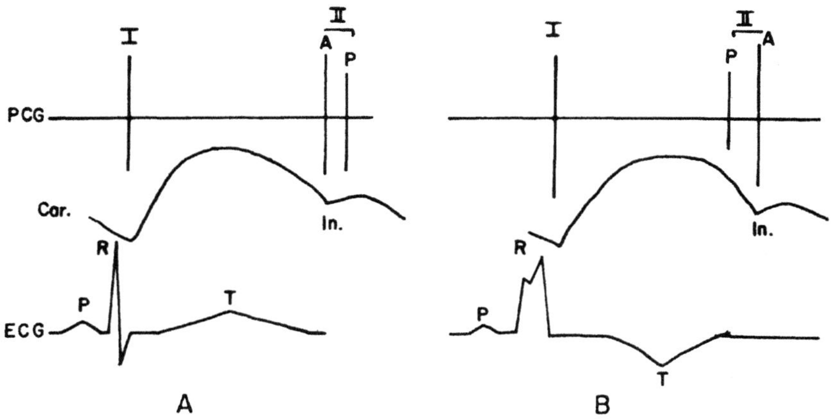

FIGURE 148. Scheme of changes of the second sound in bundle branch block.
(A) = Right. (B) = Left.

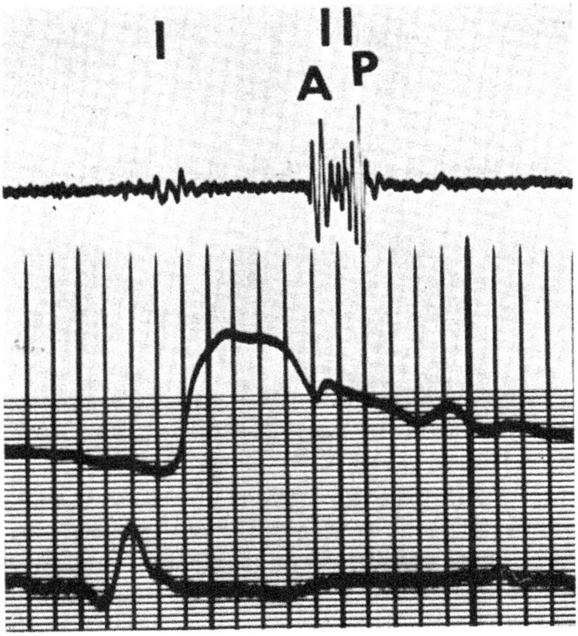

FIGURE 149. Right bundle branch block in a 58-year-old woman. PCG over the 2nd left i.c.s. and carotid tracing. No delay of the carotid rise. Coincidence of the aortic component of the second sound with the carotid incisura; delayed pulmonic component causing a wide splitting of the normal type.

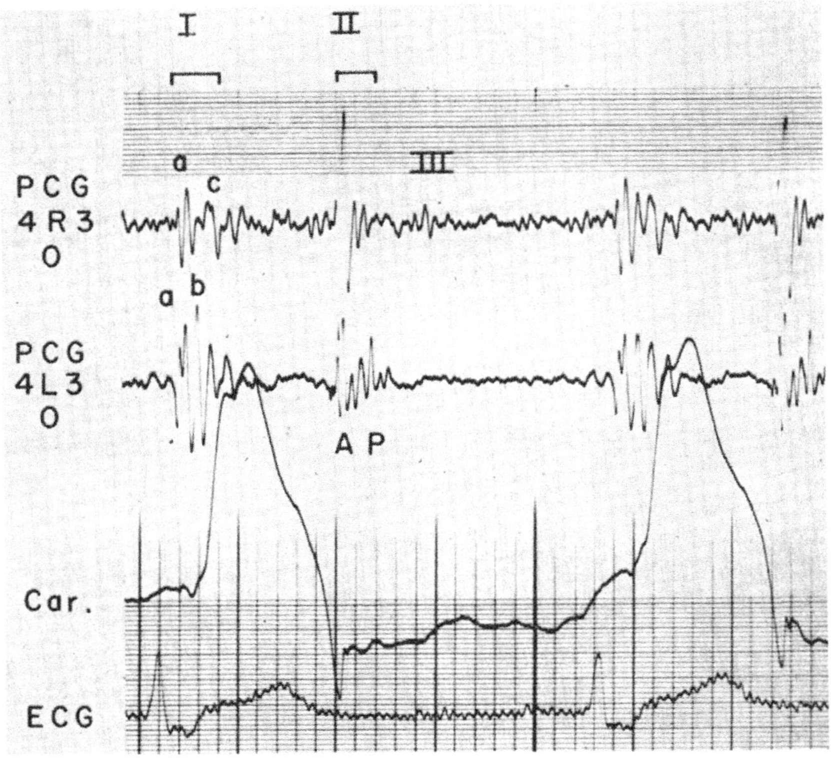

FIGURE 150. Case of right bundle branch block. The first sound at the apex is of good quality and is not delayed. The aortic component of the second sound coincides with the carotid incisura. There is a third sound. Wide splitting of the second sound of the normal type.

vated, the septum has already undergone activation and starts contracting. This might cause a certain degree of dissynergy of the left ventricle.

The interval between the first (*Ia*) and the second (*Ib*) components (*Ia* minus *Ib*) was only minimally longer in left than in right BBB (44 versus 42 msec.). The interval between the first (*Ia*) and the third (*Ic*) components (Q-Ia minus Q-Ic) was slightly longer in LBBB than in RBBB (81 msec. versus 78 msec.) but the difference was minor. The prolongation of the Q-Ic interval in LBBB was similar to that of the Q-carotid rise interval.

It was concluded, therefore, that, even though the tension

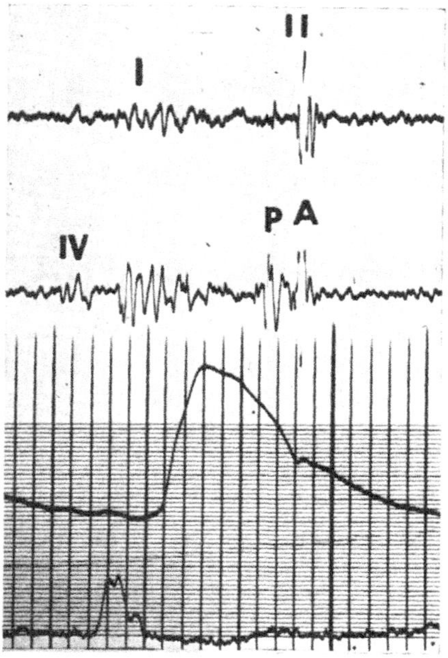

FIGURE 151. Case of left bundle branch block.
From above: PCG at the 2nd right i.c.s.
PCG at the 3rd left i.c.s.
Carotid tracing
ECG
Delay of the first sound and of the carotid rise. There is a fourth sound. The second sound is widely split but this is a reverse splitting because the second component coincides with the carotid incisura.

period is longer in LBBB than in either normal subjects or patients with RBBB, the onset of left ventricular contraction is delayed in LBBB.

The following conclusions were reached:

(1) When pressure alterations, shunts, or valvular defects causing ventricular hypertrophy or overload are excluded, ventricular asynchronism is constant in bundle branch block.

(2) The first sound, being related only to events of the left heart and aorta, shows a normal relationship to the Q wave of the electrocardiogram in right bundle branch block, a marked delay in left bundle branch block.

(3) The first sound, being related only to events of the left

heart and aorta, has three components in both types of bundle branch block.

(4) Occasional simulation of splitting of the first sound on auscultation was explained by incorrect evaluation of either a fourth sound preceding the first sound or an ejection sound. Therefore, "splitting" of the first sound is not a feature of bundle branch block.

These conclusions are similar to those already presented in regard to ventricular ectopic beats. However, a point of difference should be recognized: in LBBB, a widespread damage of the left ventricular wall is usualy present and will contribute to a decrease in amplitude of the first heart sound; in right ventricular ectopic beats, such damage is often absent, due to the thin and long structure of the right branch. Moreover, it would not modify the first sound.

Second Heart Sound. In *right BBB,* the delay in excitation of the right ventricle causes a delay in its contraction. Thus, right ventricular contraction ends late, pulmonary valve closure is delayed, and the pulmonary artery incisura is delayed. As a result, a *wide splitting of the second sound of the normal type* is constantly noted (Figs. 148-150). There still are moderate respiratory variations but the second sound is never single.

In *left BBB,* it is common to observe a *wide splitting of the second sound of a reverse type* (the aortic component follows the pulmonary componet) (Fig. 151). It has been stated that activation time of the left ventricle is still normal (Braunwald *et al.*, 1956, 1957) and that there is a prolongation of the isovolumic time of this ventricle causing a late ending of the left ventricular systole in spite of an abbreviation of the ejection time (Adolph *et al.*). However, the study of the latter, which confirmed the delay of the first sound previously found by us, was based on indirect tracings, and therefore, cannot be accepted as conclusive.

In normal man, onset of *right ventricular contraction is usually slightly delayed* over that of the left ventricle. On catheterization of 10 normal subjects, we found an average delay of 17 msec. In 10 cases of left BBB, on the contrary, the *average onset of left ventricular activation followed right ventricular activation by 11.7 msec. (Fig. 152).* This points out that a total delay of 28.7 msec. is present in LBBB in comparison with normal subjects. Additional prolongation of isovolumic contraction was often present, so that onset of the aortic pulse *followed* that of the pulmonic pulse by an average of 46.2 msec. This should be compared with a normal average of 26.5 msec. (Luisada, Kumar and Pouget). As a result of these two changes (delayed activation of LV plus prolonged isovolumic period of LV), even though LV ejection may be shortened, the aortic valve closes *after* the pulmonic, and the incisura of the aortic pressure pulse occurs *after* that of the pulmonary pressure pulse. The two components of the second sound occur at the time of the respective incisuras, and their relationship results in a *reverse* splitting of the second sound with the aortic component following the pulmonic component.

WOLFF-PARKINSON-WHITE SYNDROME

In this syndrome, pre-excitaion causes early activation of either a part of one or both ventricles or an entire ventricle. In the last occurrence, a situation similar to that of a ventricular ectopic beat would arise.

March *et al.* found early completion of contraction of both ventricles in 1 case (normal second sound), early completion of left ventricular contraction in 1 case (wide splitting), and late completion of both ventricles in 3 cases (normal second sound). Zuberbuhler and Bauersfeld studied 4 cases of the *type B* of the W.P.W. syndrome, i.e., those with early activation of the right ventricle. In 3 of the 4 cases, catheterization studies revealed early onset and early end of contraction of the right ventricle. This explained the *reverse* splitting of the second sound with widening in expiration that was presented by the same cases. In other words, cases of type B behave like cases with either right ventricular ectopic beats or right ventricular pacing.

Chapter 12

Diseases of the Myocardium

AUSCULTATION OF THE HEART and phonocardiography are of value for the diagnosis and evaluation of patients with myocardial disease. This group includes cases of myocarditis; infiltration of the myocardium by tumor or amyloidosis; endomyocardial fibrosis or endocardial fibroelastosis; metabolic and endocrine heart disease; cardiomyopathies assocated with neuromuscular disorders (Levine and Harvey, Harvey and Perloff); and conditions caused by alcohol, vitamin B deficiency, or of unknown origin.

Some of the auscultatory data are the result of either conduction disturbances or ectopic rhythms, associated pericardial disease, mitral insufficiency, or heart failure. Others are directly connected with the aspects of myocardial contraction.

THIRD OR FOURTH SOUND--TRIPLE OR QUADRUPLE RHYTHMS

A *triple rhythm (gallop rhythm),* caused by the addition of a diastolic sound, is frequent. A summation type of triple rhythm can be observed in patients with tachycardia, whereas a *quadruple rhythm* may occur in the less frequent cases with bradycardia. A ventricular type of triple rhythm (increase of the third sound) is more often found in patients with heart failure, whereas an atrial type (increase of the fourth sound) is more common with prolonged AV conduction and in hypertensive patients (Fig. 33-47).

Occasionally a quadruple rhythm may simulate on auscultation an apical diastolic rumble or there may be a real *diastolic rumble* (more seldom a diastolic-presystolic rumble) of a functional nature (Luisada *et al.*, 1959, 1972;/Rushmer *et al.*, Harvey and Perloff) (Chap. 27—Fig. 75).

Second Sound. The pulmonary component of the second sound is usually louder and of greater magnitude if there is left ventricular failure that causes a secondary pulmonary hypertension.

Bundle Branch Block. In bundle branch block, various abnormal data can be incountered, such as (1) right BBB with poor magnitude of the first sound and wide splitting of the second sound; or (2) left BBB with poor magnitude of the first sound and either single second sound or reverse splitting of the second sound (Figs. 148-152).

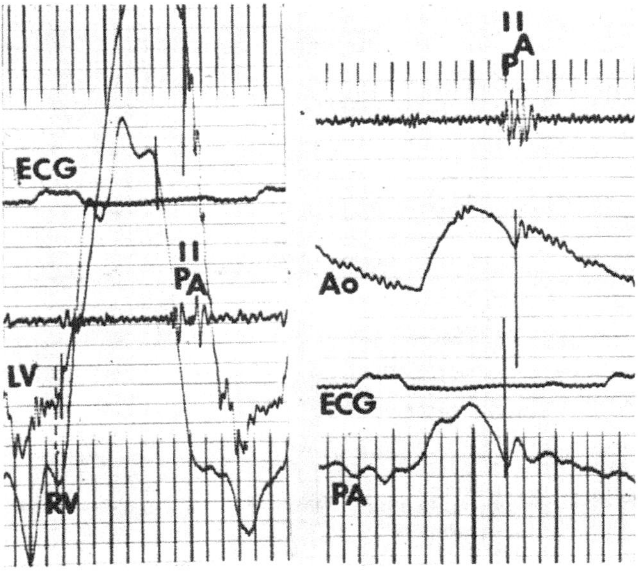

FIGURE 152. Left bundle branch block in a 31-year-old woman.
At left — Pressure tracings of the two ventricles at same amplification.
PCG, ECG
RV = 45/6; LV = 94/15
The left ventricular pressure rise starts slightly after the right (10 msec); its end is much after the right. At right — Arterial pressures. The aortic incisura is delayed over the pulmonic incisura.

In either type, but more ofter in left BBB, seeming *splitting of the first sound* can be heard. This is usually an incorrect observation because such impression is due to the existence of an atrial type of triple rhythm. On the other hand, the phonocardiogram can reveal the existence of a wide splitting. In such cases, the second audible group of vibrations usually corresponds to the third component of the first sound, which has greater magnitude because of aortic dilatation (*ejection sound*), a fact that was demonstrated by van Bogaert *et al.* (1963). Triplication of the first sound has been described in BBB, a fact that was incorrectly interpreted by some authors and was probably caused by the graphic transcription of the three normal components of the first sound.

Systolic Murmur. A systolic murmur is frequently found in primary myocardial disease (Harvey and Perloff); it may be either a flow murmur of the pulmonary artery or a murmur of relative mitral or tricuspid insufficiency.

Blowing, Early Diastolic Murmur. The blowing, early diastolic murmur of aortic insufficiency, sometimes heard and recorded during *acute rheumatic carditis,* may disappear later. It is not known whether the murmur is caused by dilatation of the aortic ring or reversible edema of the aortic leaflets.

Middiastolic and Presystolic Murmurs. These are usually noted in cases with myocarditis or cardiomyopathy in failure. An example was presented in Fig. 75. Others, caused by rheumatic carditis, will be discussed below.

Murmurs in Endomyocardial Fibrosis. Murmurs of mitral and tricuspid regurgitation and triple or quadruple rhythm have been described in endomyocardial fibrosis (Somers and Williams).

Friction Rub. The friction rub of "dry" pericarditis or the dull, faint heart sounds of pericarditis with effusion are observed whenver a pericardial involvement is associated with the myocardial disease. This is more common in both viral myocarditis and rheumatic myocarditis.

Alternans. If there is alternans, the phonocardiogram shows smaller first and second sounds every other beat (Fig. 153). The apex cardiogram and the carotid tracing also show smaller pulsations every other beat. The murmurs have the same be-

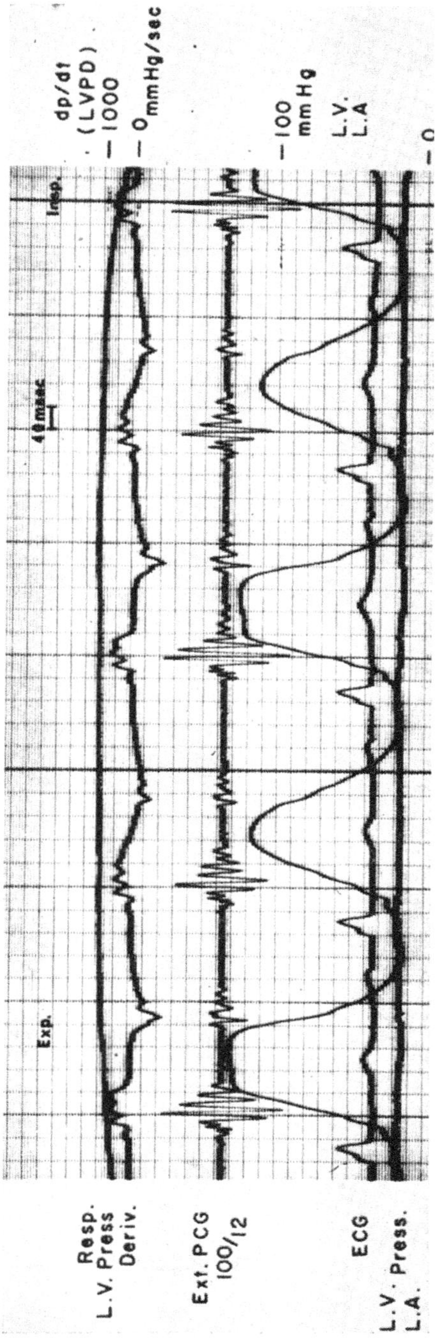

FIGURE 153. Experimental alternans in a dog.

havior with one exception—that of muscular subaortic stenosis, where the smaller beat may be accompanied by a louder murmur.

Rheumatic Carditis

During acute or subacute rheumatic fever, all structures of the heart are involved by inflammation. Thus, endocardial inflam-

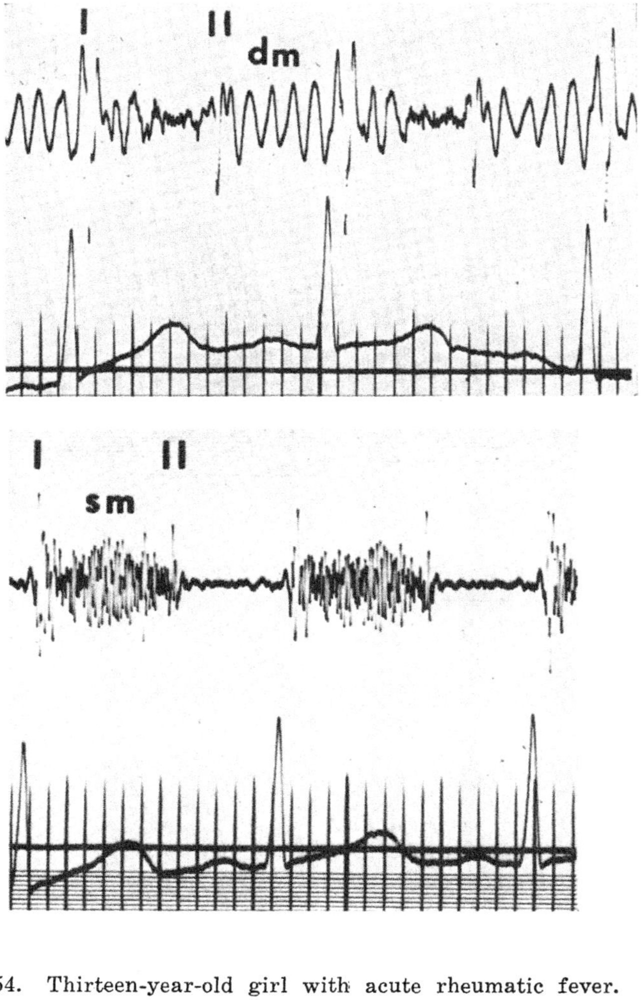

A

B

Figure 154. Thirteen-year-old girl with acute rheumatic fever.
(A) Low frequency tracing at apex (displacement, 20-300 Hz) showing large diastolic vibrations.
(B) High frequency tracing at apex (acceleration, 10-400 Hz) showing a large pansystolic murmur.

mation of the valves and chordae, dilatation of the chambers caused by myocarditis, and pericardial inflammation may occur at the same time or in rapid succession. Quite often it is difficult to decide if a certain murmur is caused by valvular or myocardial damage.

A frequent occurrence is that of a young patient with marked cardiac enlargement and both *systolic* and *diastolic murmurs* (Figs. 74 and 154). All murmurs, or some of them, may persist after recovery indicating that evolution of valve damage has occurred. On the other hand, frequently most of the murmurs later disappear, and this is evidence that they were caused by cardiac dilatation and abnormal cardiac function plus possible edema of valve structures.

At times, one can hear the diastolic murmur of aortic insuf-

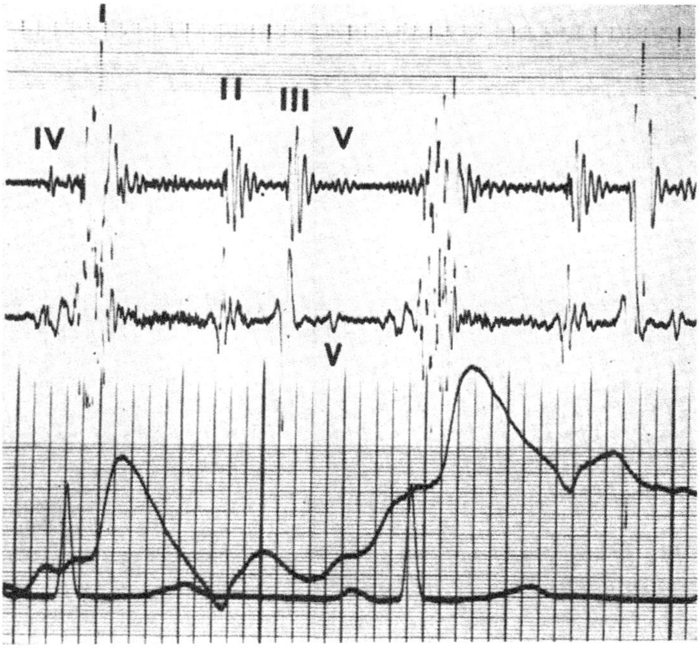

FIGURE 155. Twenty-one-year-old woman with sickle cell anemia. Tracings at apex.
From above: displacement tracing (20-200 Hz), velocity tracing (unfiltered), carotid tracing, ECG.
Large first sound, gigantic third sound, fourth sound, fifth sound.

ficiency, and this may disappear after two or three months indicating that it was caused by either dilatation of the aortic ring or edema of the cusps.

A *rumbling middiastolic and presystolic murmur* may vary from a simple increase of the third and fourth sounds to a murmur that simulates that of mitral stenosis (Figs. 74 and 154).

Hyperthyroidism-Vitamin B Deficiency

The increased circulatory activity of *hyperthyroidism* is accompanied by alterations of the heart sounds and frequently by murmurs (so-called *hyperkinetic syndrome*). The first sound is loud and snapping, sometimes simulating that of mitral stenosis (Ueda *et al.*, 1963) but the Q-I interval is normal. This fact can be explained by a more rapid rise of left ventricular pressure related to the endocrine disturbance.

The Q-II interval is abbreviated more than accounted for by the heart rate; the second sound is often widely split and has a large pulmonary component (Renfro *et al.*; Ueda *et al.*, 1963).

It is common to observe an atrial or ventricular type of *triple rhythm* (Renfro *et al.*; Ueda *et al.*, 1963).

A *diamond-shaped systolic murmur* with an early peak (obviously a *flow murmur*) is often observed. A *rumbling diastolic murmur* at the midprecordium or apex can be observed; it starts soon after a large third sound, has a higher pitch than that of the murmur of rheumatic carditis, and is favored by cardiac dilatation (Ueda *et al.*, 1963).

A *thyroid bruit* was found by Ueda *et al.* (1963) in about one-half of their patients, and a *venous hum* was recorded in nearly all cases by Renfro *et al.* Vitamin B deficiency may cause a picture similar to that of hyperthyroidism.

Anemia

Patients with anemia frequently have cardiac murmurs, especially the so-called "innocent" or "flow murmurs." Even though these murmurs are *not* related to valvular lesions, they are definitely related to changes in cardiovascular dynamics. Tachycardia, higher cardiac output, and dilatation of the cardiac chambers result from the low hemoglobin level. The myocardium is stressed while its oxygen supply tends to decrease. Thus, either diffuse or tiny and widespread areas of necrosis can develop in the subendocardial layers of the left ventricle, also involving the papillary muscles. It is obvious then that myocardial damage may add one more cause of murmurs, especially in chronic anemias, and that the damage is not reversible.

In a recent study from our laboratory (Argano), the *heart sounds* were found normal in anemia. The most common finding was a crescendo-decrescendo murmur at the left of the sternum (2nd left i.c.s.), obviously a *pulmonary flow murmur*. A *high-pitched, early-diastolic murmur* in decrescendo (aortic or pulmonary insufficiency) was recorded in 4 out of 20 cases. A *functional mid-diastolic rumble* was recorded in 2 cases.

In sickle-cell anemia, a *diamond-shaped murmur* was present in 9 out of 10 cases. A *third sound* was present in 9 out of 10 cases and assumed a prolonged configuration (*mid-diastolic rumble*) in 6. Five cases had a *fourth heart sound*. A *pansystolic murmur* of mitral insufficiency was present in 1 case.

Thus, patients with sickle-cell anemia have more often both diastolic sounds and diastolic murmurs (Figs. 155-158) than other patients with anemia.

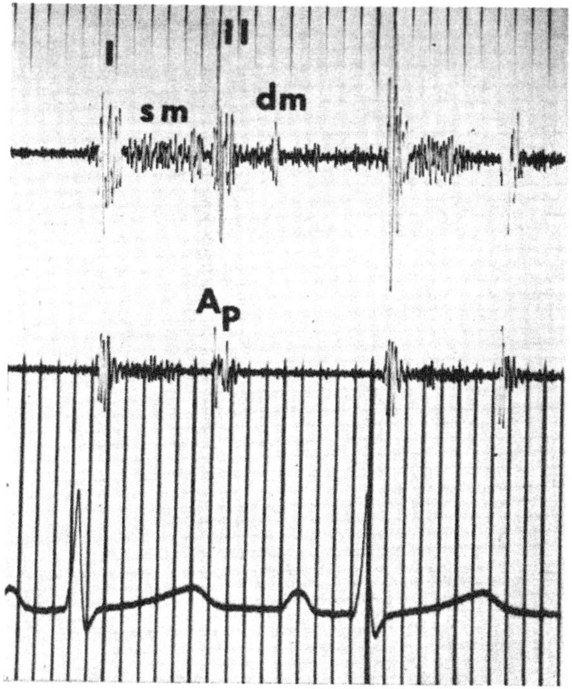

FIGURE 156. Twelve-year-old girl with sickle cell anemia.

From above: velocity tracing at apex, velocity tracing at apex, velocity tracing at 2nd left i.c.s. (both with a 200 Hz high pass filter). Diamond-shaped systolic murmur, large III sound within a series of diastolic vibrations (dm).

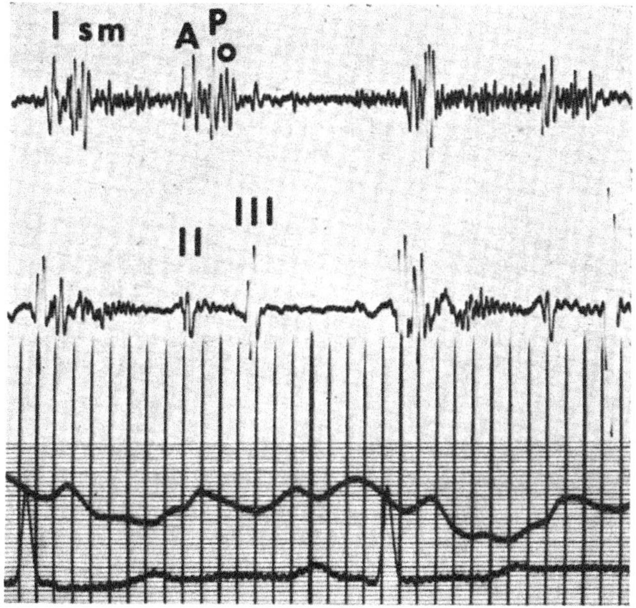

FIGURE 157. Same case as in Figure 154, on a different date. Tracings over 2nd left i.c.s. (above) and apex (center) with high pass filter at 100 Hz. Low grade systolic murmur. Split second sound. Opening sound. Large third sound.

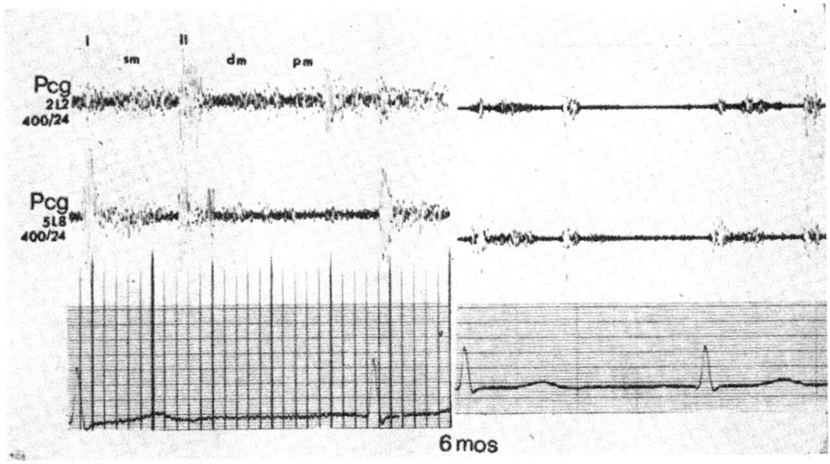

FIGURE 158. Ten-year-old boy with sickle cell anemia. Tracings over 2nd left i.c.s. (above) and apex (center).
At left: during an acute crisis (systolic and distolic-presystolic murmurs).
At right: six months later, after recovery (diamond-shaped systolic murmur, minimal early diastolic murmur).

Chapter 13

Coronary Heart Disease

MYOCARDIAL INFARCT

FIRST HEART SOUND. In acute myocardial infarct, the first heart sound is decreased on auscultation only in 24% of the cases (Master and Friedman). In phonocardiograms, a reduction of the first sound is again found only in some of the cases. However, more frequently, there is a downward shift in frequency so that this sound may be poorly audible even though it may have the same magnitude. Decrease in amplitude was found more commonly in inferior than in anterior infarcts (Price and Brown) ; this fact might be related to a greater size of the infarcted area, as well as to more frequent involvement of a papillary muscle, in the inferior infarct.

Decrease in the amplitude of the first sound in experimental myocardial infarct was found by Sakamoto *et al.* in our laboratory. A slower rise of left ventricular pressure was noted, revealed by a decrease of the dp/dt. This fact was explained by the elastic distension of the infarcted area. Of course slower rise would cause both a downward shift in frequency and a decrease in amplitude (Fig. 159).

Second Heart Sound. The second sound may be *single* or may show a reverse type of *splitting,* greater in expiration. When this phenomenon is noted in the absence of left BBB, it is

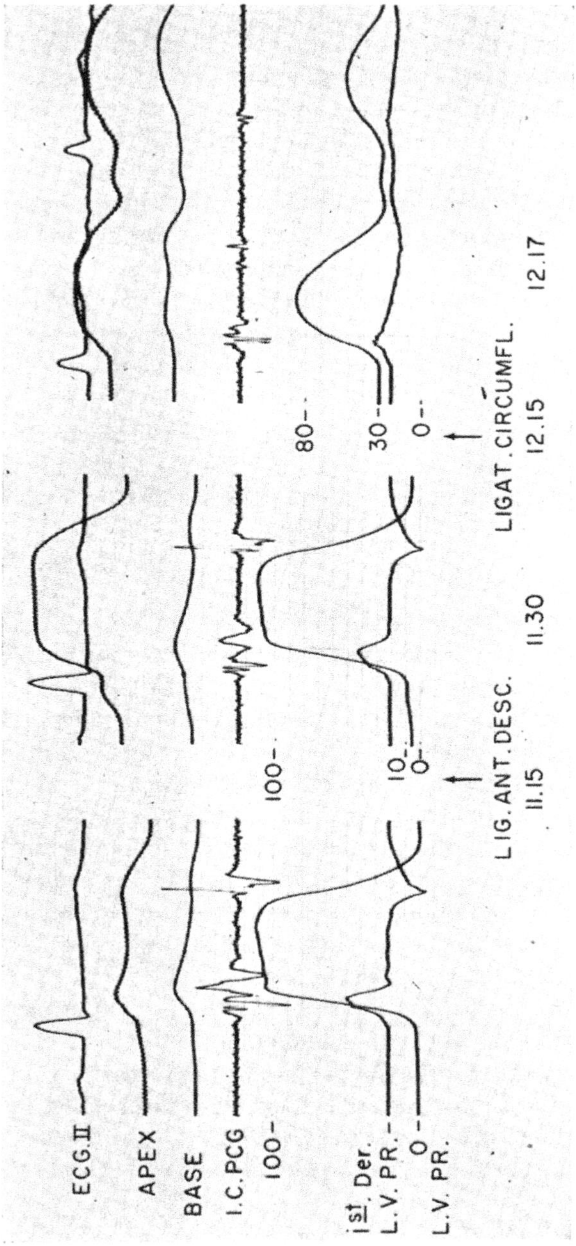

FIGURE 159. Effect of experimental ligature of coronary arteries in a dog on the pulsations of the apex and base (upper two tracings), on the first heart sound, and on the left ventricular pressure as well as its first derivative (dp/dt).

significant, and reveals the effect of prolonged left ventricular systoles, related to left heart failure.

Third and Fourth Sounds. An increase of the third or fourth sound (or their summation) is frequent in the acute stage of an infarct. This is most likely related to an increase in the diastolic pressure of the left ventricle, be this caused by left heart failure or the mechanical effect of a soft area of the wall (*dynamic aneurysm*). According to Giorgetti *et al.*, the third sound progressively becomes more distant from the second (up to 200 msec.) while the fourth gradually approaches the first until it may fuse with it (in the absence of changes in the AV conduction time). Other studies (Gola and Vitols) have revealed an abbreviation of the P-IV interval without, however, explaining it.

Friction rubs are typical in acute myocardial infarction. However, they may last only a few hours and are usually heard and recorded best in the anteroseptal, frank anterior, or anterolateral infarcts.

ACUTE COMPLICATIONS OF MYOCARDIAL INFARCT

(1) *Murmur of mitral insufficiency*

According to Heikkila, 55% of his patients with an acute infarct developed a murmur of mitral insufficiency within the first 5 days. This murmur exhibited a remarkable variability in regard to type, intensity, and permanence even in the same patients. It was apical and *pansystolic* in 91%; *ejection type* initially, then changing to *pansystolic* later, in the others. Whenever the murmur was ejection type or decrescendo, the insufficiency was mild. The determinant cause of the murmur seems to be papillary muscle dysfunction, especially in regard to the posterior papillary muscle.

A *functional diastolic rumble,* at times with presystolic accentuation, may occur (Scherf and Brooks).

Another cause of a *large pansystolic* murmur is *rupture of a papillary muscle.* The murmur is then loudest at the midprecordium (Fowler, 1969) and radiates to the left axilla. Survival is exceptional but possible, for some time at least.

(2) *Perforation of the ventricular septum*

This is relatively more common in anteroseptal infarcts. The murmur is *pansystolic* and loudest at the 4th and 5th i.c.s., close to the sternum (Fowler, 1969). If perforation is small and survival is possible, the differential diagnosis with a ruptured papillary muscle is difficult.

(3) *Rupture of the ventricular wall*
This causes *tamponade* and may be revealed by *small heart sounds* and the presence of *friction rubs*.

(4) *AV block, bundle branch block*
The typical changes of the heart sounds caused by these disturbances of conduction are noted.

(5) *Ectopic beats, ventricular tachycardia, atrial fibrillation or flutter*
Again, the typical changes related to the arrhythmia or pathological tachycardia are observed.

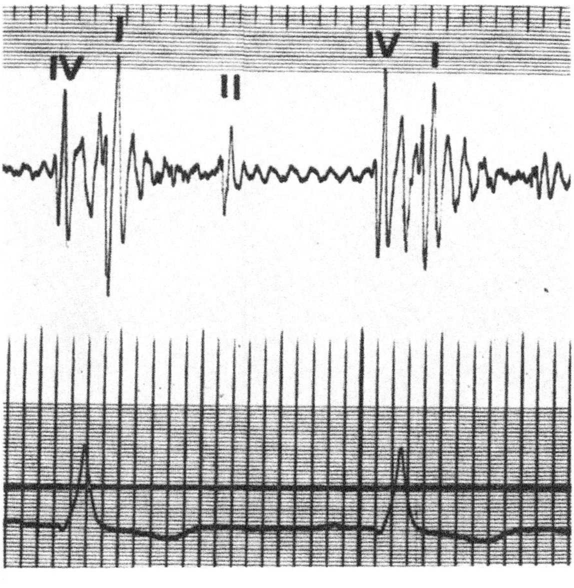

FIGURE 160. Large fourth sound in a patient with coronary heart disease and a history of hypertension.

APEX CARDIOGRAM

A systolic bulge (*pansystolic plateau*) can be recorded in one or more precordial areas or at the epigastrium (Davie *et al.*).

The precordial pulsation is often outside the point of maximal impulse.

Vakil (1956) found an abnormal pulsatory pattern above and within the apical impulse in 5.7 per cent of his patients during the acute stage of myocardial infarction, usually appearing within the first week of illness. This pattern was transient in 9 patients and permanent in 2 patients. This abnormal pattern, which corresponds to that of the "dynamic aneurysm" described by Luisada and Fleischner, makes it difficult to decide, by means of this method, whether a permanent ventricular aneurysm has developed.

Chronic Changes Following an Infarct

Diastolic Sounds. Following a myocardial infarct, it is common to hear and record diastolic sounds, the most common being a *large third sound.* Occasionally a *large fourth sound* (especially in old hypertensive patients) can be observed (Fig. 160). And, of course, both a *third and a fourth sound* (or a summation sound) can be recorded (Fig. 161). These sounds can also be recorded in patients with a ventricular aneurysm or a papillary muscle syndrome.

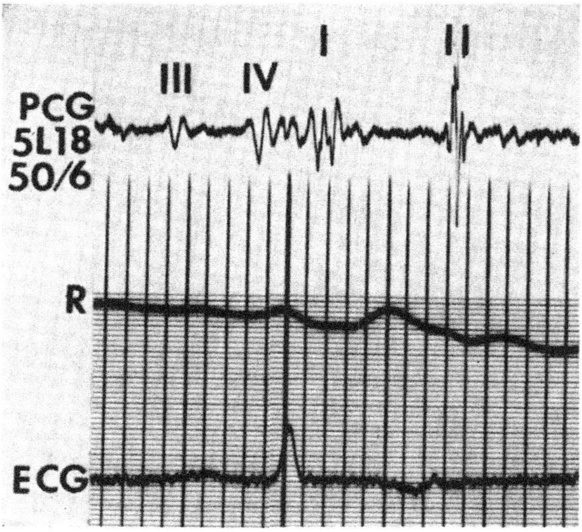

FIGURE 161. Large third and fourth sounds in a patient with coronary heart disease and history of an old infarct.

Murmurs. A *pansystolic murmur* (mitral insufficiency) and a *diastolic-presystolic* murmur of functional nature can be recorded (Figs. 162 and 163).

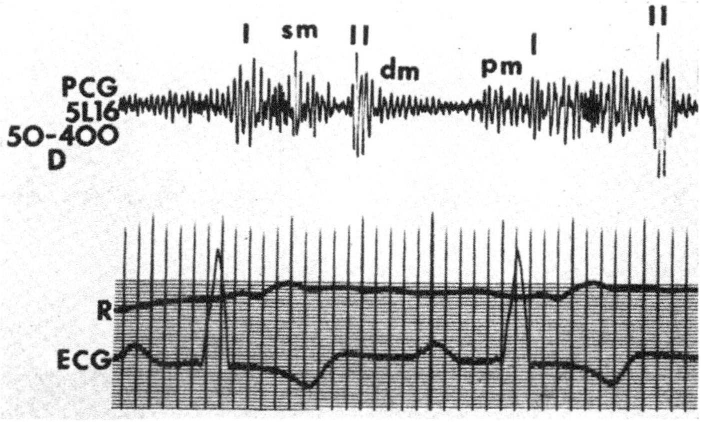

FIGURE 162. Pansystolic murmur recorded in a 47-year-old patient with coronary heart disease, history of infarct, and evidence of papillary muscle syndrome.
Tracings recorded over the 2nd left i.c.s. and the apex with high pass filter at 100 Hz.

(1) *Papillary muscle dysfunction.* Described by Burch *et al.* (1963), this complication may occur even as a result of an intramural infarct. The murmur may be *diamond-shaped* (Burch and Di Pasquale, 1968) but may be *pansystolic* (Fig. 162), occasionally with a diamond-shape configuration (Fig. 164). A *late-systolic murmur in crescendo,* often preceded by a *midsystolic or late-systolic click* (Barlow *et al.*), can also be recorded (Figs. 165 and 166).

(2) *Ventricular aneurysm.* While a ventricular aneurysm may not be revealed by typical auscultatory and phonocardiographic findings, occasionally unusual phenomena may be present: musical systolic murmur (Scherf and Brooks, Vakil, Fisher and Oo), short apical systolic murmur (Parkinson *et al.*, McGinn *et al.*), early-diastolic murmur or diastolic-presystolic murmur (Scherf and Brooks, Fisher and Oo). Twenty-five per cent of Vakil's

cases had musical murmurs in both systole and diastole. The diastolic murmur usually starts immediately after the second sound (or even before it), and seems caused by the blood of the aneurysm that flows back into the main chamber of the left ventricle.

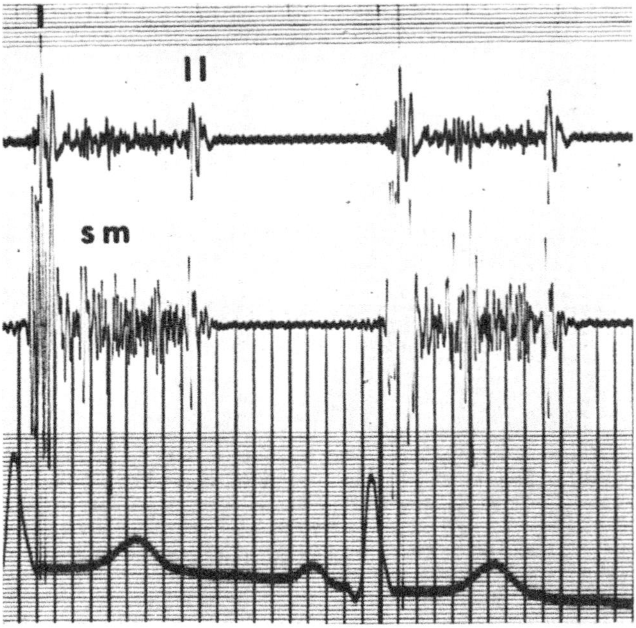

FIGURE 163. Tracings recorded at the apex in a 63-year-old man with hypertensive and coronary heart disease. Displacement tracing at 50-400 Hz. Systolic and diastolic-presystolic murmurs.

A *large fourth sound* or a *quadruple rhythm* are recorded in certain cases (McGinn *et al.*).

A singular case was studied by us. While auscultation failed to reveal unusual murmurs, phonocardiography revealed both a bizarre, high-frequency presystolic murmur and a bizarre, high-frequency, early-diastolic murmur (Figs. 168-170). They were obviously too faint for auscultation.

APEX CARDIOGRAM

(1) *Ventricular aneurysm without thrombosis.*

Slow upstroke, delayed peak of a sustained systolic wave,

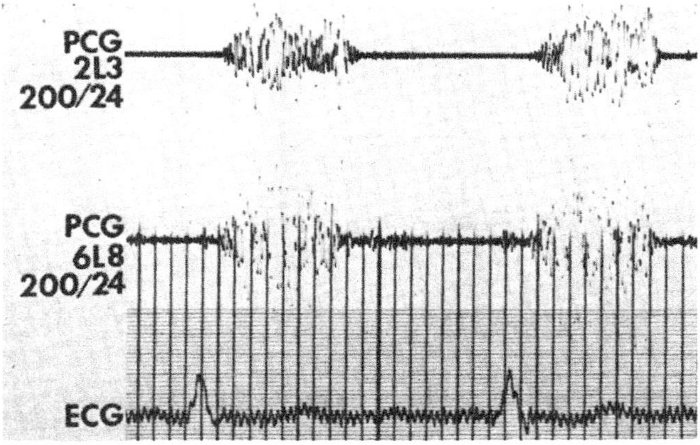

FIGURE 164. Pansystolic murmur with diamond-shaped appearance in a 61-year-old patient with a history of infarct and a shortened mitral chorda (autopsy). Small diastolic vibrations.

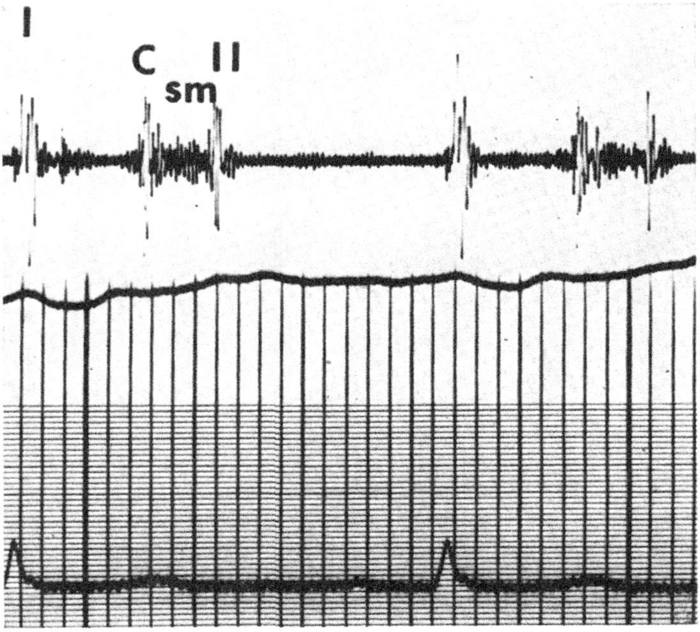

FIGURE 165. Tracing recorded at the apex in a 60-year-old man with a history of myocardial infarct and a papillary muscle syndrome. Late-systolic click (C) and late-systolic murmur. Velocity tracing with a high pass filter at 200 Hz.

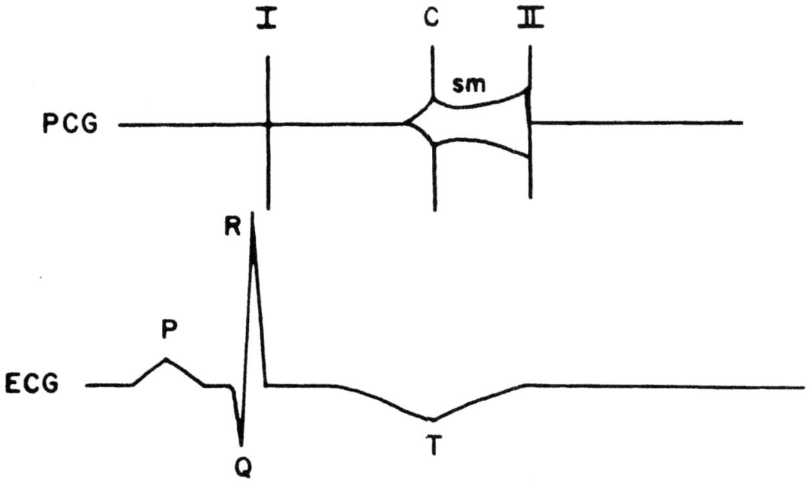

FIGURE 166. Scheme of the systolic click and of the systolic murmur in patients with papillary muscle dysfunction.

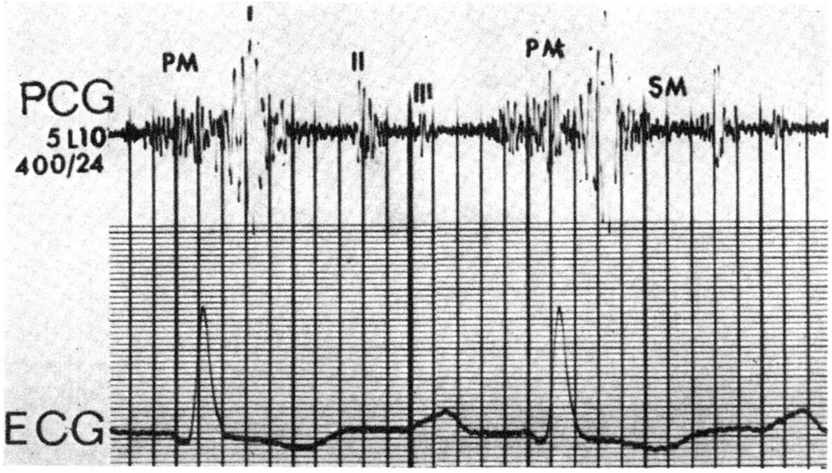

FIGURE 167. Patient with ventricular aneurysm. Large presystolic murmur, small systolic murmur, high frequency third sound (or opening snap).

and absence of the rapid filling wave have been described (Ahuja *et al.*).

(2) *Ventricular aneurysm with thrombosis.*

A double systolic peak, a midsystolic bulge, and a rapid wave of filling can be recorded (Ahuja *et al.*).

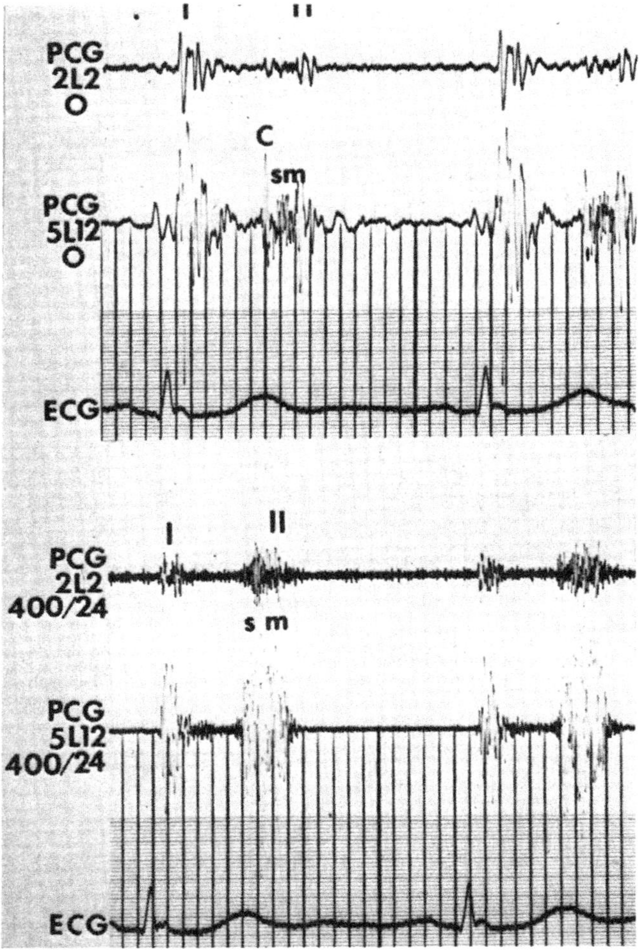

FIGURE 168. Tracings recorded over the 2nd left i.c.s. and the apex in a 58-year-old patient with ventricular aneurysm.
Low frequency tracings (upper series) show a fourth sound, a late-systolic click (C) and a late systolic murmur (sm). High frequency tracings show that there is a pansystolic murmur at the apex (with some transmission to the base) and with marked late-systolic accentuation. A papillary muscle may also have been involved.

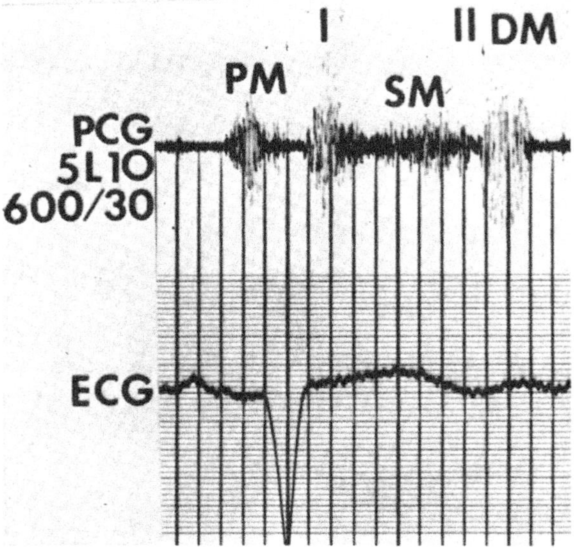

FIGURE 169. Tracings recorded at the apex in a 66-year-old woman with a ventricular aneurysm. High frequency tracings record a presystolic murmur (not in crescendo), a pansystolic murmur with late accentuation, and a very large early-diastolic murmur, well separated from the second sound and not in decrescendo. (From Luisada and Argano: *Dis Chest*, 1969).

CORONARY HEART DISEASE WITHOUT INFARCT, ANEURYSM, OR LEFT BBB

PHONOCARDIOGRAM

The phonocardiogram of these patients may show a *reverse splitting of the second sound* and a *large third sound.* These changes may either occur acutely during an attack of precordial pain or be persistent (MacDonald; Seymour *et al.*; Crampton; Agnew *et al.*). The change of the second sound has been attributed to prolongation of left ventricular systole.

APEX CARDIOGRAM AND CARDIOGRAMS OVER OTHER AREAS

There is an *increased height of the A wave* (often correlated with an atrial type of triple rhythm) (Benchimol and Dimond), and there is a systolic bulge (*pansystolic plateau*) in one or more precordial or epigastric areas, more frequently in anterior infarctions. The maximum of this paradoxical impulse is most often recorded above or within the point of maximal impulse (Davie *et al.*; Suh and Eddleman; Roervik).

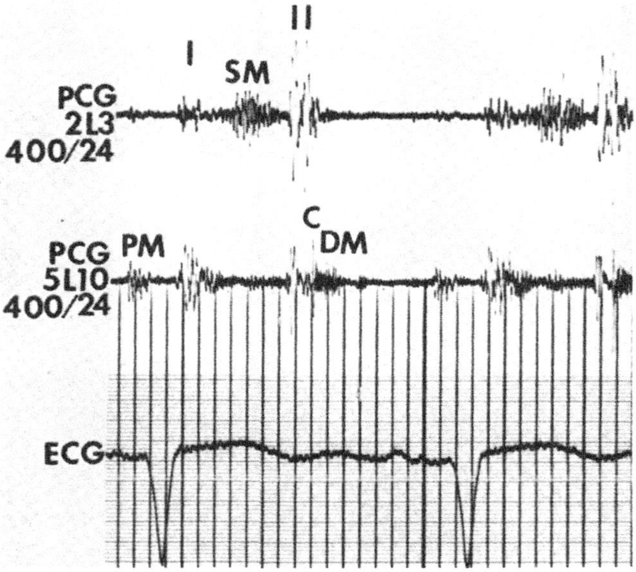

FIGURE 170. Same patient as in Figure 169. Tracings recorded with less high filtration reveal that the early-diastolic murmur is initiated by a click (C).

Of great interest is the fact, ascertained by Harrison, that several abnormalities that are found at rest in patients with old infarcts may also appear acutely during the episode of precordial pain in patients with angina pectoris. This seems to indicate a temporary weakening of part of the left ventircular wall during the transient episode of ischemia.

The "accelerogram" has been extensively studied in coronary or ischemic heart disease. Rosa *et al.* (1962 A) found abnormal tracings* in over 94 per cent of their patients with old or recent myocardial infarct and in 66.6 per cent of their patients with angina pectoris.

---

*Rosa *et al.* evaluated the tracing according to (1) number of distorted phases, (2) pulsatory abnormalities in diastole or systole, and (3) presence or absence of major respiratory variations. None of these data were considered specific. However, while the diastolic abnormalities (obviously connected with the various types of gallop rhythms) were also found in patients with hypertension, they were more common in coronary patients. In particular, the presystolic abnormality was found in a high percentage of patients.

Chapter 14

Pericarditis

PERICARDITIS MAY DEVELOP ACCORDING to three stages: "dry" pericarditis, pericarditis with effusion, and constrictive pericarditis.

"DRY" PERICARDITIS

In this stage, fluid accumulation in the pericardial sac is minimal and the typical manifestation is that of *pericardial friction rubs.*

Pericardial rubs are caused by the slashing noise of thick sero-fibrinous exudate due to the cardiac motion. They are typically found in "dry" pericarditis but can be heard also in pericarditis with effusion.

Pericardial rubs are easily audible as groups of high-pitched vibrations that seem to be "superficial" or "extraneous," i.e., not caused by intracardiac noise. This psychologic interpretation of the vibrations is not reflected in the phonocardiogram. The auditory characteristics of a rub are variable, giving the impression of soft, grating, or scratching noises (Sahli).

Pericardial rubs may be transient and may be localized, but the last finding is not constant. Inspiration may increase the loudness of rubs or leave them unchanged. On the other hand, *pleuropericardial rubs* (pleural rubs with a cardiac cycle) are deeply affected by respiration, may disappear in one of its phases

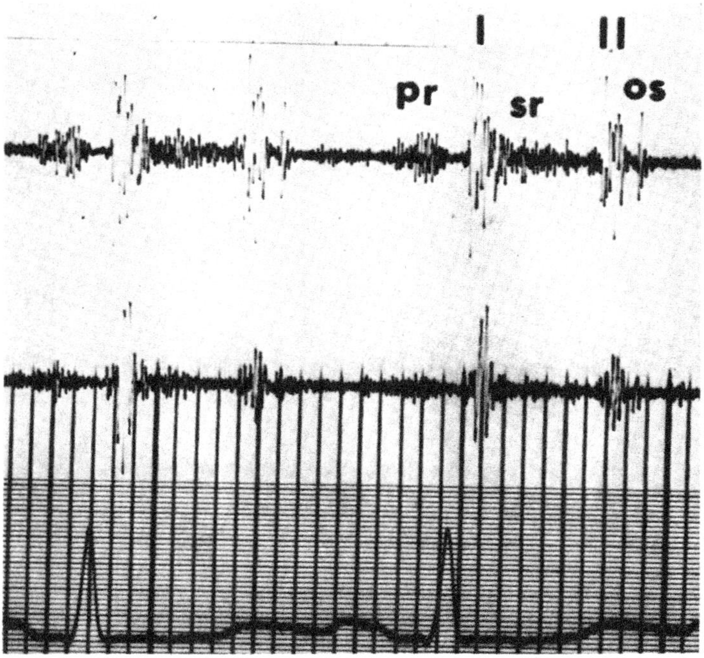

FIGURE 171. Tracings recorded over the 2nd left i.c.s. (upper tracing) and the apex (lower tracing) in a patient with mitral stenosis after commissurotomy. Friction rubs in presystole (pr), systole (sr) and early-diastole (dr).

or with held respiration, and are usually present along one of the margins of the heart, usually the left.

A rapidly changing graphic picture of the pericardial rub has been observed by Holldack *et al.* (1959). They also noted sudedn changes in amplitude, even from beat to beat.

The graphic characteristics of the friction rubs have been described in Chapter 5 (Figs. 53-55).

A less common cause of rubs is *pericardiotomy* performed in occasion of a cardiac intervention (Fig. 171).

PERICARDITIS WITH EFFUSION

In this stage, the rubs decrease or disappear, and *the heart sounds* may become fainter on account of the transmission through the fluid. Weaker cardiac action, caused by either inflammation of the heart muscle (myocarditis) or encroachment on venous return, may be another cause of fainter heart sounds, especially the first sound.

If the pericardium is tapped and air is introduced, a typical "squashing" noise may develop (*bruit de moulin*) (Fig. 172).

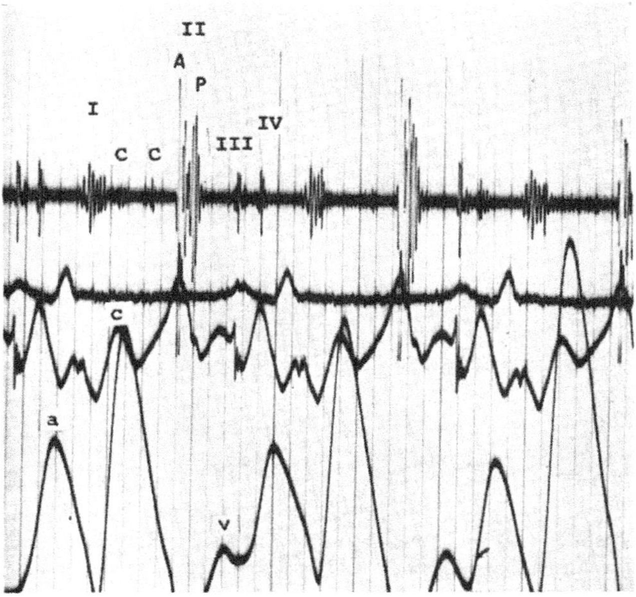

FIGURE 172. Tracings recorded in a 63-year-old man with tuberculous pericarditis following paracenthesis and injection of air in the pericardial sac. Auscultation revealed a clicking, thrashing rhythm *(bruit de moulin)*. From above = phonocardiogram recorded at the midprecordium with a 120-240 Hz filter (24 db/oct. slopes); electrocardiogram; apex cardiogram; jugular tracing.
There is a close splitting of the 2nd sound; there are two small clicks (C,C) in systole and two large ones (III, IV) in diastole.

Friction rubs may be still present in this stage due to the posterior accumulation of the fluid and the "floating" of the heart with contact of the apex with the pericardial sac, at least in some phase of the cycle.

A *late-systolic sound,* occurring in the very last part of systole, was described by Kay *et al.* and was attributed to a forward thrust of the heart, suspended in fluid, against the pericardial sac and the chest wall.

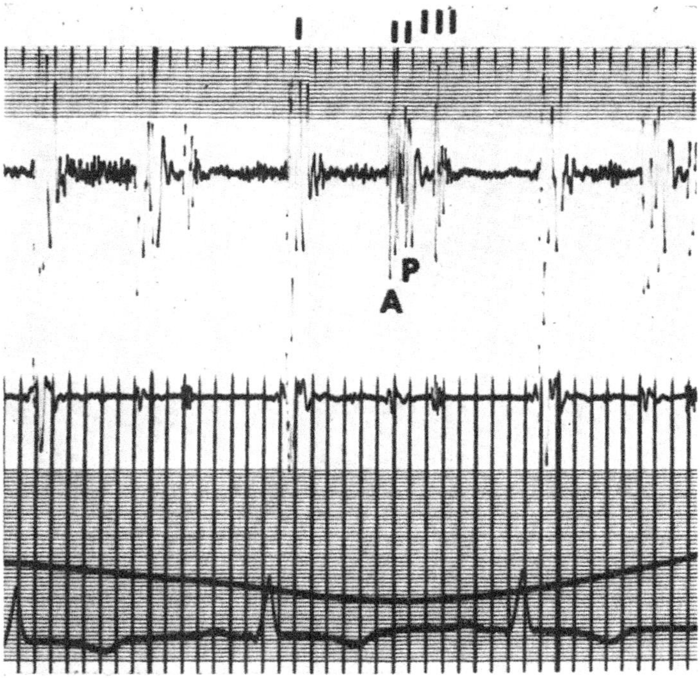

FIGURE 173. Tracings recorded in a 54-year-old woman with pure mitral insufficiency and extensive pleuropericardial adhesions. There is a wide splitting of the second sound and a large, snapping vibration in diastole (130-140 msec. after the aortic component), which should be considered as a pericardial knock. Diastolic and systolic murmurs.

ADHESIVE AND CONSTRICTIVE PERICARDITIS

Systolic Click. A *midsystolic* or *late-systolic* click of high frequency is often found in adhesive pericarditis, especially the form involving discrete bands of fibrous tissue that can be tensed by cardiac dynamics. This was termed *"bruit de triolet."* The significance of this click (without systolic murmur) has already been discussed (Chapter 4) (Fig. 48).

Pericardial Knock. The pericardial knock (or click) is found in early diastole in patients with calcific or constrictive pericarditis (Potain, Smith). This sound, which is snapping and high-pitched, may be both heard and palpated.

In graphic tracings, this sound appears as a vibration of high frequency (50 to 200 Hz), often higher than that of a third sound, even when the latter has the characteristics of "gallop."

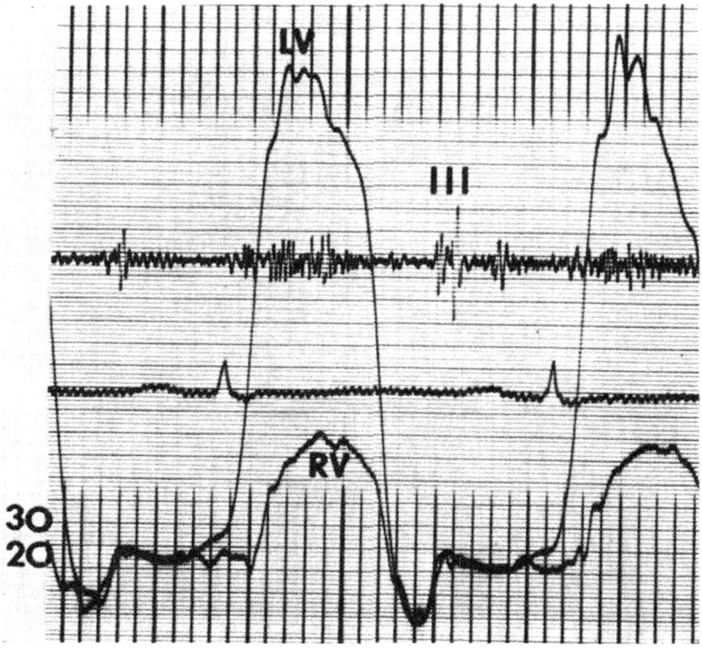

FIGURE. 174. Tracings recorded on catheterization in a 52-year-old patient with constrictive and calcific pericarditis and right bundle branch block.
Both ventricles have a typical early-diastolic dip followed by a diastolic plateau.
The external phonocardiogram shows a double pericardial knock (III) and a fourth sound (IV). The end-diastolic pressure was 20 mm Hg in the right ventricle, and 30 in the left.

This knock, moreover, occurs earlier than the average third sound. It still coincides with the peak of the wave of rapid filling because this phase ends earlier than in a normal heart (Figs. 173 and 175). It is likely that the pericardial knock has the same mechanism of the third sound, even though sudden checking of filling by the adherent pericardium contributes to an earlier appearance and a higher pitch.

Chest Pulsations. In constrictive pericarditis, several types of pulsations of the chest wall have been described (Williams, Broadbent, Skoda, Friedreich, Dressler). The most typical was considered a "retraction of the apex," i.e., a *negative pulsation.* However, the variability of pattern of the apical impulse does

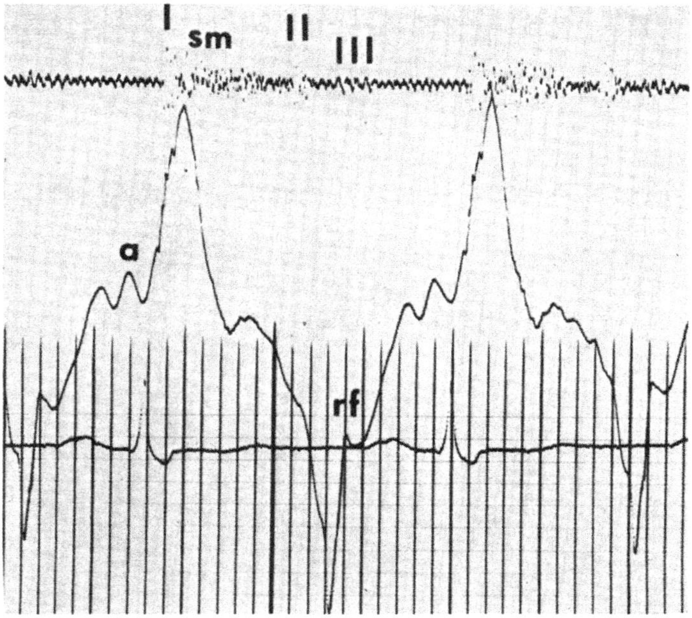

FIGURE 175. (A) Same patient as in Figure 174. The apex cardiogram shows a large presystolic wave (a) and an extremely rapid wave of filling (rf). The pericardial knock is poorly visible in this record.

not allow us to consider this as pathognomonic unless it is quite obvious and marked, and severe right ventricular hypertrophy can be excluded.

Another sign was first described by means of electrokymography and cardiac catheterization and was then confirmed by apex cardiography. It consists of a rapid drop in early diastole (*diastolic dip*), an *early rebound,* and a subsequent flat course (*diastolic plateau*). This pattern has been called the V-type of ventricular contraction (Fig. 174). When these data are present and there is a loud diastolic sound, the diagnosis of constrictive pericarditis is confirmed.

The *apex cardiogram* confirms the abnormal diastolic filling (Fig. 175 A).

Murmurs. Murmurs are not part of the usual picture. However, in certain cases, fibrous or calcific bands penetrating into the heart, or constricting bands circling the atrioventricular grooves or compressing the outflow tracts, may cause either *diastolic* or

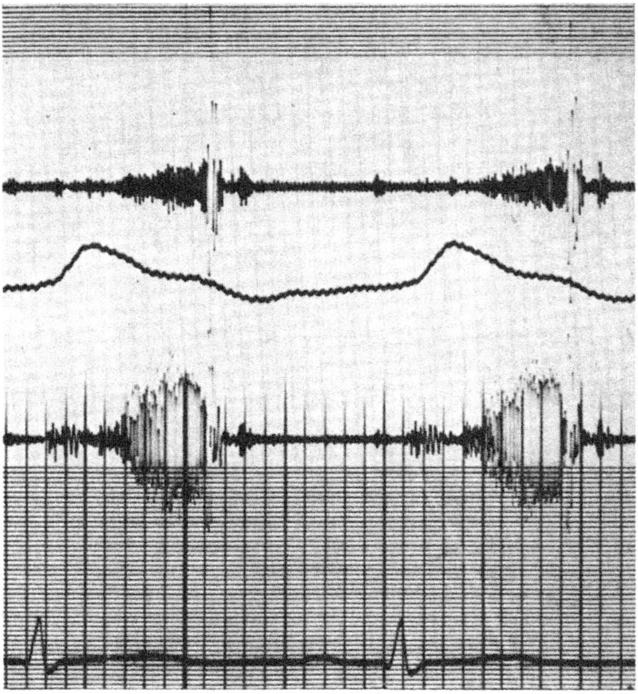

(B) Other case of constrictive pericarditis confirmed by catheterization (age 18—diastolic dip). There is a large systolic murmur in crescendo, also recorded at the 2nd right i.c.s. (above) but chiefly developed at the apex (center). Tracings with high pass filter at 200 Hz.

systolic murmurs similar to those more commonly caused by valve lesions. A *late-systolic murmur in crescendo* can be observed in some of the cases, and one can only speculate that the abnormal dynamics caused by the compression of fibrous tissue may impair the closure of the mitral valve. A *mid-diastolic rumble* of mitral or tricuspid stenosis or a systolic murmur of aortic or pulmonic stenosis can be recorded (Gouley, McGaff *et al.*, Schrire *et al.*). However, before accepting that these lesions are present, one should exclude *functional diastolic rumbles* and *flow murmurs* caused by altered cardiac dynamics. These functional rumbles have been repeatedly documented. Of course, if valvular lesions, caused by previous rheumatic endocarditis, are present, the murmurs of these lesions will be recorded.

Chapter 15

Hypertensive Heart Disease

THE CHANGES OF THE heart sounds and the possible occurrence of murmurs in systemic hypertension have been repeatedly investigated. However, few studies correlated these changes with the severity of the hypertension on the basis of systematic phonocardiographic documents.

The following phenomena have been described in previous publications:

(1) A large *fourth or third sound* (Luisada, 1955; Kontos *et al.*; Sakamoto *et al.*, 1960);

(2) An *ejection sound* (Puchner *et al.*; Kincaid-Smith and Barlow; Moyer *et al.*);

(3) A *diastolic rumble* at the apex (Luisada, 1955; Kincaid-Smith and Barlow; Moyer *et al.*);

(4) An *ejection systolic murmur* (Moyer et al.);

(5) A *single second sound* or a prolongation of both Q-IIA and Q-IIP with extremely *close splitting* (Shah *et al.*, 1964).

These phenomena have been studied more recently by Luisada and Argano (1971) in a series of 88 cases from 5 to 80 years of age, and were correlated with the severity of the hypertensive state.

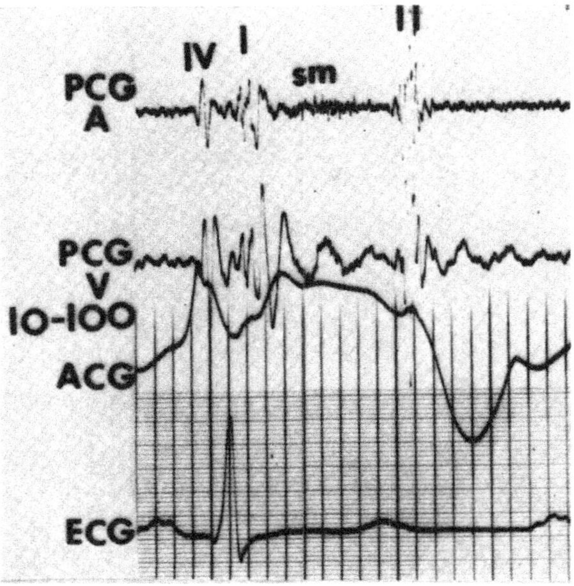

FIGURE 176. Tracing recorded over the 3rd left i.c.s. in a 36-year-old man with severe hypertension. The tracing reveals a large presystolic wave in the apex cardiogram (ACG) coinciding with a large fourth sound. There is also a small systolic murmur. The apex cardiogram has the plateau-like pattern of left ventricular hypertrophy.

THIRD SOUND—FOURTH SOUND

The IV sound was found in 42 per cent of moderate versus 83 per cent of severe cases (Figs. 43, 176-178). This should be compared with the III sound that was found only in 19 per cent of moderate versus 27 per cent of severe cases. Thus, the IV sound was far more common than the III, and was particularly frequent in severe hypertension.

Ejection Sound. An ejection sound was present in 23 per cent of moderate versus 36 per cent of the severe cases (Fig. 178).

Ejection Murmur. An ejection murmur at the base was present in 63 per cent of moderate and in 100 per cent of severe cases (Figs. 176-178).

Apical Pansystolic Murmur. An apical pansystolic murmur was presnt only in 4 per cent of moderate and 14 per cent of severe cases.

Early-diastolic Murmur. An early-diastolic, high frequency

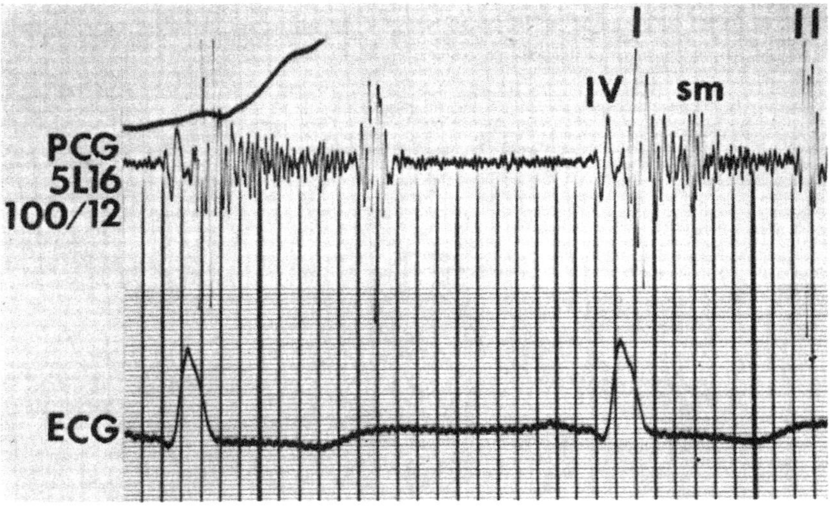

FIGURE 177. Tracings recorded over the apex in a 47-year-old woman with severe systemic hypertension.
Large fourth sound. Systolic murmur.

murmur was found in 38 per cent of the moderate versus 60 per cent of the severe cases (Figs. 178).

Other Findings. Either a *late systolic click* or a *mid-diastolic rumble* was an occasional finding.

SECOND HEART SOUND

The second heart sound was most often *closely split* or *single* (34 per cent in moderate cases versus 60 per cent of severe cases) but it was not unusual to find it either *normally* or *reversely split.*

From the above data, an auscultatory and phonocardiographic picture clearly emerges (Fig. 179). This should be known because the association of a fourth (presystolic) sound with an aortic ejection sound and both an ejection systolic and an early-diastolic blowing murmur may be erroneously interpreted as resulting from rheumatic valvular disease or the effects of atherosclerosis.

The fourth sound is now interpreted as resulting from the more powerful left atrial contraction in the presence of a decreased compliance of the left ventricular wall. Both the ejection sound and the ejection systolic murmur are probably caused by dynamic dilatation of the ascending aorta.

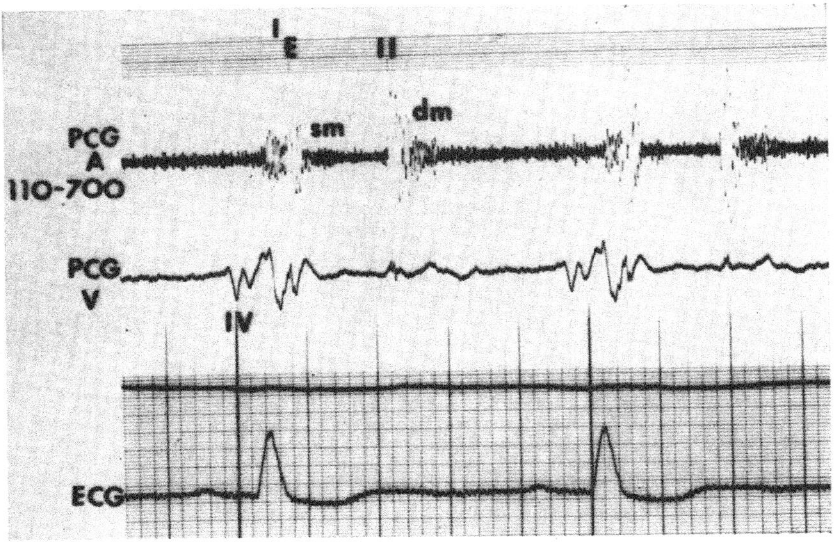

FIGURE 178. Tracings recorded at the apex in a 62-year-old woman with severe systemic hypertension. The high frequency tracing (A) reveals an ejection sound (E), a small systolic murmur, and an early-diastolic murmur in decrescendo. The low frequency tracing (V) shows a large, complex fourth sound.

The early-diastolic blowing murmur, usually a short one, can be explained by minimal aortic insufficiency caused by stretch of the aortic ring. As it was present even in young patients, it cannot be caused in all cases by aortic atherosclerosis with secondary fibrosis of the aortic valve.

In regard to the second sound, no correlation was found between its various types and the age of the patient. Thus, the finding of Slodki and Shah, who described prolongation of both the Q-IIA and the Q-IIP intervals with increasing age, cannot be considered as an obligatory phenomenon. The two above authors thought that a single second sound was the rule in normal subjects above age 50. However, a less constant behavior of the second sound of normal individuals was described by Adolph and Fowler, and, even in normal subjects in the old age group, various behaviors of the second sound can be expected (Slodki *et al.*, 1969).

The third sound has been found only in a minority of the cases in the present series. It is our present impression that

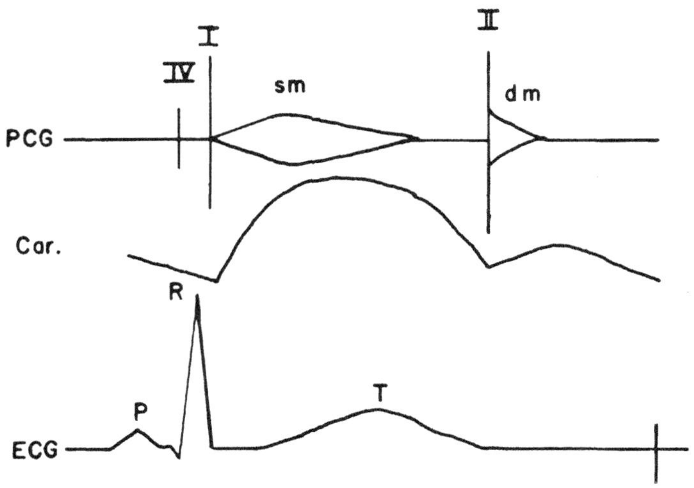

FIGURE 179. Scheme of the phonocardiogram in systemic hypertension.

either a III sound or a short diastolic rumble occur only in cases with definite left ventricular failure.

APEX CARDIOGRAM

The apex cardiogram was studied by Luisada and Magri (1952), Beilin and Mounsey, Ueda *et al.* (1962A), and Roervik. Abnormal prolongation of the apical impulse up to the second sound was found in patients with marked left ventricular hypertrophy (so-called *pansystolic plateau*). A *tall atrial wave* in presystole was also common in such cases (Fig. 176). A *midsystolic bulge* was observed by Ueda *et al.* (1962A) in patients with left ventricular hypertrophy and myocardial damage.

An abnormal pattern of the midprecordial "accelerogram" was found by Rosa *et al.* (1962) nearly ten times more often than in nonhypertensive, noncardiac subjects. The percentage of abnormalities sharply increased in patients with a diastolic pressure above 100 mm Hg and with electrocardiographic evidence of left ventricular strain; it further increased if there was heart failure or malignant hypertension.

Patients with severe uremia may have a *wide splitting of the second sound* (Fig. 180) as pointed out by Gibson. Peritoneal dialysis may cause a marked decrease of the width of the splitting in such cases.

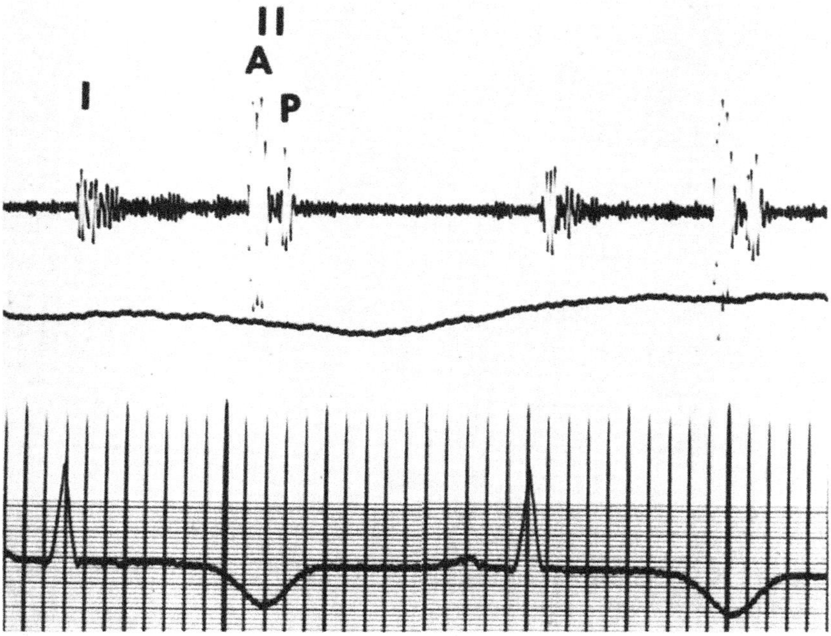

FIGURE 180. Phonocardiogram in a case of uremia with severe electrolyte and fluid retention. The patient was treated by means of hemodialisis on the following day. Small systolic murmur. Minimal presystolic murmur. Wide splitting of the second sound.

A possible interpretation is that uremic patients with severe retension of nitrogen products and water may have a greater distension of the pulmonary artery than of the aorta. This would cause a delayed rebound of the pressure pulse in the pulmonary artery with delay of the incisura and of the pulmonic component of the second sound. This interpretation seems to be confirmed by experimental data (Fig. 31).

Chapter 16

Pulmonary Hypertension and Pulmonary Heart Disease

ACUTE PULMONARY EMBOLISM

IT IS TRADITIONAL TO mention three auscultatory aspects of this condition, and they have been consistently confirmed by phonocardiography.

(1) *Wide and fixed splitting of the second sound with large pulmonary component* (Shapiro *et al.*, 1965; Cobbs *et al.*, 1966). This sign has great diagnostic value in cases of *massive embolization* with scarcity of other signs, especially if one has the documentation that the second sound was normal at a prior examination. In less severe cases, the width of the splitting was found to be about 30 msec. in expiration, greater in inspiration. In severe cases, it was 75 msec. in both expiration and inspiration. This behavior of the pulmonary component cannot be attributed to the pulmonary hypertension *per se*, as the latter would tend to cause an opposite modification. Therefore, two possible mechanisms can be considered: either right heart stress or failure prolonging right ventricular systoles and delaying pulmonary valve closure, or the effect of the high pressure on the timing of the incisura and the re-

bound of pulse over the closed pulmonary valve. The former seemed more likely because elevated right ventricular end-diastolic pressure is commonly found in these cases (Cobbs *et al.*).*

(2) *A pulmonary flow murmur.* This is important if it occurs in a patient who previously did not have such a murmur. It can be explained by the dilation of the pulmonary artery and its main branches.

(3) A *scratching sound* in systole over the 3rd left interspace. This has been explained as a *friction rub,* caused by the irritation of the pericardium in the small section that covers the root of the pulmonary artery.

PULMONARY HYPERTENSION CAUSED BY CHRONIC PULMONARY DISEASE OF THE OBLITERATIVE TYPE

Contrary to previous belief, wide splitting of the second sound, not disappearing in expiration, is a constant finding in chronic pulmonary heart disease (Figs. 181-185).

In the study of Cobbs *et al.* (1966), the hemodynamic factors most closely correlated with the degree of splitting were the level of the pulmonary artery systolic pressure and the presence of an elevated right ventricular diastolic pressure. With PA systolic pressure at the systemic level and clear-cut right ventricular failure, the degree of splitting at rest was 50 msec. or more, and increased with exercise. In the cases studied by

*Recent experiments conducted in our laboratory have demonstrated a new mechanism for the wide splitting of the second sound in acute pulmonary embolism. When the late stage of the picture is reached, the right ventricle still ends its contraction at the same time as the left. However, its relaxation occurs very slowly so that pulmonary valve closure is markedly delayed. The only possible explanation is that, with severe dilatation of both the right ventricle and the pulmonary artery, end of ventricular contraction is followed by a regurgitant jet into the ventricle. When right ventricular pressure finally is approaching the diastolic level, the pulmonic valve closes and no further regurgitation takes place. Thus a murmur is recorded only between aortic and pulmonary components ending with the latter. Whether a similar mechanism also occurs in chronic pulmonary hypertension, we still do not know. Of course, in some of the experiments, right ventricular failure takes place (with later ending of RV pressure) or a more conventional type of pulmonary insufficiency occurs (causing an early-diastolic murmur) (Luisada *et al.*, 1972).

Sutton *et al.* (1968), the second sound was 20-30 msec. in expiration and 30-50 msec. in inspiration.

Another finding that can be observed is a *diastolic rumble* over the mid-precordium and often the apex* (Figs. 182-184, 186). This was recorded in several cases from our Hospital including one further documented by autopsy (Freilich and Szatkowski). This rumble unfortunately may be misinterpreted as the result of mitral stenosis.**

Tricuspid insufficiency is not unusual in the advanced stage, and is revealed by a pansystolic murmur, often at the "apex," and by typical changes of the *jugular* and *hepatic tracings* (Fig. 185). *Pulmonary insufficiency* is less frequent but may occur (Fig. 184).

PULMONARY HYPERTENSION CAUSED BY CONGENITAL SHUNTS

These situations have been studied in detail by Sutton *et al.* Cases of *atrial septal defect* maintain the wide splitting of the second sound in the presence of pulmonary hypertension but the amplitude of the pulmonary component becomes greater. In cases of *ventricular septal defect* with pulmonary hypertension, the splitting of the second sound remains as in cases without hypertension but there is a tendency for a single sound in expiration. In cases of *patent ductus* with left-to-right shunt and *hyperkinetic pulmonary hypertension,* splitting of the second sound is either normal or reversed. In cases of patent ductus with bi-directional shunt (*so-called Eisenmenger situation*), the second sound is single or nearly single in expiration but still often has a 20-40 msec. splitting in inspiration.

Actual single second sound is documented only in cases with large ventricular septal defect or single ventricle and pulmonary resistance at the systemic level (Eisenmenger situation) (Fig. 123). Clinical auscultation may be misleading because, in some

*In severe cases, the "apex" of the heart is formed by the dilated right ventricle.

**This rumble has been called "the Austin Flint murmur of the right heart," based on the belief that it is secondary to pulmonary insufficiency. We believe that it would be more appropriate to emphasize the similarity between this diastolic rumble and that found in cases of mitral insufficiency. This because pulmonary insufficiency is seldom severe in these cases while tricuspid regurgitation may be massive.

cases there is a fainter aortic component followed by a louder pulmonary component, and auscultation reveals only the latter.

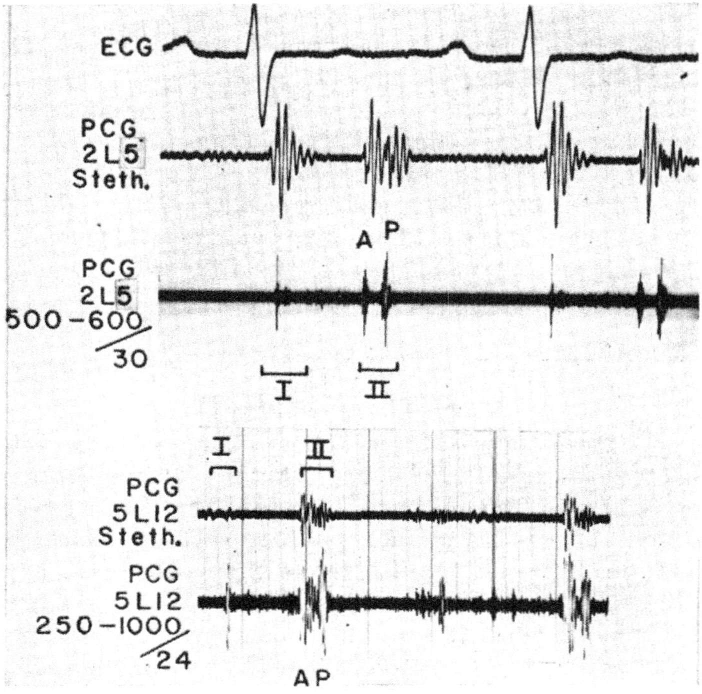

FIGURE 181. Tracings of a 50-year-old man with primary pulmonary hypertension (RV pressure = 75/5 mm Hg).
Top = tracings on the 2nd left i.c.s. showing a widely split second sound with a large, high frequency pulmonary component.
Below = Tracings at the apex: the pulmonary component is well recorded at the apex.

PRIMARY PULMONARY HYPERTENSION

According to Cobb *et al.*, splitting of the second sound is narrow in these cases. However, Sutton *et al.* found *wider splitting in expiration* and a 20-60 msec. splitting in inspiration. Usually Q-IIA remains constant while Q-IIP increases in inspiration.

In conclusion, pulmonary hypertension tends to increase the degree of splitting of the second sound. An exception is represented by cases with the Eisenmenger's situation, especially

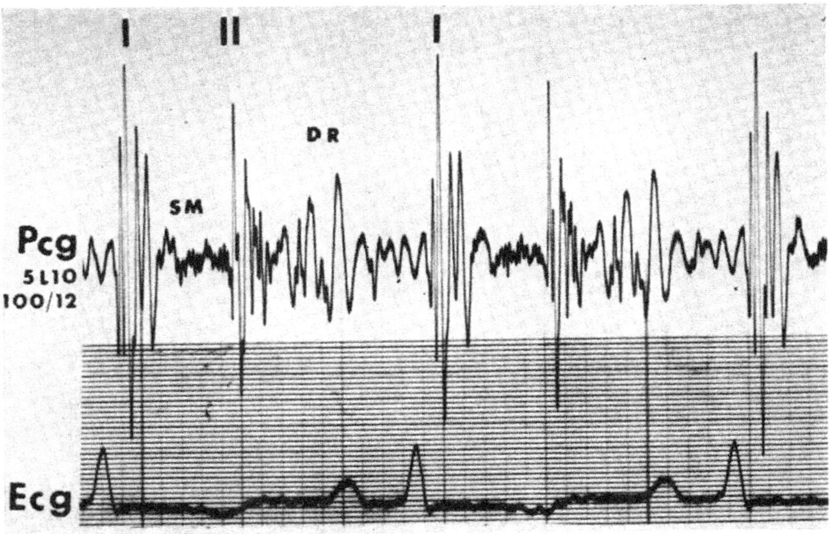

FIGURE 182. Tracings recorded in expiratory apnea in a 38-year-old woman with chronic obstruction of the pulmonary vessels due to multiple and repeated embolizations. The tracing is recorded in the low-frequency range over the midprecordium. Large diastolic rumble.

if there is a large ventricular septal defect or a single ventricle and the pulmonary resistance is at a systemic level. The common cause of prolongation of the Q-IIP interval seems to be a *prolongation of right ventricular systoles related to strain and probably to failure.* This seems confirmed by the clinical observations of Bose and Das that studied these cases both during obvious failure and following improvement. (See also footnote on page 000.)

In all conditions of pulmonary hypertension, the magnitude of the pulmonary component is increased, and its transmission is wider, so that it can be recorded both at the right of the sternum and at the apex. Clinical confusion with an opening snap is then common.

IDIOPATHIC DILATATION OF THE PULMONARY ARTERY

In this condition, *wider and poorly variable splitting of the second sound* is recorded. The cause of the phenomenon seems to be the dilatation of pulmonary artery, which causes a delay of both the incisura and the rebound of the pulmonary pressure pulse, like in atrial septal defect.

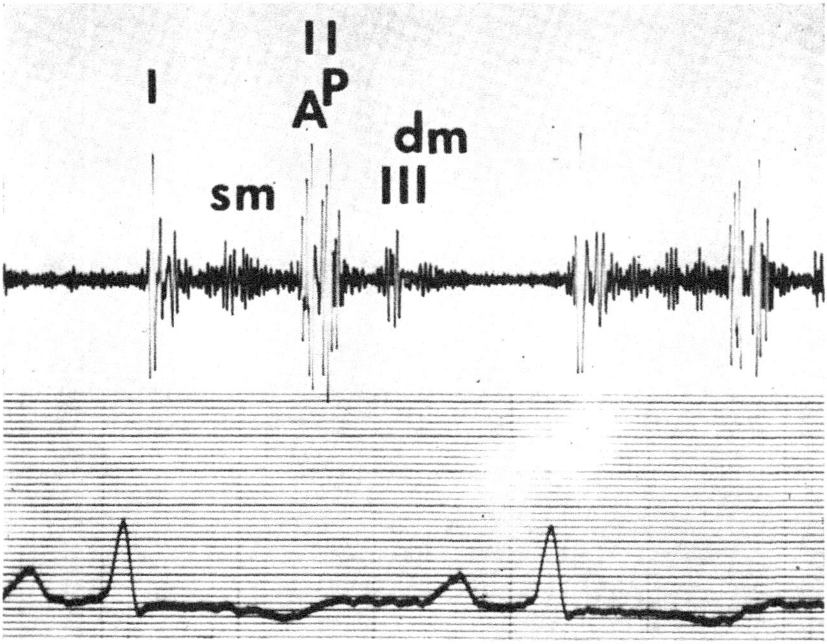

FIGURE 183. Tracings recorded in expiratory apnea over the 2nd left i.c.s. in the same patient of Figure 182 in the high frequency range (high pass filter at 400 Hz, slope 24 db/oct.).
Diamond-shaped systolic murmur, clearly split 2nd sound, mid-diastolic murmur with large 3rd sound.

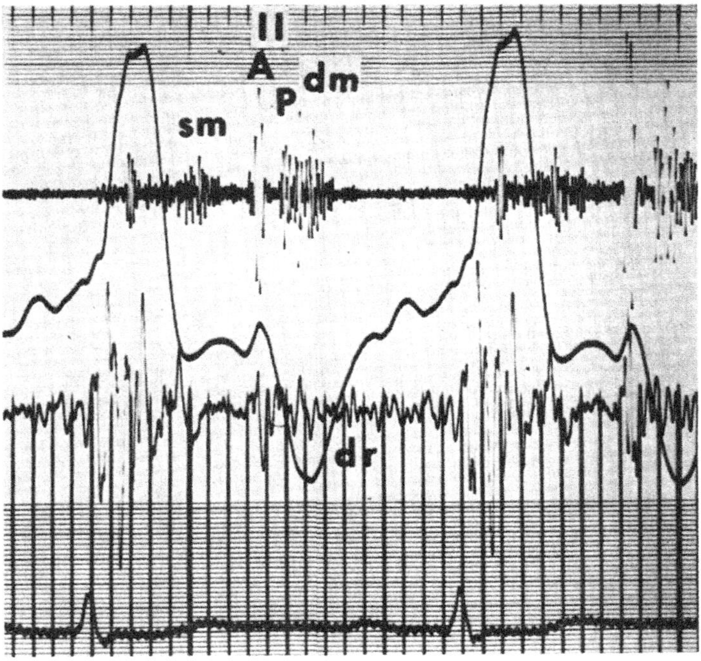

FIGURE 184. Tracings recorded in expiratory apnea in a 63-year-old woman with extensive pulmonary fibrosis caused by therapeutic radiation.
Above = high frequency tracing (high pass filter at 200 Hz) over the 2nd left i.c.s.
Center = low frequency tracing (high pass filter at 50 Hz) over apex. Apex cardiogram.
Diamond-shaped systolic murmur; widely split 2nd sound; early diastolic murmur of pulmonary insufficiency. Low frequency apical diastolic rumble.

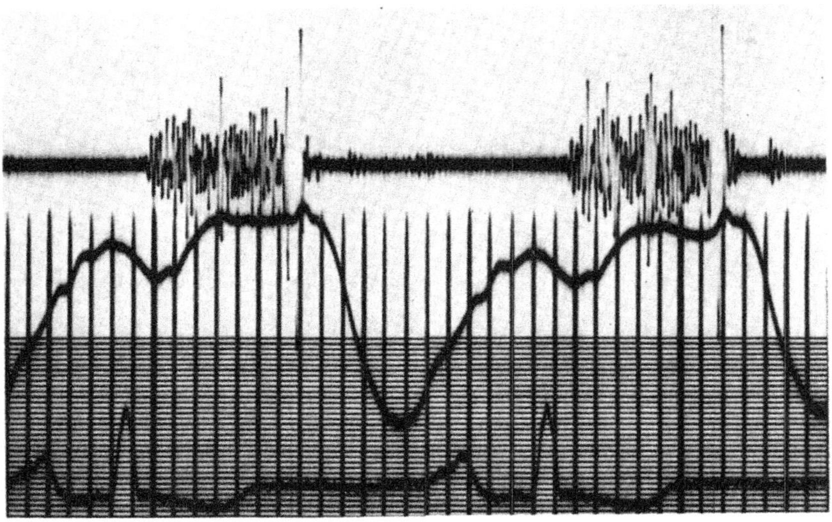

FIGURE 185. Tracings recorded at the apex in a 51-year-old patient with pulmonary emphysema (catheterization, autopsy). There is a large pansystolic murmur of tricuspid insufficiency and a systolic plateau in the jugular tracing.

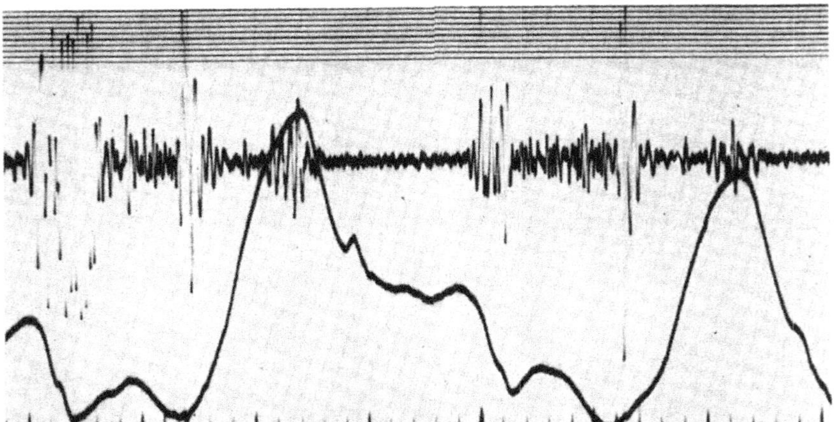

FIGURE 186. Same patient as in Figure 185. Tracings over the 3rd left i.c.s. reveal a mid-diastolic rumble. The jugular tracing at this time fails to show a plateau but shows, on the contrary, a large positive wave coinciding with the rumble. Actually the jugular tracing is the mirror image of the previous, and this may be due to technical reasons.

Chapter 17

Aortic and Pulmonary Artery Diseases

AORTIC DISEASES

DILATATION

THE AORTA MAY BECOME dilated as a result of atherosclerosis, syphilitic aortitis, or congenital weakness (like in Marfan's syndrome). Systemic hypertension, on the one hand, and aortic insufficiency, on the other, will cause dynamic stretching, which, in the long run, may result in permanent dilatation of the aorta.

Low Frequency Tracings. In normal subjects, a tracing recorded in the 10 to 40 Hz range shows a relatively brief series of oscillations that start with the first sound and continue until about midsystole (vascular oscillations) ; the second sound is marked by a small deflection (Fig. 187 A).

When the aorta is dilated, these oscillations are larger and continue much longer, so that they may reach the end of systole (Fig. 187 B,C,D). The vibrations of the second sound become larger, especially in the presence of hypertension (187 B,D), and may be followed by a series of vibrations in decrescendo if there is aortic insufficiency (Fig. 187 D).

High Frequency Tracings. These tracings constantly show a series of systolic vibrations that may either appear as an *early- and mid-systolic murmur* (Fig. 188) or, more commonly, as a

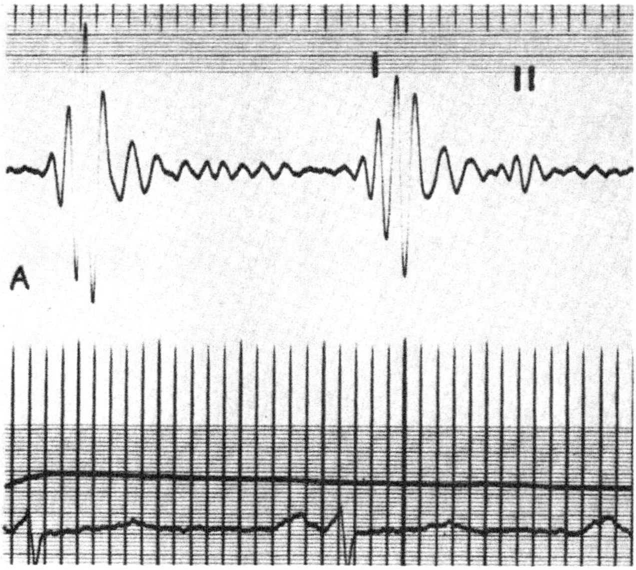

FIGURE 187. Low frequency tracings recorded at the apex in the 10-40 Hz range in clinical cases and comparison with normal vibrations.
(A) 54-year-old man with normal aorta. Early-systolic vibrations. Compare this tracing with the following (B, C, and D).

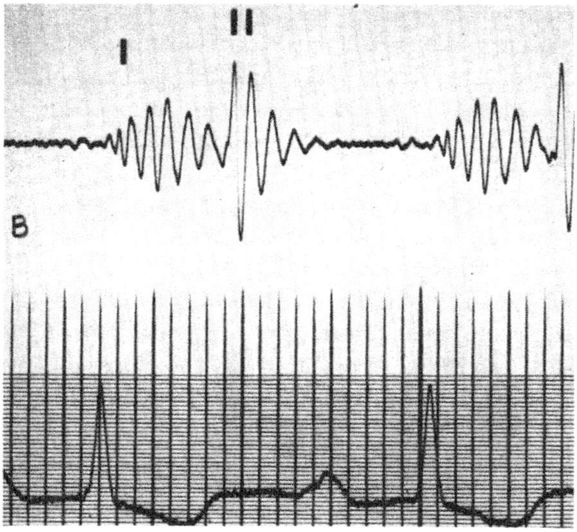

(B) 43-year-old hypertensive woman with dilatation of the ascending aorta on chest film. Pansystolic crescendo-decrescendo series of vibrations. Large II sound. Small early-diastolic vibrations.

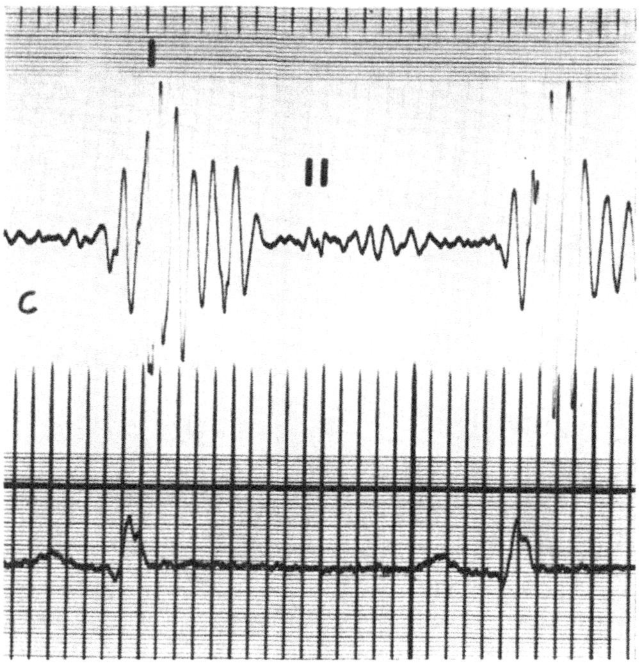

(C) 69-year-old man with syphilitic aortitis, dilatation of the ascending aorta on x-ray and minimal aortic insufficiency. Presystolic vibration (IV sound); early and mid-systolic vibrations; mid-diastolic vibrations.

diamond-shaped murmur (Figs. 190, 191 A,B,C). In the last case, the peak of the murmur occurs early, in contrast with that of the murmur caused by marked aortic stenosis. These vibrations are the graphic expression of the "aortic flow murmur", which is so frequent in such cases. Occasionally we can record either *one systolic click* (Fig. 188) or, more rarely, *two systolic clicks* (Fig. 189). One can only surmise that atherosclerotic plaques may be the cause of these clicks through their higher frequency of vibrations.

Aortogram, Carotid and Subclavian Tracings. These ultra-low frequency tracings, recording the transmitted pulsations of the arteries, frequently show an *anacrotic depression* (Figs. 191 A,B). This depression is usually high in the ascending branch of the pulse curve in contrast with that noted in the pulse tracings of aortic stenosis. One can only speculate about the part played by abnormal function of the left ventricle and by an

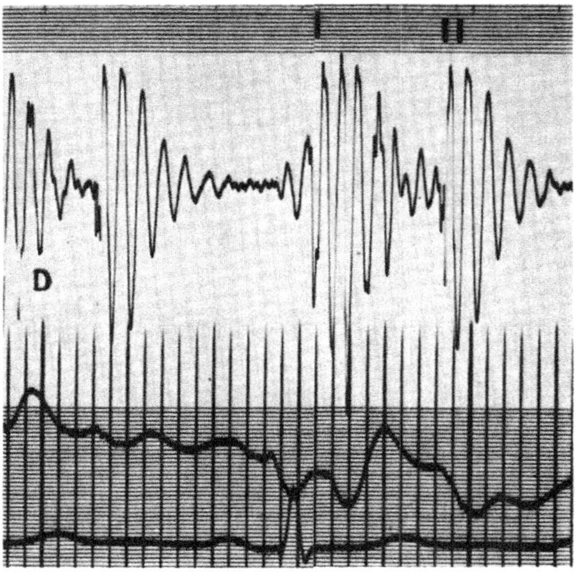

(D) 50-year-old man with chronic hypertension, syphilitic aortitis and medium-severe aortic insufficiency. Pansystolic, crescendo-decrescendo vibrations; early-diastolic decrescendo vibrations.

abnormal aorta, respectively, in causing this notch. However, when the dilatation is severe and there is a marked loss of the elastic structures of the aorta, a rounded pulse wave is often noted with a small incisura, similar to the pulse of an aneurysm Fig. 191 C).

Aneurysms

Graphic documents of aneurysms are scanty due to the fact that treatment of syphilis has markedly reduced their incidence.

A high frequency tracing of an aneurysm constantly shows a *systolic diamond-shaped murmur,* which has an earlier peak if the aneurysm is in the ascending aorta, and a later peak if it is in the abdominal aorta. An *early-diastolic murmur* is frequent but is much longer in the presence of aortic insufficiency (Fig. 192) than when it is merely due to the altered aortic dynamics (Fig. 193).

The *apex cardiogram* occasionally shows a double apical impulse (*sign of Hope*) if the aneurysm is located in the arch or early descending aorta, and causes a vascular rebound of the

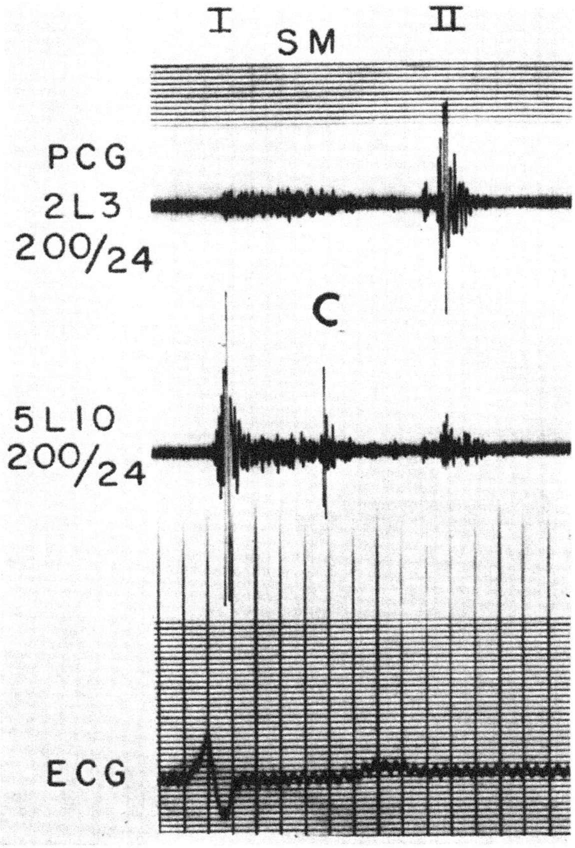

FIGURE 188. Sixty-four-year-old patient with syphilitic aortitis. Tracings over 2nd left i.c.s. (above) and apex (center). Early and mid-systolic murmur with mid-systolic click (C). Large second sound.

heart against the chest wall (Fig. 192).

The *pulse of the aneurysm* usually has a slow rise and a delayed peak (Fig. 194 B), in contrast with that of a dilated, non aneurysmal vessel (Fig. 194 A). However, occasionally the pulse rise is rapid even over an abdominal aneurysm (Fig. 193).

PULMONARY ARTERY DISEASES

Dilatation of the pulmonary artery invariably causes a *systolic flow murmur.* This is usually diamond-shaped and has an early peak. Examples were given in Chapter 16. When the

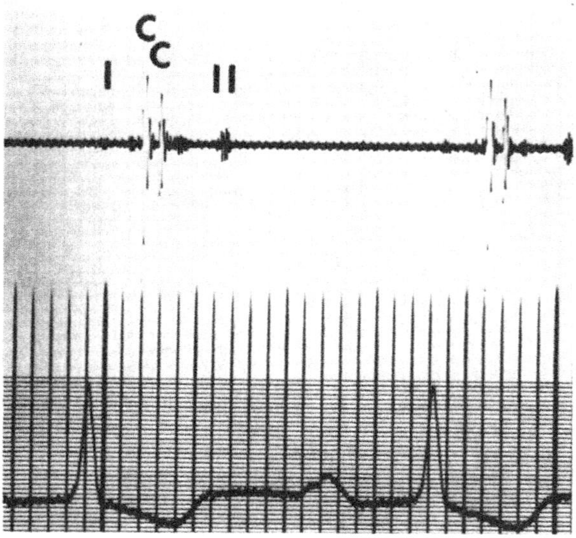

FIGURE 189. Same patient as in Figure 187 B. A high frequency (D—150-500 Hz) at the apex reveals a double click (C,C) at mid-systole.

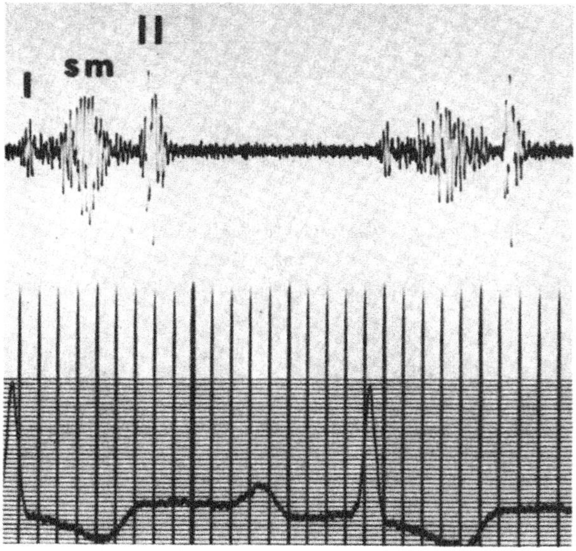

FIGURE 190. Same patient as in Figures 187 B and 188. Velocity tracing at the apex (100-600 Hz). Systolic, diamond-shaped murmur. Large second sound.

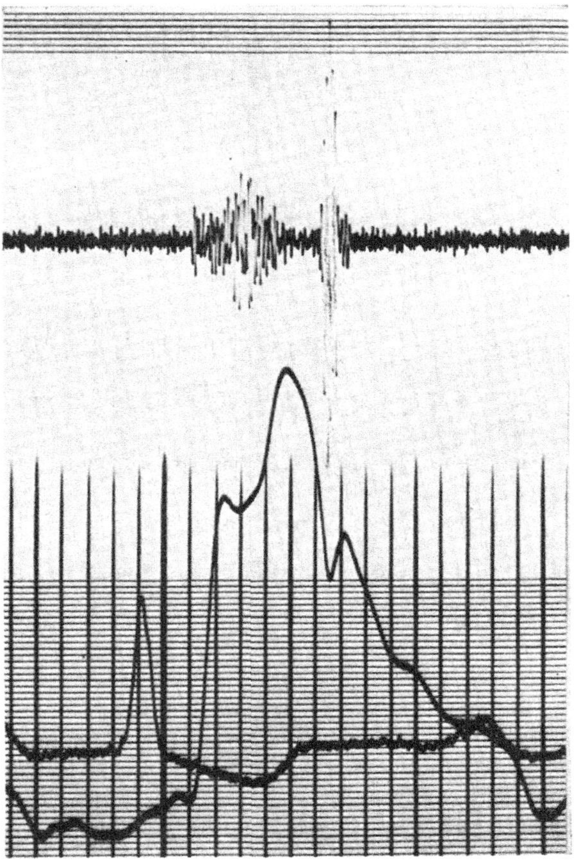

FIGURE 191. Carotid tracings in three cases of aortic dilatation. (A) Same patient as in 187 B and 190.

dilatation is severe and caused by severe pulmonary hypertension, an *early-diastolic murmur* is noted (Fig. 194 C) (*Graham Steell murmur,* if there is mitral stenosis). In cases with an *aneurysm of the pulmonary artery,* a typical pulse and a double murmur are recorded (Fig. 194 C).

THE SECOND HEART SOUND IN AORTIC AND PULMONARY DISEASES

In general, dilatation of the pulmonary artery delays the incisura and the rebound of the pulmonic pulse over the previously closed pulmonic valve. This will cause *an increase in the*

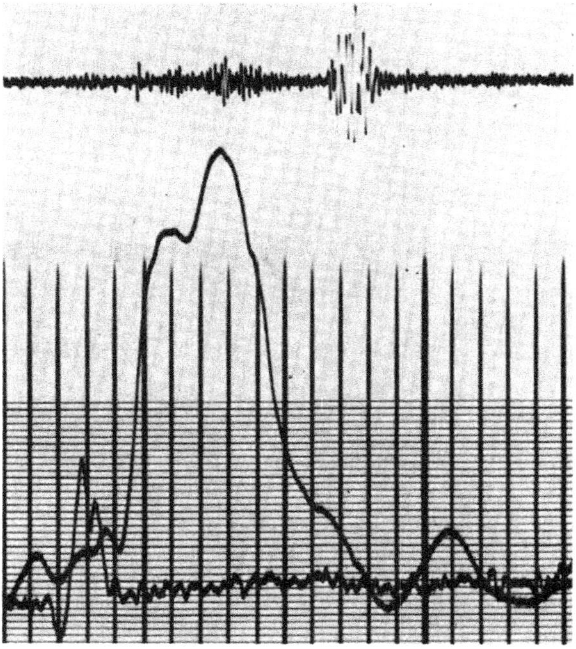

(B) Same patient as in 187 C.

interval between pulmonic valve closure and pulmonic component of the second sound. Necessarily, *wider splitting of the second sound* will take place.

This was documented in atrial septal defect and, to a lesser extent, in ventricular septal defet (Chapters 9 and 16). Due to the greater compliance of the pulmonic wall in comparison with the aortic wall, this phenomenon may occur even in situations in which there is an increased blood volume or an increased output distending both vessels, like in uremia (Fig. 180).

A reciprocal change, i.e., delay of the aortic component, does *not* take place in aortic dilatation because, while dilatation of the pulmonary artery caused by greater flow extends to the smaller branches, diatation of the aorta due to wall lesion is limited to the aorta itself, and peripheral resistance is still high, due to vasoconstriction.

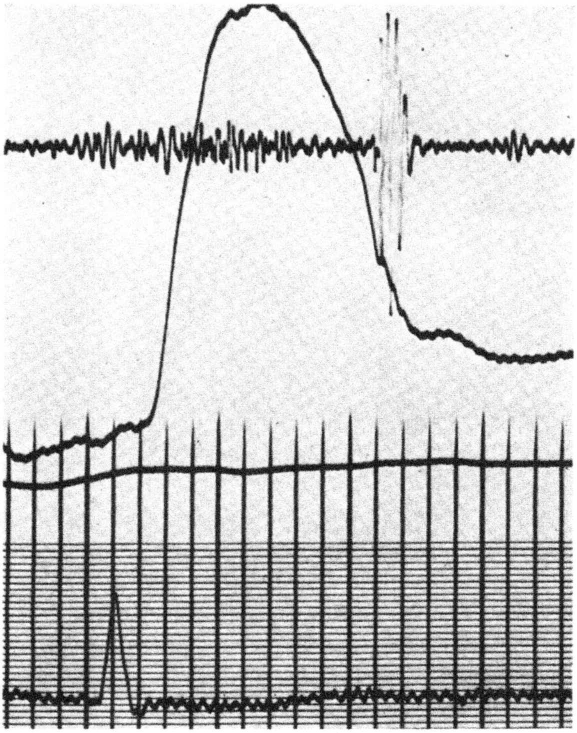

(C) Seventy-eight-year-old woman with aortic dilatation.

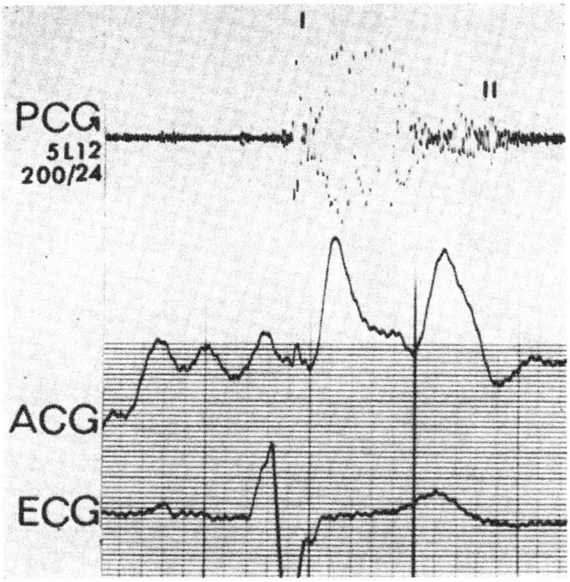

FIGURE 192. Tracings recorded in a 64-year-old man with aortic aneurysm and moderate aortic insufficiency. There is a very large systolic murmur and a smaller early-diastolic murmur. The center tracing is an apex cardiogram which shows a double apical impulse *(sign of Hope).*

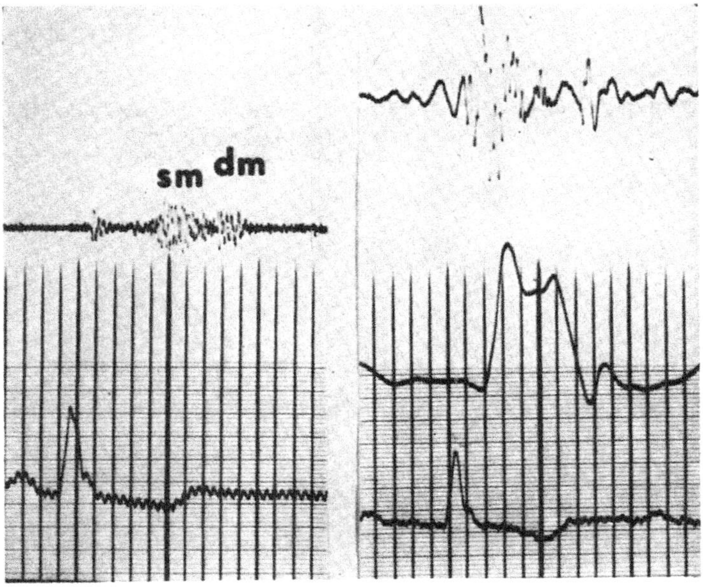

FIGURE 193. Tracings recorded in a 65-year-old woman with an aneurysm of the abdominal aorta.
At left = phonoarteriogram of the aneurysm (recorded with high pass filter at 200 Hz). It shows a diamond-shaped, systolic murmur with late peak and an early-diastolic murmur.
At right = above: phonocardiogram at the apex (recorded with high pass filter at 50 Hz). below: pulsation of the aneurysm with rapid rise and rapid drop.

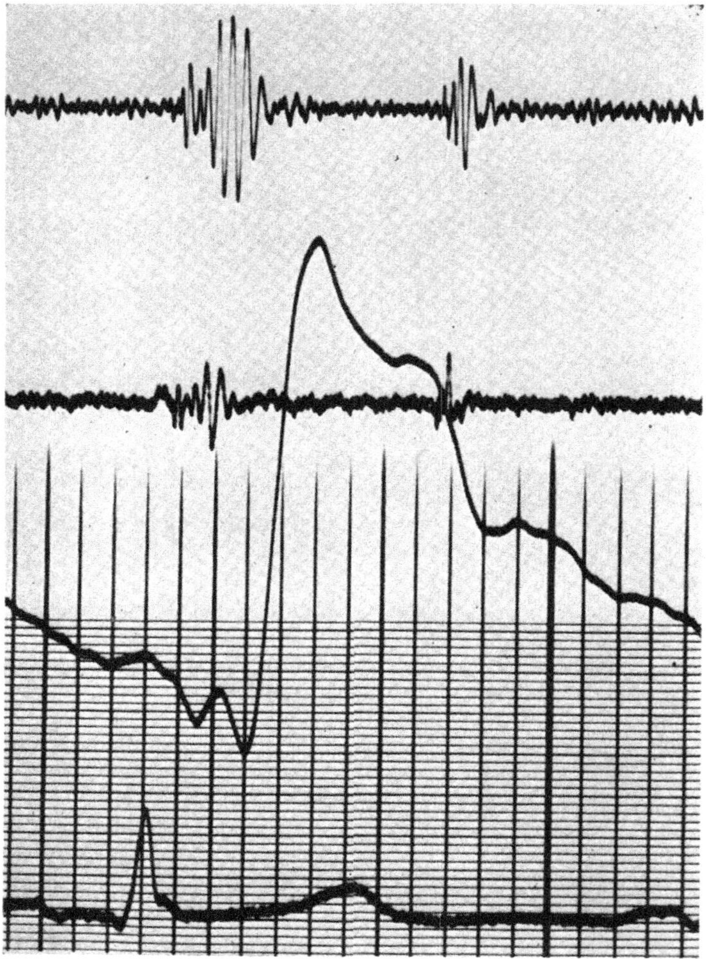

FIGURE 194. (A) Phonocardiograms at apex and pulse tracing at the epigastrium in a 45-year-old man in whom an aneurysm of the abdominal aorta had been suspected. The pulse is normal and excludes an aneurysm.

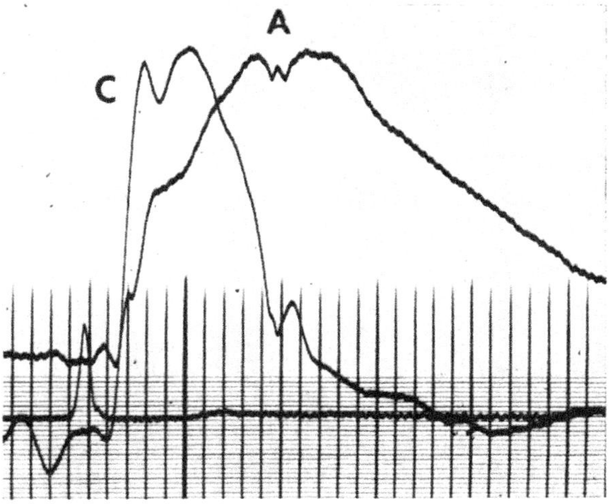

(B) Pulse tracing at the right carotid (C) and over an aneurysm of the abdominal aorta (A) in an 80-year-old woman. Both the rise and drop of the pulse are much slower in the tracing of the aneurysm.

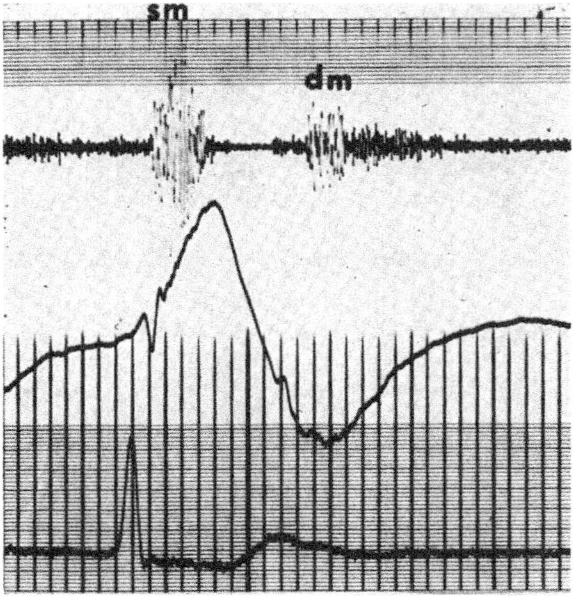

(C) Sound tracing and pulsation recorded over the 3rd left i.c.s. near the anterior axillary line. Sixty-year-old woman with severe pulmonary hypertension and an aneurysm of the pulmonary artery. The patient had been previously operated on (commissurotomy) for mitral stenosis but still had a pulmonary arterial pressure at systemic level when this tracing was recorded. Short and large, early-systolic murmur (flow murmur); early-diastolic murmur (pulmonary insufficiency).

Chapter 18

Syphilitic Heart Disease

SYPHILITIC HEART DISEASE MAY cause different lesions of the heart and vessels, and thus various changes of cardiovascular dynamics. These are often accompanied by abnormalities of the heart sounds, appearance of murmurs, and changes of the precordial and arterial pulses.

AORTITIS

This process is usualy maximal in the ascending aorta and thus will cause dilatation of this section of the vessel, as well as a *systolic flow murmur*. Changes of both the low and the high frequency tracings will be noted in conjunction with the systolic murmur (Chapter 17). The aortic component of the *second heart sound* is frequently large and high pitched.

AORTIC INSUFFICIENCY

The typical murmurs caused by this lesion wil be present.

(a) Early-diastolic or pandiastolic murmur, usualy of high frequency.

(b) Mid-diastolic, and occasionally also presystolic murmur of lower frequency (*A. Flint murmur*). Due to the massive dilatation of the left ventricle that can be found in some of these cases, the functional diastolic rumble is more common than in rheumatic aortic insufficiency.

(c) Typical changes of the apex cardiogram and of the carotid pulse. These data have been already discussed in Chapter 8.

Aortic Aneurysms

In Chapter 17, we have reviewed some of the data in regard to aortic aneurysms.

The combination of aortitis with aortic insufficiency (and possibly also with aneurysms) is typical of syphilitic heart disease.

Chapter 19

Bacterial Endocarditis

WITH THE EXCEPTION OF rare cases of "malignant" or acute bacterial endocarditis, this process usually develops over either valves that are already damaged by rheumatic fever or congenital narrowing, on the rim or wall of congenital shunting lesions, or on a bicuspid aortic valve. Therefore, one is always confronted by the problem whether certain murmurs were preexisting or whether some of them has developed as a result of the disease.

Changing murmurs have been considered typical of bacterial endocarditis. However, changes of cardiac dynamics always cause changes of the existing murmurs, whatever their cause. Therefore, such term should be employed only in those cases, which have a certain murmur (i.e., the murmur of mitral insufficiency or stenosis) and develop an additional murmur (usually that of aortic insufficiency).

Even though thrombotic masses occasionally cause the murmur of mitral or aortic stenosis, the most common murmur caused by bacterial endocarditis is that of *aortic insufficiency.* This develops as a result of perforation of one or more valve cusps and frequently asumes a musical tonality. *Mitral insufficiency* may result from a similar process or rupture of a chorda.

Anemia caused by S.B.E., may contribute additional murmurs through dilatation of the mitral or aortic ring, dilatation of the ventricles, and acceleration of the circulation ("functional murmurs"). It also would cause a well developed third heart sound.

Chapter 20

The Continuous Murmur

A CONTINUOUS MURMUR IS A murmur that is present in the entire cardiac cycle, though with accentuation in a certain phase. It is typically caused by blood flowing in an artery (occasionally a vein) or through a communication between either two arteries or an artery and a vein.* The most commonly observed and best known continuous murmur is that of patent ductus arteriosus. However, a similar type of murmur can be found in several other conditions.

Aorto-pulmonary Window

The murmur is recorded best in the third and fourth left interspaces and is similar to that of P.D.A. It is accompanied by a thrill.

Rupture of a Sinus of Valsalva

An aneurysm of the sinus of Valsalva can rupture into either the *right atrium* or the *right ventricle*. The acoustic phenomena are similar in both cases. The murmur is best recorded in the third and fourth left interspaces and is accompanied by a thrill. The murmur has a wide radiation and its diastolic component may radiate to the right of the sternum. The shunt causes a large pulse pressure. The pulmonary component of the *second*

*From a dynamic standpoint, the pulmonary artery can be considered like a vein on account of its low pressure.

sound is often loud due to increased pulmonary arterial pressure. A *functional mid-diastolic rumble* can be observed (Morch and Greenwood).

PULMONARY ARTERIOVENOUS FISTULA

The continuous murmur is usually recorded on one side of the chest.

COARCTATION OF ONE PULMONARY ARTERY

Again the murmur is best recorded on one side of the chest. However, in spite of the often mentioned characteristics of a "continuous murmur", the latter may be only ssytolic (W. P. Harvey, 1968). According to Perloff and LeBauer, *proximal pulmonary artery stenosis* causes only a systolic gradient and hence a *systolic murmur* while an occasional continuous murmur may be related to associated broncho-pulmonary collaterals. *Amyl nitrite* increases the stenotic murmur but decreases the bronchial collateral murmur. The pulmonary component of the second sound is of normal intensity and thus allows for differentiation from both pulmonary valvular stenosis and pulmonary hypertension. On the contrary, *distal pulmonary artery stenosis* causes a *continuous murmur.*

CORONARY ARTERIOVENOUS FISTULAS

Origin of a coronary artery from the pulmonary artery (a low pressure vessel) usually causes blood to flow from a normal coronary artery, through collaterals, to the abnormal coronary and then to the pulmonary artery. This is why it is termed "arteriovenous fistula." The murmur is usually recorded best above the apex (Fig. 196 B). This murmur should be differentiated from the murmur of *stenosis of the anterior descending coronary artery* (Dock and Zoneraich, Cheng). The latter is more often diastolic than continuous and is typically decreased by amyl nitrite inhalation.

COARCTATION OF THE AORTA

The continuous murmur is recorded in the back between the spine and the left scapula (Fig. 196 A). Collateral arteries of the thoracic and abdominal walls usualy present *no* murmurs. However, they may do so (Chapter 9).

Small Atrial Septal Defect (Congenital or Following Transseptal Catheterization) in Cases with Left Atrial Hypertension

Contrary to the average atrial septal defect, here a *continuous murmur* over the precordium has been observed (Ross *et al.*, 1963). On the other hand, cases of atrial septal defect with large shunt and no pulmonary hypertension may present a *systolic murmur* over the lung fields, caused simply by the large flow (W. P. Harvey, 1968).

Bronchopulmonary Collaterals or Surgical Fistulas of Blalock or Potts

These can be found in patients with pulmonary stenosis, pulmonary atresia, or tetralogy of Fallot. The continuous murmur from a surgical fistula tends to mar the baseline and mix with other murmurs but is greatest on the site of the operation. The murmur caused by collaterals is faint but bilateral and diffuse.

Arteriovenous Fistulas of the Neck or the Extremities

The continuous murmur is typically recorded over the area presenting large venous pulsations. However, compression of the artery by scar tissue, and arterial pulsations transmitted to the vein by proximity may simulate the fistula, so that phonoarteriography shows a continuous murmur in both instances. Thus, it has no diagnostic value.

The Venous Hum

This particular type of murmur is typically found in, but is not exclusive of children, and is usualy limited to the right side of the neck. However, it may be occasionaly heard and recorded in the first, second, and third right interspaces. It is more common inthe "hyperkinetic syndrome," especially when it is found in adults. It is usually a faint, soft, high frequency murmur that continues throughout the cardiac cycle but may be larger and louder in systole (Fig. 197). This murmur is increased by either *inspiration or amyl nitrite,* and may disappear with either rotation of the head toward the right or pressure on the jugular vein.

In contrast with this typical description, the venous hum may be loud and "roaring" with predominance of low frequency vibrations and with accentuation in diastole (W. P. Harvey, 1968).

Three types of venous hums have been described by Veyrat *et al.*:

(a) A *late-systolic and early-diastolic murmur* encompassing the second sound, followed by a *mid-diastolic murmur;* the two phases of accentuation correspond to two phases of increase of venous flow;

(b) A *continuous murmur* with accentuation in the same two phases of (a);

(c) A *continuous murmur* without accentuation.

Various maneuvers may cause the change from one type to another.

The Mammary Souffle

This murmur is found over the breasts of pregnant or lactating women (Tabatznik *et al.*).

The Uterine Souffle

This type of murmur is found over the abdomen of pregnant women.

Intracardiac and intravascular phonocardiography may be of help in differentiating PDA from either A-P window, AI with VSD, coarctation of the pulmonary arteries, or rupture of a sinus of Valsalva. The differentiation is based on the fact that the largest murmur is recorded when the catheter tip is at the location of the shunt.

Chapter 21

Thoracic and Extra-Thoracic Vascular Murmurs

PULSELESS DISEASE (TAKAYASU'S SYNDROME — OBSTRUCTION OF VESSELS ARISING FROM THE AORTIC ARCH)

THE DIAGNOSIS OF THIS syndrome, that is found both in elderly patients and in young women, is helped by auscultation, phonocardiography, and pulse tracings of the various arteries. In particular, one should explore the two carotids and the two subclavian arteries, the abdominal aorta, and the two femoral arteries. Angiocardiographic studies will complete the preliminary examination.

Various murmurs or bruits can be heard over the different arteries. Apparently the arterial bruit of each involved artery is fairly localized and is not widely transmitted, especialy proximally (Ueda *et al.*, 1966). No correlation exists between severity of the narrowing and loudness of the murmur. On the other hand, diffuse narrowing of several arteries seems to increase the loudness of the murmur in other, less severely narrowed, vessels (Ueda *et al.*, 1966).

Whenever a phonocardiogram is recorded over a *normal artery* (necessarily with a certain degree of compression), a

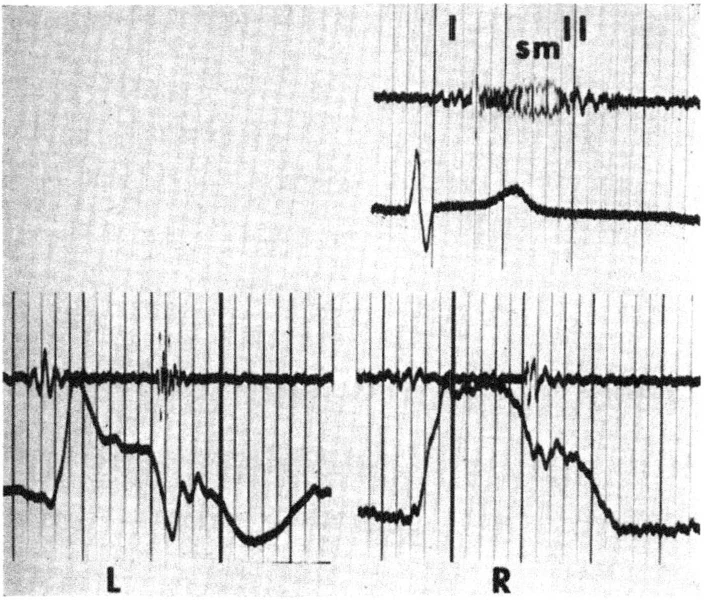

FIGURE 195. Tracings recorded in a 25-year-old woman with Takayasu's syndrome (no femoral pulses; very small brachial and radial pulses; smaller right carotid pulse accompanied by a bruit).
Above: phonoarteriogram of right carotid artery showing a systolic, diamond-shaped murmur with very late peak.
Below: at left = pulse of the left carotid with rapid rise and rapid drop. at right = pulse of the right carotid with slower rise and slower drop. The phonocardiograms of the two lower tracings are precordial tracings, recorded for timing. The two pulses, due to the indirect technique, may not have the same amplification.

short systolic murmur is recorded. If there is *narrowing* (whether from atherosclerosis or other process), the murmur is *systolic* and becomes *diamond-shaped;* it has a peak that occurs progressively later with increased narrowing (Fig. 195). With severe stenosis, a *continuous murmur* is often recorded (Ueda *et al.* 1966). Complete occlusion of course will abolish the murmur (Fig. 196).

A *kinked artery* will give a murmur similar to that of mild arterial stenosis, but the *onset of the murmur is delayed* (Holldack, 1949; Ueda *et al.*, 1966).

A similar graphic picture is often observed in cases with localized *subclavian* or *carotid atherosclerosis.* Again the peak

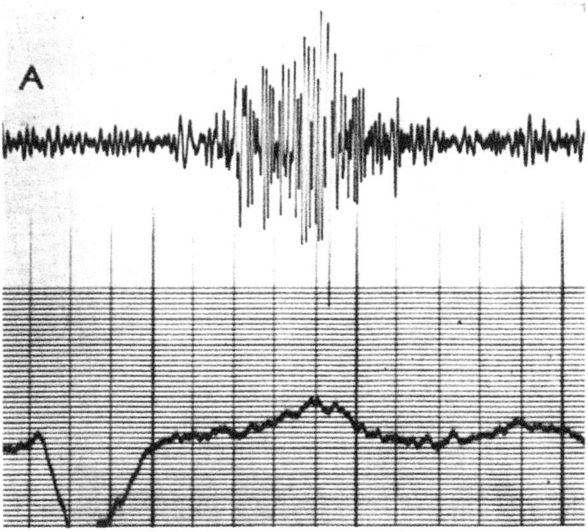

FIGURE 196. (A) Coarctation of the aorta in a 39-year-old man. Tracing recorded in the back. There is a continuous murmur with late-systolic and early-diastolic accentuation. The ECG is broader because the tracing was recorded at a speed of 200 mm/sec.

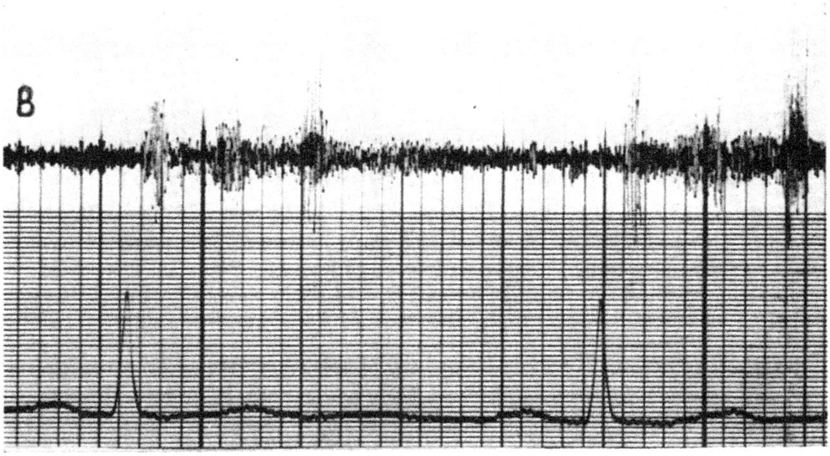

(B) Murmur of a coronary AV fistula, recorded at the apex with high pass filter at 400 Hz. Forty-year-old patient.

of the murmur will occur later, becoming mid-systolic or late-systolic with increasing severity of the obstruction. This mur-

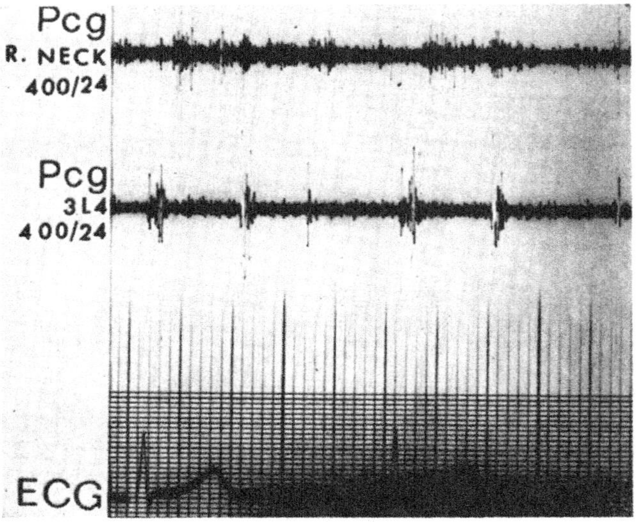

FIGURE 197. Venous hum recorded in a 13-year-old boy. The upper tracing, recorded on the right side of the neck, shows clearly the high-pitched vibrations. The center tracing, recorded over the 3rd left i.c.s., shows some transmission of the murmur to the precordium.
Tracings recorded at 50 mm/sec.

mur is abolished by ipsilateral compression and is accentuated by contralateral compression.

The slope of the *pulse* of the narrowed artery tends to rise more slowly, and will be followed by a more rounded contour.

Functional drug tests may help in the evaluation of the lesions. In *moderate stenosis of an artery, amyl nitrite increases the murmur; in severe stenosis, it decreases it,* probably due to a "steal" phenomenon (Ueda *et al.,* 1966).

BRUITS OF THE HEAD AND NECK

Intracranial bruits can be observed in cases with *angiomata* of the *cranial vessels.*

Aneurysms, arterial stenosis, arterovenous fistulas, neoplasms compressing an artery, or inflammatory processes can cause *bruits in the neck.* A *carotid glomus tumor* may give a late-systolic and early-diastolic murmur, which sounds like a continuous murmur with systolic accentuation.

Systolic or continuous murmurs in one or both carotids and orbital regions occasionally occur in *vertebral-basilar artery oc-*

clusion. It becomes therefore important to correctly study the orbital regions in the presence of a carotid bruit. The microphone should lie over the patient's upper eyelid, and care should be taken that good contact is made without excessive compression. The patient should keep the other eye open in order to avoid sound artefacts related to muscular activity..

Carotid artery-cavernous sinus murmurs are continuous, loud, and maximal over the orbits; they are never abolished by compression of a single carotid. Graphically, they assume maximal peaking in late systole and early diastole.

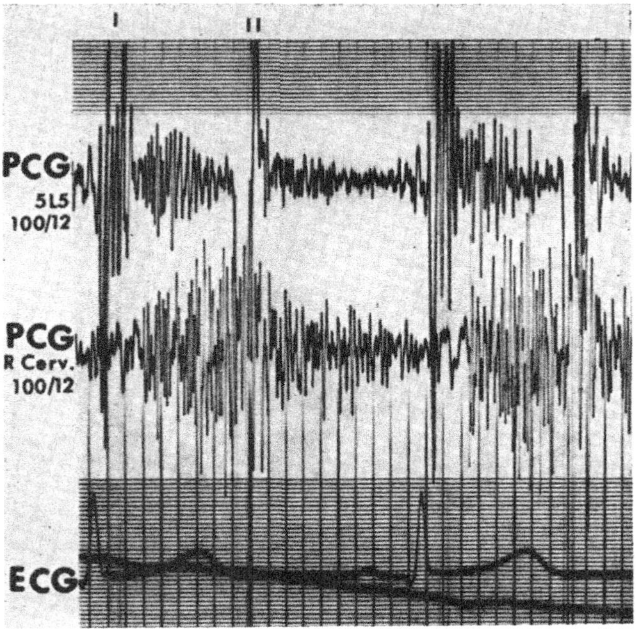

FIGURE 198. Carotid bruit recorded in a 6-year-old child with kinked right carotid artery. The upper tracing is recorded at the apex. The center tracing is recorded over the right side of the neck. The latter shows a continuous murmur with late-systolic, early-diastolic accentuation. The apical tracing shows a diamond-shaped systolic murmur and an early-diastolic murmur.

Supraclavicular arterial bruits are found in many children and young adults due to vicinity of the subclavian artery with bone structures. Shoulder extension commonly results in the disappearance of such murmurs and will help in differentiating them from those caused by organic disease of the vessels. *A carotid bruit* can be heard and recorded in children or young persons due to tortuosity or kinking of a carotid artery (Fig. 198).

Bruits of the Extremities

Arteriovenous fistulas of the arms or legs are either congenital, traumatic, or iatrogenic (A-V shunts in patients on chronic hemodialysis programs). Patency in the latter cases may be confirmed by the palpation of a thrill and auscultation over the area. The high frequency tracing recorded over a *radial A-V fistula* shows a *continuous murmur* with a late systolic-early diastolic peak.

In the lower extremities, bruits and thrills are usually absent in congenital A-V fistulas because the communications are either multiple, tortuous, or narrow. This topic has been already discussed in Chapter 20. A murmur heard over the femoral and iliac arteries may be significant in correlation with a history of claudication, especially if it increases in magnitude and extends into diastole following exercise (P. W. Harvey, 1968).

Bruits of the Abdomen

Auscultation and recordings of the abdomen may provide a clue to several possible processes. A ssytolic bruit may be associted with *atherosclerosis* or *aneurysm of the aorta*. It is generally of medium-low pitch (Fig. 193).

Whenever a continuous murmur arises in a patient known to have an aneurysm of the aorta, *rupture of the aneurysm into the inferior cava* must be suspected. In such cases, the murmur becomes loud, high-pitched, and harsh; it is frequently accompanied by a thrill, and is transmitted to the lumbo-sacral areas.

The most common abdominal murmur in liver disease is the *venous hum of portal hypertension*. This arises from the venous channels connecting the umbilical or paraumbilical veins with the veins of the abdominal wall. This venous murmur (in contrast with the most common occurrence in the case of a venous hum of the neck) is low-pitched and is altered both by changes in posture and by respiration. It is frequently obliterated by firm pressure. While this *venous hum* merely indicates the presence of a large portal collateral circulation, *arterial murmurs* over the liver are most often caused by either a hepatoma or hepatitis. Such murmurs may be either continuous with systolic accentuation or only systolic.

Peritoneal friction rubs have been reported in acute cholecystitis, gonococcal perihepatitis, and metastatic carcinoma of the liver.

The rare finding of a *continuous murmur in the pelvic area* in women should lead to the suspicion of an arterovenous fistula of the ovarian or uterine vessels. More commonly, a systolic murmur or bruit in this area is transmitted from the aorta or the iliac vessels.

Bruits of the Flanks

Auscultation over both flanks should never be omitted in considering renal diseases. *Renal artery stenosis* generally gives rise to a systolic murmur both in atherosclerotic disease and in congenital narrowing of the renal arteries. According to Ueda *et al.* (1968), the murmurs found in renovascular hypertension, apart from their lateral localization, are high-pitched and either pansystolic or continuous. In contrast, the murmur caused by the dilated abdominal aorta is median, is of a lower pitch, and has an early peak and a long decrescendo phase in systole. Amyl nitrite decreases both types of murmurs.

A *continuous murmur* over either flank is most comomnly associated with congenital stenosis, atherosclerosis or fibromuscular hyperplasia of a renal artery, an A-V fistula, or vascular tumors of the kidneys.

Chapter 22

Heart Failure

HEART FAILURE MAY CAUSE the appearance of extra-sounds and murmurs that will be superimposed over those caused by the underlying heart disease.

A *third sound* or a *summation type of triple rhythm* is commonly recorded.

A *pansystolic murmur* of mitral or tricuspid insufficiency may appear in either right or left heart failure, due to dilatation of one or both ventricles.

A *diastolic rumble,* occasionally with *presystolic accentuation,* may appear in heart failure (functional diastolic rumble, caused by ventricular dilatation and favored by AV valve insufficiency).

An *early-diastolic murmur* of pulmonary insufficiency may appear in cases of pulmonary hypertension with heart failure, due to dilatation of the pulmonic ring.

It is unfortunate that, as the borderline between ventricular stress (often lasting for many years) and ventricular failure (also of long duration) is still rather vague, it is often difficult to attribute diagnostic value to these data.

Part III

Borderline Conditions

Chapter 23

Deformity of the Chest

STRAIGHT BACK SYNDROME

THIS IS A SYNDROME caused by the absence of the physiologic kiphosis of the spine (Rawlings). The reduction in the postero-anterior diameter of the chest tends to cause compresson of the heart and may result in a "false enlargement" of the cardiac shadow, seen in the chest films.

The *second heart sound* can be widely split and may not fuse in expiration. A grade 1-3 *diamond-shaped systolic murmur* at the 3rd left i.c.s. has been often demonstrated.

Distortion of the outflow tract of the right ventricle and kinking of the great vessels have been considered as possible causes of he systolic murmur. A gradient of from 4 to 11 mm Hg has been demonstrated in the infundibulum of the right ventricle of some of these cases (Siegel and Schechter). Pressure on the chest wall may increase the magnitude of the murmur (Rubenstein and Johnson). However, Wennewold and Stage found no difference in either the externally or the internally recorded murmur between subjects with straight back and others, also presenting an ejection type of systolic murmur but having no abnormality of the spine. Others failed to observe any pressure gradient. For these reasons, a final interpretation of this syndrome is needed. One should keep in mind that a pulmonary flow murmur and occasionally also a widely split second sound can be found in normal young subjects *without* a straight back.

PECTUM EXCAVATUM

Subjects wth this deformity of the chest often have a *pulmonary flow murmur*. The chest films disclose a leftward shift of the cardiac shadow. The phonocardiogram is similar to that of other types of pulmonary flow murmurs.

KYPHOSCOLIOSIS

These patients present a severe deformity of the spine and tend to have rigidity of the chest cage. As a result, respiration is somewhat impaired, and a chronic type of "pulmonary heart disease," caused by hypoxia, may develop. The auscultatory and phonocardiographic findings are similar to those of other forms of chronic pulmonary diseases (Chapter 16).

Chapter 24

The Senile Heart

AUSCULTATION HAS AN IMPORTANT place in the physical examination of aged persons because it may supply clues about possible alterations of the cardiovascular system. However, auscultatory data in the aged are often misinterpreted so that a serious prognosis may be made in persons with mild or no heart disease. For this reason one should become familiar with the possible data that can be obtained through auscultation in this age group and with their meaning.

Phonocardiography has helped in clarifying some of the problems, and comparison of phonocardiographic and catheterization data has led to a new outlook in regard to the senile heart.

THE "NORMAL" SENILE HEART

A fairly large number of persons between 60 and 100 years of age have no obvious alterations that can be detected by auscultation. However, both *heart sounds* tend to become shorter and seem to have a snapping quality (Aravanis and Harris). The *second sound* is often single, but in some subjects splitting of the second sound is audible. A "normal" type of splitting and a "reverse" type have been clarified by phonocardiographic documents (Slodki *et al.*, 1969). The *normal type* is most marked at the end of inspiration, and the aortic component of the second sound is followed by the pulmonary component, as is true in younger individuals. The *reverse type* of splitting is most marked at the end of expiration; in this situation, the pulmonary component occurs first and is followed by the aortic component.

These two alterations, if isolated and unaccompanied by murmurs, have no serious meaning and probably represent the result of a prolongation of systole in both ventricles but more in either the right (normal splitting) or the left (reverse splitting). Both types of splitting are similar to those caused by bundle branch block, although they are less marked, so that the second sound often becomes single in some phase of respiration.

Atherosclerosis of the Aorta and Normal Blood Pressure

A number of old people present a minor degree of dilatation of the ascending aorta caused by the formation of atheromatous plaques. This causes a disproporation between the caliber of the outflow tract of the left ventricle and that of the ascending aorta, which results in minor turbulence. As a consequence there will be an *aortic flow murmur*, best heard at the third left and second right interspaces (aortic area—Shah *et al.*, 1964). The

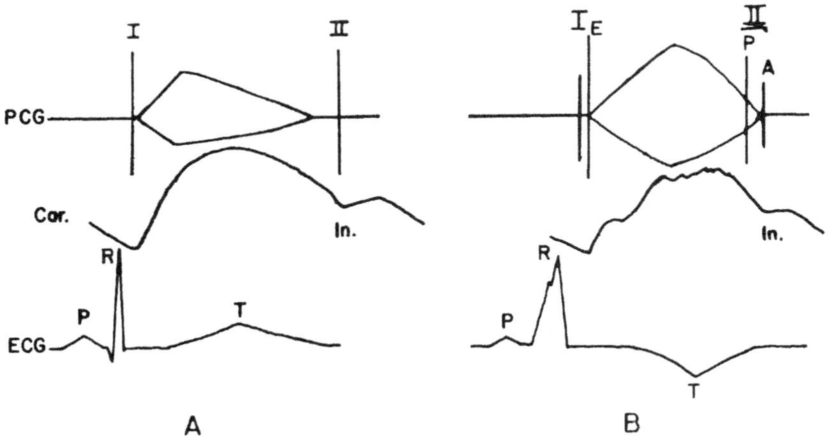

FIGURE 199. Scheme of the phonocardiographic picture of atherosclerosis of the aorta in old age (A), and, for comparison, the picture of aortic stenosis (B). The latter shows reverse splitting of the second sound II with marked alteration of the carotid pulse. I, first sound; E, Ejection sound; P, pulmonary component of second sound; A, aortic component of second sound.

diagnosis of aortic stenosis is sometimes erroneously made. However, when one observes that there is no thrill, that the murmur is grade 1 or 2, that the second sound is either single or normally split, and that the electrocardiogram shows no evidence of left ventricular hypertrophy, a correct evaluation will be made. A most significant aid is the tracing of the *carotid pulse* because it shows a normal configuration that contrasts with the severely deformed pattern of the pulse in aortic stenosis (Fig. 199).

Hypertension

Elderly hypertensive subjects may have a variety of auscultatory changes: (1) a third or fourth sound (gallop sound), or both, audible in diastole; (2) a short flow murmur, present over the third left and second right interspaces; and (3) a short, decrescendo, blowing murmur of aortic insufficiency audible over the same spaces (Fig. 200).

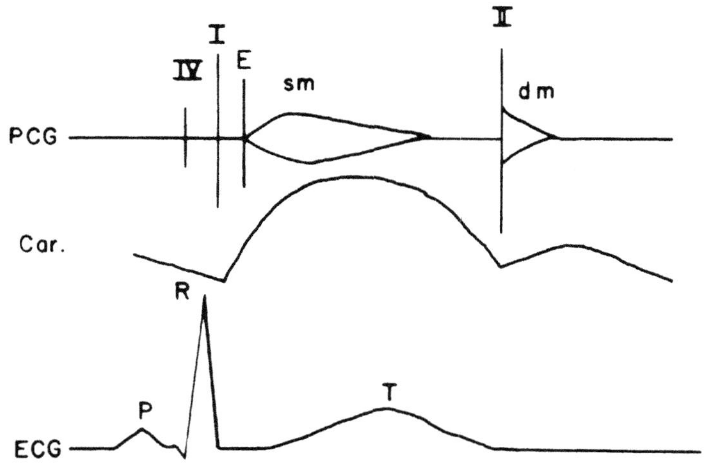

FIGURE 200. Scheme of the phonocardiographic picture of hypertension. IV, fourth sound; sm, systolic murmur; dm, diastolic murmur.

These findings are caused by left ventricular strain (gallops), dynamic aortic dilatation (systolic murmur), and dilatation of the aortic ring (short diastolic murmur). These changes have no serious significance, but a hard gallop sound would signify a more severe left ventricular embarrassment, and a *long aortic diastolic murmur* would signify lowering of one or more aortic leaflets as a result of alterations of the aorta above the valve. These undoubtedly would have a more serious meaning. Also a *long pansystolic murmur* at the apex and left axilla would be evidence of mitral insufficiency, caused by dilatation of the left ventricle with stretching of the chordae and widening of the mitral ring. A similar phenomenon may be the result of a previous myocardial infarct, often with damage to a papillary muscle.

Bundle Branch Block

In older as well as in younger persons, bundle branch block alters the spread of activation of the heart so that one ventricle is activated slightly after the other. Due to the thinness of the right branch, a discrete lesion is sufficient for the occurrence of *right* bundle branch block. In this condition, the right ven-

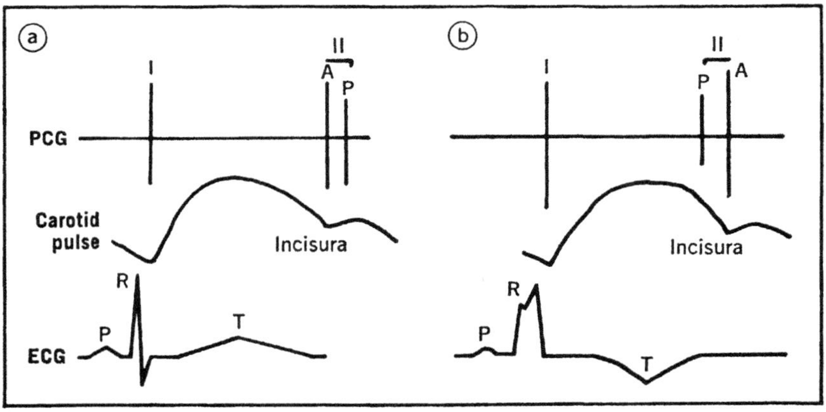

FIGURE 201. Scheme of the typical alterations of the second heart sound in bundle-branch block. In both right (a) and left (b) bundle branch blocks there is a wide, fixed splitting of the second sound. Normal type of splitting (a); reverse type of splitting (b).

tricle starts to contract after the left. The first sound, being related only to the left side of the heart, is still normal. The second sound is widely split. This splitting is of the *normal type,* so that the aortic component is followed after a wide gap by the pulmonary component (Fig. 201).

Left bundle branch block requires a more widespread lesion because the left branch splinters into several divisions. Left ventricular hypertrophy of fibrosis, or both, is usually present. Activation of the left ventricle is slightly delayed and is also prolonged (due to a prolongation of the tension period). The first sound is often soft and poorly audible, whereas the second sound exhibits a *reverse* type of splitting (the aortic component follows the pulmonary component).

In either right or left bundle branch block, the gap between

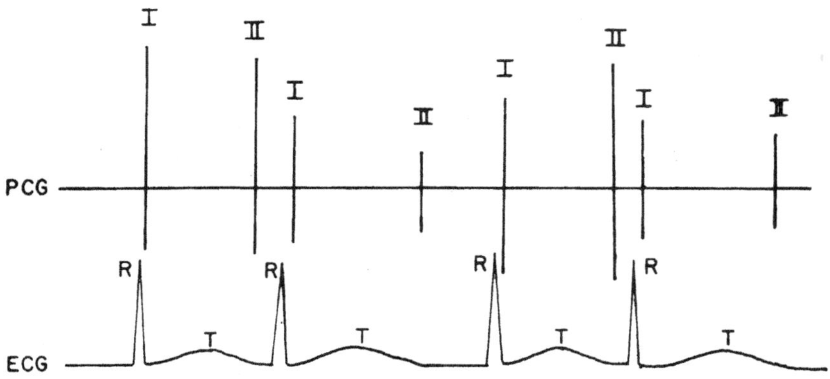

FIGURE 202. Atrial fibrillation in old age. Irregularity of the ventricular rhythm. Variable amplitude of heart sounds (scheme).

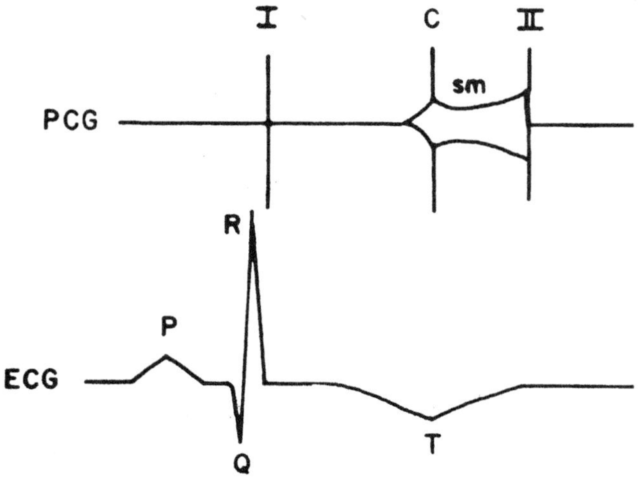

FIGURE 203. One possible aspect of the "papillary muscle syndrome." C, late-systolic click; sm, late-systolic murmur (scheme).

the first and second components of the second sound varies slightly with respiration but is never obliterated (fixed splitting of the second sound). This contrasts with the senile type of splitting, in which respiration usually causes a single second sound in one of its phases.

In either type of block, the second component of the second sound may be misinterpreted as an opening snap, a fact that would lead to the erroneous diagnosis of mitral stenosis. An *electrocardiogram* showing the existence of bundle branch block should alert the physician to the possibility of the aforementioned auscultatory changes. A *carotid tracing,* recorded with the phonocardiogram, gives the best proof of interpretation by showing which component of the second sound coincides with the incisura of the arterial pulse, and is therefore the aortic component.

Atrial Fibrillation

Some old persons have an iregular cardiac rhythm due to atrial fibrillation. This arrhythmia is caused by coronary heart disease and may persist for decades without endangering life. On auscultation, the typical complete irregularity of the heart sounds is heard (Fig. 202), and one may even hear a diastolic rumble caused by the tumultuous filling that occurs with short diastoles (Luisada and Montes, Luisada, et al., 1955, 1959). This functional rumble should not be misinterpreted as evidence of mitral stenosis.

Previous Infarcts

In coronary patients who have had infarcts, auscultation can give a variety of findings. The most common ones are a soft first sound, a faint systolic murmur, and a normal second sound. The second sound, however, may have a reverse type of splitting. In some instances, a loud third or fourth sound is audible in diastole, giving rise to a triple rhythm. This is more common if the left ventricle has some degree of failure. In other cases (the minority), either a pansystolic murmur, a late systolic murmur in crescendo, or a midsystolic click followed by a late-systolic murmur, well transmitted to the left axilla, is audible at the apex. These phenomena are evidence of mitral insufficiency caused by fibrosis of a papillary muscle, which in turn is related to the previous infarct (*papillary muscle syndrome*) (Fig. 203).

The presence of a *ventricular aneurysm* might cause additional murmurs (Chapter 13).

Conclusions

Changes in both the first and second heart sounds may be audible on auscultation in older persons whose hearts are "normal". A common phenomenon in aged persons with aortic atherosclerosis is an ejection murmur. Bundle branch block or atrial fibrillation, if present, would change the heart sounds. Whenever there has been a myocardial infarct, alteration of the first and second sounds occurs; there may be a third or fourth sound, and sometimes there is evidence of mitral insufficiency.

Chapter 25

The Heart in Pregnancy

IT IS WELL KNOWN that pregnancy may cause extra-sounds and murmurs related to the increased blood volume, typical of this condition. Physicians should be aware of them in order not to diagnose heart disease. On the other hand, a careful study is indicated in pregnant women with murmurs because they may have heart disease, the effect of which is either masked or increased by the overload of pregnancy.

In a study from our laboratory (Meyer *et al.*), the following data were found in a group of 200 pregnant women.

(1) *Increased magnitude of the third sound* was found in 58.5% (51% between the 6th and the 15th week, and 59% between the 24th and the 36th week).

(2) *Increased magnitude of the fourth sound* was found in 22% (10% between the 6th and the 15th week, and 24% between the 24th and the 36th week) (Fig. 204).

(3) An *ejection sound* was found in 37.5% (again more frequent in the last part of pregnancy).

(4) A *great vessels flow murmur* was found in 70.5% (51% between the 6th and the 15th week, and 86% betwen the 24th and the 36th week) (Fig. 205).

In regard to the *second heart sound,* subsequent studies showed a *narrow splitting* of this sound in most cases (Fig 204)

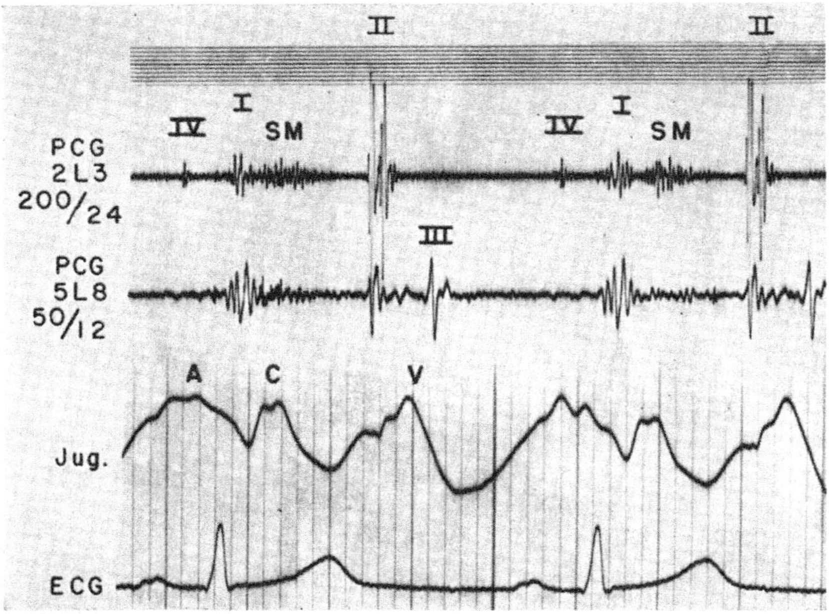

FIGURE 204. Tracings recorded in a 19-year-old pregnant woman. The upper tracing is recorded at the 2nd left i.c.s. and shows a high frequency fourth sound, a small flow murmur, and a closely split second sound. The second tracing is recorded at the apex and shows a very large, medium-frequency third sound. The third tracing is a jugular tracing with normal but prominent waves.

but there were occasional exceptions, where a *wider splitting of the normal type* was recorded.

Similar findings are reported by Cutforth and MacDonald. They also found:

(5) A *functional diastolic rumble* in 18% of the cases.

In regard to the "wide splitting of the first sound" reported by the same authors, it is likely that it was caused by the first component followed by an ejection sound.

Heart disease is the greatest single cause of maternal mortality as reported by many obstetrics clinics throughout the world. Estimation of its frequency among gravid women ranges between 1.5% and 3%. The maternal mortality in this group is five times the normal, representing 25% of the entire maternal mortality rate. The analysis of the type of heart disease seen among pregnant women, as reported by many authors, shows the following distribution: 95% rheumatic heart disease; 2% congenital malformations; and 3% hypertensive, luetic, thyrotoxic, infectious, and rare types of heart disease. Therefore, patients in this group with heart disease should have abnormal sound phenomena, different or greater than those of normal pregnancy. They may include a pansystolic murmur at the apex or midprecordium, a diastolic-presystolic murmur at the apex, an abnormal splitting of the second sound, or a diastolic murmur at the base.

The well-established physiological changes of pregnancy which combine to produce a significant volume load, are: (1) a 30%-

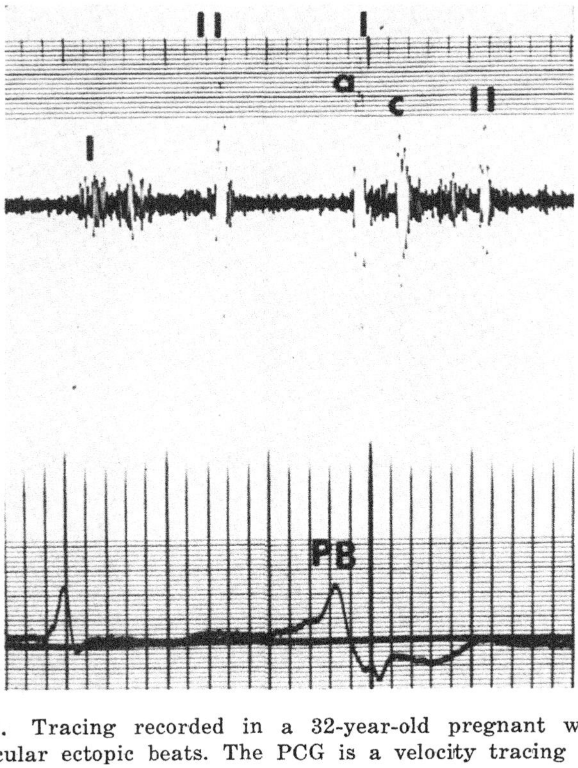

FIGURE 205. Tracing recorded in a 32-year-old pregnant woman with left ventricular ectopic beats. The PCG is a velocity tracing recorded at the apex with a 200-500 Hz band pass filter.
There is a well developed flow murmur in the normal beat. The premature beat (PB) shows a larger first sound, a smaller second sound, and a large ejection click (C).

40% rise in blood volume (40%-50% increase in plasma volume with only 20%-30% increase in red blood cells); (2) a moderate rise in the basal metabolic rate, usually accompanied by tachycardia; (3) an arterio-venous fistula effect in the placenta, where the arterial blood flows into venous "lakes" with no capillary phase. The increased cardiac load becomes evident in the 16th week; its effect then increases gradually until the 24th week, and more rapidly to a peak of hemodynamic load by the 36th week.

Tortuosity of the uterine veins may be responsible for a decrease in the arteriovenous shunt effect in the last month of gestation. The increased blood volume tends to produce "flow murmurs" in the great vessels, usually best heard in the second left intercostal space. Increased ventricular filling during early diastole is indicated by a large third sound. Vigorous atrial contraction against an overfilled and, therefore, less compliant ventricle may cause a larger fourth sound. Pulmonic and, more rarely, aortic ejection sounds complete the list of the sound phenomena encountered during pregnancy. These findings can be expected to increase in frequency as the peak of hemodynamic load is approached. However, Goldberg and Uhland find a remarkable variability in the intensity of the murmur from week to week without any apparent cause.

It should be kept in mind that a *uterine souffle* can be recorded over the central part of the abdomen, and a *mammary souffle* can be recorded over the breasts. Both of them are continuous murmur.

If there are *ectopic beats* or other arrhythmias, they can cause additional sound abnormalities (Fig. 205).

Chapter 26

Functional and Innocent Systolic Murmurs

THE TERM "FUNCTIONAL MURMUR" was created about a century ago to separate murmurs not caused by obvious valvular deformity from the others. The most common instances were systolic murmurs of the apex or base of the heart, encountered in severe anemia, pregnancy, or congestive heart failure, which subsequently disappeared.

MECHANISMS

In the last thirty years, some of the mechanisms causing functional murmurs have been clarified and can be summarized as follows.

1. Severe *tachycardia* may cause a moderately loud systolic murmur. The mechanism is probably multiple (acceleration of flow, incomplete valvular closure, etc.).
2. Severe *anemia* usually causes a loud apical and midprecordial systolic murmur. The mechanism is probably multiple, including acceleration of flow, decreased viscosity, and dilatation of the tricuspid and mitral rings (relative tricuspid and mitral insufficiency). It may also cause an apical and midprecordial diastolic rumble.
3. Conditions associated with *increased blood volume* (pregnancy), *greater rapidity or quantity of systemic or sectional blood flow* (hyperthyroidism, septal defects), or *dilatation of*

one of the large arteries (aortitis or atherosclerosis of the aorta, the Marfan syndrome, idiopathic dilatation of he pulmonary artery) frequently present a systolic murmur at the base. The mechanism is disproportion between normal ostium and dilated vessel, as well as trigonoidation of the valve (relative aortic or pulmonic stenosis).

Unfortunately physicians gradually have extended the meaning of the word and applied the term *functional* to the following murmurs:

(1) Any murmur of persons who have no history of rheumatic fever, have no obvious clinical cardiac enlargement, and are able to develop a normal physical activity without exertional dyspnea.

(2) Any murmur persisting for many years and unaccompanied by obvious changes of cardiac size or physical deterioration.

(3) Any murmur of children or adolescents up to Grade 3.

The reasons for the extremely wide application of the term are obvious and frequently based on emotional factors. It is apparent that the term "functional murmur" has tended to change its meaning through abuse and misuse. Created to depict a murmur caused by mechanisms other than permanent valvular deformity, it now tends more and more to become synonymous with innocent murmur.

The American Heart Association has suggested the term *innocent murmur,* a better term than "functional murmur." Unfortunately this term is based on a complex interpretation and has a negative meaning: *An innocent murmur is a murmur that is not going to be associated with any serious cardiac disease.* The future developments are simply surmised. Therefore, this term represents a *prognosis,* not a diagnosis. This would not be serious if proof of such innocence were available. Unfortunately this proof frequently is either not sought or unavailable. Moreover, psychologic and social elements have become so utterly mixed with the term that it has lost its original meaning.

A murmur that is definitely functional may *not* be innocent. Murmurs may be caused by rheumatic carditis; severe, chronic anemia; vitamin deficiency; heart failure; severe dilatation of either the right ventricle (chronic cor pulmonale) or the left ventricle (coronary or hypertensive heart disease). Murmurs may be *indirectly* caused by a congenital malformation through dilatation of either or both ventricles or the pulmonary artery.

On the other hand, an innocent murmur may *not* be functional. A murmur may be caused by a small septal defect, minimal pulmonary or aortic stenosis, a minor abnormality of the great vessels or a minimal rheumatic deformity of the mitral valve.

The fact that a murmur increases after exertion or when the patient lies supine does not exclude its being caused by a valvular lesion. Exertion and recumbency increase the blood flow to the right heart and accentuate the murmurs of many congenital or acquired cardiac lesions.

The Systolic Murmur in General

The occurrence of systolic murmurs without definite evidence of heart disease has been recognized since Laennec. Certain patients, after an attack of rheumatic fever, present a systolic murmur caused by deformity of a valve, which then persists unchanged for twenty or thirty years with no evidence of cardiac enlargement or failure (Wilson). Therefore, the three criteria—no failure, no enlargement, no change of the murmur—are inadequate for excluding a valvular deformity of minor or moderate severity.

Even the disappearance of a murmur is not sufficient proof that it is a "physiologic" murmur. Bland *et al.* (1936) collected several cases of rheumatic heart disease in which the murmur disappeared in spite of minor valvular lesions. This was proved by autopsy in one case.

It is interesting to note that extremely sensitive phonocardiographic recordings show *a systolic murmur in practically all normal adults* (Groom *et al.* 1960; Ueda *et al.*, 1962). This systolic murmur usually has the graphic characteristics of an ejection (or flow) murmur, and is often musical. On the other hand, a nonmusical, aortic or pulmonic, systolic murmur is more common among *elderly subjects* (Ueda *et al.*, 1962).

The Systolic Murmur of Children and Adolescents

If an audible systolic murmur is not rare in apparently normal adults, it is even more common in children and adolescents.

Painstaking attention has been devoted to a possible clinical differentiation between harmless (or innocent) systolic murmurs and significant systolic murmurs of children or young people. The data collected from medical literature in regard to innocent murmurs can be summarized as follows.

Area. The murmur is heard more often over the pulmonic area but may be heard at the apex, within the apex, or over the entire precordium. Friedman *et al.* found 39 per cent of murmurs in children to be pulmonic, whereas Messeloff found 22 per cent of them in this area.

Type of Murmur. The murmur over the pulmonic area has been defined as groaning, twanging-string, vibratory, musical, or soft and blowing.

Loudness. The murmur may vary from Grade 1 to Grade 3 but seldom is louder than this.

Duration. The murmur is usually short.

Recumbency. The murmur is usualy louder in the recumbent position and may be inaudible when sitting.

Respiration. The murmur is usually louder at the end of expiration but occasionally may be louder at the end of inspiration.

Exertion. Exertion usualy increases the loudness of the murmur; on the contrary, it may cause it to disappear-

Transmission. It is claimed that the murmur has a limited transmission.

Even though based upon the experience of numerous clinicians, the aforementioned criteria are by no means safe. They would have a much greater value if the subjects had been followed throughout their lifetime* or if autopsy data were available. Moreover, these data are not basically different from those that can be observed in certain organic murmurs. For example, organic murmurs originating in the tricuspid or pulmonary valve are increased by recumbency, whereas most organic murmurs are increased by exertion.

Since a high percentage of adults with valvular disease of obvious rheumatic etiology have no knowledge of ever having suffered from rheumatic fever during childhood or adolescence, a negative history has a limited value.

*Marienfeld et al. followed for twenty years 139 subjects diagnosed as having an innocent (vibratory) murmur. Only 2 patients had evidence of heart disease at the end of the period. There were 2 others who had heart disease that was not related to the murmur found in childhood. The murmur had disappeared in 80 per cent of the subjects, whereas it was still present in the others.

Graphic Studies of Systolic Murmurs in Children

Bass *et al.* ascertained by the use of several filters that the innocent murmur contained only high-pitched vibrations in 86 per cent of 64 cases, and occupied either the first half or the major part of ventricular systole. McKee recorded phonocardiograms of normal children and observed a faint, low-frequency systolic murmur in 90 per cent. Mannheimer (1940) made a systematic study of the phonocardiograms of children by means of different filters. Systolic murmurs of comparatively low frequency were commonly recorded. In 50 per cent of the cases, the murmur included vibrations of 100 to 250 Hz, and in 72 per cent of the cases the murmur included vibrations of 50 to 175 Hz (135 children were studied in all). Murmurs below 1 mm of amplitude were discounted.

Evans (1947) and then Wells (1957) considered as typical a systolic pulmonic murmur that starts late in systole and is spent before the second sound (Fig. 206). Harris *et al.* studied

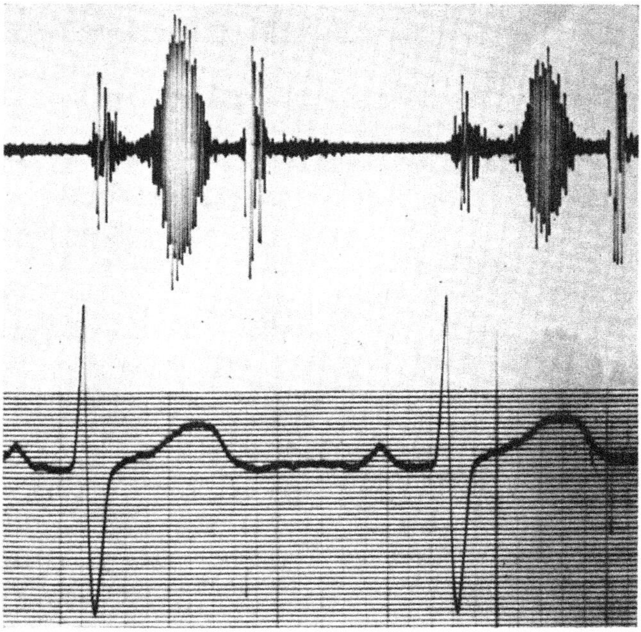

FIGURE 206. Typical mid-systolic, vibratory murmur in a 4-year-old child. Tracing recorded over the second left i.c.s. with a high pass filter at 100 Hz.

13 children with mitral insufficiency and 30 children with a non-blowing, systolic apical or precordial murmur (25 of them of a musical type). The musical (or vibratory) murmur was shorter, had regular vibrations, and was separated by a short pause from the first heart sound. The same observation was repeated by Harris and Friedman in a group of 28 children, and by Harris in 40 children. Harris and Needleman reported graphic studies in normal children and in children with rheumatic or congenital heart disease.

Graphic studies of Luisada and Magri (1953) and Zilli and Gamna revealed large vibratory or musical pulmonic murmurs during the acute stage of rheumatic fever and their attenuation or disappearance when clinical evidence of activity had ceased.

Luisada *et al.* (1958) studied 500 unselected children between the ages of 4 and 17 and were unable to differentiate the innocent murmurs from the others.

Humphries and McKusick recognized two different types of innocent murmurs as follows: (1) the *ejection systolic murmur* having a groaning or vibratory quality, which corresponded to the description of Wells; and (2) the *circumscribed late-systolic murmur*.

Three types of innocent systolic murmurs have been described by Fogel.

(1) The *parasternal murmur* is most frequent in children, is vibratory in type, and is of brief duration.
(2) The *pulmonic murmur* is most frequent in adolescents, is short, blowing, occasionally groaning, and radiates along the sternal border.
(3) The *cardiorespiratory murmur* is short and high-pitched and is best recorded within and below the apex, with marked respiratory variations. This is an infrequent finding.

It is obvious that the first and second murmurs can be easily confused; they are separated mostly because of the different age group.

At present the problem can be summarized in terms of the configuration and duration of the various systolic murmurs that can be observed and recorded.

Pansystolic Murmur. A pansystolic murmur over the midprecordium or apex is usually pathologic, being caused by either

VSD or mitral insufficiency. Magnitude, duration, frequency of vibrations, and area of transmission are sufficient for the differentiation between the two types and for their separation from other murmurs.

Late-systolic Murmur. A late systolic murmur at or within the apex, once considered physiologic by Leatham (1960), is now considered evidence of either mitral insufficiency (as also confirmed by Barlow *et al.*) or pericardial adhesions.* In the first instance, its vibrations start earlier. The effects of pressor drugs and respiration are of some help for the differentiation of the two types. A *midsystolic or late-systolic click* favors the interpretation of chordal or papillary muscle dysfunction.

Midsystolic Murmur. A midsystolic murmur with musical quality (twanging-string or groaning), well separated from both heart sounds, is usually innocent (Wells, 1957) (Figs. 206 and 207 A). It is probably caused by pulmonary artery dynamics, as proved by functional and pharmacodynamic tests,** even though it may be loudest in the third or fourth interspace. In contrast with his earlier study (1967), Wennewold (1968) was able to record this murmur in the pulmonary artery or its branches in 54 out of 80 subjects. In 11, it was only present in the branches; in 17, it was also recorded in the outflow tract of the right ventricle. On the other hand, 15 subjects with an external murmur had no murmur within the chambers of the right side, so that it was accepted that it originated in "aortic or subaortic area."

Early-systolic Murmur. The murmur may occur only in early systole (Fig. 207 B) and may have different origins.

Among the possible causes of musical, systolic murmurs, one should keep in mind the possibility of an *anomalous left ventricular band.* This may become taut when the left ventricle enlarges, either during exercise or because of heart failure (Roberts, 1969).

Diamond-shaped Murmurs. A diamond-shaped murmur of the outflow tract of the right heart can be caused by numerous dynamic reasons, not always easy to recognize (Fig. 207A). This murmur may be grunting or groaning in quality, and is increased by exertion, excitement, bradycardia, inspiration, or amyl nitrite.

* Other possibilities have been discussed in Segal's book.
**Amyl nitrite typically increases this murmur.

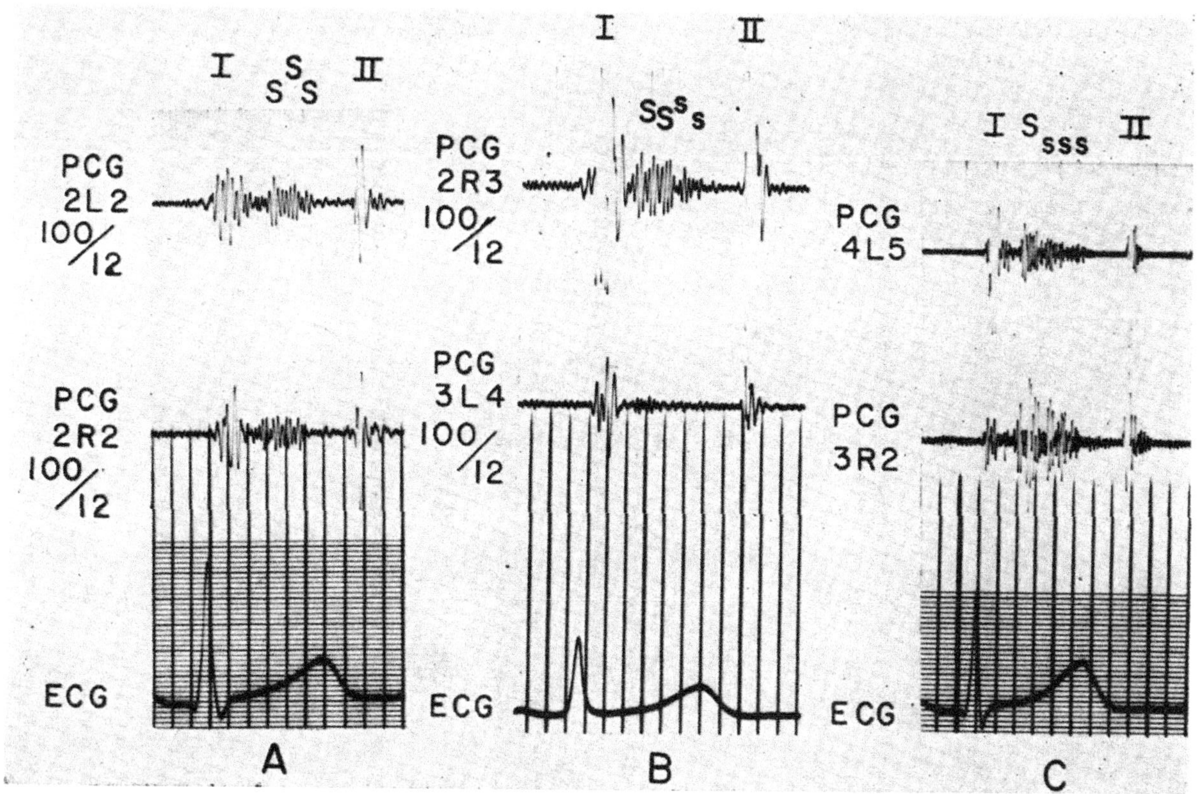

FIGURE 207. (A) A mid-systolic, somewhat musical, pulmonic murmur in an 8-year-old boy. No gradient of pressure was found on catheterization. However, angiocardiography revealed a slightly narrow outflow tract of the right ventricle. (B) An early-systolic murmur originating in the large arteries in a 6-year-old child. Documentation of the origin was given from the area of greatest loudness (at right of sternum and intravascular phonocardiography.
(C) A mid-systolic murmur caused by a kinked innominate artery in a 2-year-old boy. Normal data were obtained on right and left heart catheterization, and demonstration was given by angiocardiography.

A diamond-shaped murmur of the outflow tract of the left heart can be caused by harmless abnormalities (*kinking or tortuosity of an artery*), as we have been able to demonstrate by angiocardiography in several of our cases (Fig. 207 B and C). It is often louder over the right supraclavicular area and has been called *supraclavicular arterial bruit* (Segal, 1964). Its differentiation from other murmurs of more serious meaning may require a complex cardiac work-up.

In addition to these murmurs, *a venous hum* can be heard and recorded over the right supraclavicular fossa but often also over the second right interspace (Fig. 197).

Many of these murmurs are hemodynamically insignificant because no shunts or pressure gradients can be easily demonstrated. Sometimes the exact point of origin of the murmur can be only determined by intracardiac and intravascular phonocardiography and at times it is impossible to exclude a minor valvular lesion. Leatham *et al.* correctly stressed the importance of the study of the second sound. Abnormalities in the degree of splitting, respiratory variations, and amplitude of the two components of this sound may be essential clues for the diagnosis of valvular lesions or congenital shunts. However, minor lesions and even small shunts may not affect the second sound. Therefore, in the final analysis, the clinical evaluation of all data will be decisive in doubtful cases.

The difficulty encountered in the evaluation of some of these murmurs are demonstrated by the following cases.

CASE 1. M.W.

This is an 8-year-old female in whom a heart murmur was discovered at birth. She had frequent sore throats and occasional joint pains during her earlier years. There was no reduction of physical activity and no history of cyanosis.

Physical Examination. BP/65 mm Hg. Pulse 88/min/reg. No cyanosis, clubbing, or edema. No venous distension. Arterial pulsations are normal. The lungs are clear to percussion and auscultation.

Heart. The first heart sound is loud at the apex. The second sound is loud at the second left i.c.s. and closely split. A grade III, midsystolic, medium-pitched murmur in crescendo-decrescendo is heard over the apex, mid-precordium, and pulmonic areas. Abdomen, extremities are normal.

ECG. Sinus arrhythmia. Normal juvenile pattern.

Chest X-ray. Normal.

Phonocardiogram. Large first sound. Loud and split second sound in the pulmonic area. High pitched, diamond-shaped, systolic murmur best recorded at the midprecordium and second left i.c.s. This systolic murmur was increased by inspiration as well as by inhalation of amyl nitrite and was considered as a pulmonic "flow murmur".

Cardiac Catheterization. No evidence of shunts. No gradient across the valves of the right side. Gradient of 15 mm Hg between left ventricle and aorta. Cardiac index, 7.3 L/min/m^2

In this case, the clinical impression was that of an atrial or ventricular septal defect. Large cardiac output and the absence of shunts led to the conclusion of a "high putput murmur." The demonstration of a small gradient across the aortic valve is interesting and may be evidence of a left-sided origin of this "innocent" murmur (minor aortic stenosis, probably congenital).

CASE 2. L.B.

This is a 12-year-old female, born cyanotic. A murmur was heard before the age of 3. She developed well and is asymptomatic. No history of rheumatic fever.

Physical Examination. BP 110/70 mm Hg. Pulse 90/min. regular. Neck veins, peripheral pulsations, lungs: normal. Abdomen, normal.

Heart. The first heart sound is normal. The second heart sound is normal in intensity and normally split. Grade III, early-systolic murmur, best heart over the midprecordium. This murmur is also heard with less intensity at the pulmonic area and at the apex.

ECG. Normal tracing.

Chest X-ray. Small vertical heart. Narrow chest in lateral position.

Phonocardiogram.

1957. Medium amplitude, high frequency diamond-shaped murmur over the third left i.c.s. The second heart sound is widely split and "fixed."

1960. The second heart sound has a "fixed" splitting. There is a third heart sound at the apex. High frequency, musical systolic murmur at 2nd-3rd left i.c.s.

1965. Close and variable splitting of the second heart sound.

The third heart sound is still present. Diamond-shaped systolic murmur at the apex and 2nd left i.c.s. with an early peak. Amyl nitrite markedly increases the systolic murmur.

Cardiac Catheterization. Intracardiac phonocardiography shows both an early-diastolic murmur and a systolic murmur in the left ventricle.

The pressures are normal in both the right and left cardiac chambers. Pulmonary artery angiography is normal and there is no recirculation of the contrast material. No evidence of shunts. No aortic root injection was done.

In this case, the history seemed to favor a congenital lesion. The clinical and phonocardiographic data favored for some years the diagnosis of atrial septal defect. When catheterization was performed, no abnormalities were found. On the other hand, at that time, even the phonocardiographic data had become less significant.

The conclusion was: "pulmonary flow murmur." The cyanosis at birth may have been caused by a shunt lesion that subsequently closed. This lesion may have been the cause of the murmur detected in early childhood and may not be related to the present murmur. On the other hand, the congenital shunt may have caused some dilatation of the pulmonary artery that persisted until the time of the examination.

CASE 3, B.E.

This is a 12-year-old female with a history of "rheumatic fever" at age 2. A heart murmur was heard at age 5. She has been asymptomatic all the time.

Physical Examination. BP 100/60. P. 88/min.

Heart. The pulmonic component of the second sound is louder than the aortic component. Grade II systolic murmur at apex transmitted to left axilla. Grade I-II systolic murmur in the pulmonic area in the supine position that almost disappears upon standing.

ECG. Normal tracing.

Chest X-ray: Normal lung field. Normal heart. Minimal dilatation of pulmonary artery.

Phonocardiogram. Normal splitting of second sound. Diamond-shaped, systolic murmur with early peak at second right and second left i.c.s. extending to apex.

Cardiac Catheterization. There is a 10 mm gradient between right ventricle and pulmonary artery. Left ventricular angiography is normal. Aortic root angiography is normal.

Intracardiac Phonocardiography. Minimal systolic murmur in the left ventricle, larger than in the right.

Serial Oxygen Studies. No evidence of shunts.

In this case, the diagnosis of rheumatic fever may have been incorrect. Still the murmur could have been caused by minimal mitral insufficiency of rheumatic etiology. The phonocardiogram showed the existence of a "flow murmur." The intracardiac phonocardiogram revealed the murmur to be larger in the left ventricle. Therefore, these possibilities can be considered: minimal mitral insufficiency, minimal pulmonic stenosis, or an aortic flow murmur. A minimal gradient of pressure was found across the pulmonic valve. Mitral insufficiency was still not excluded because no LV injection was made. In addition, there are two possible locations (pulmonary valve or aorta) for the origin of the murmur.

CASE 4, K.C.

This is a 13-year-old female. A heart murmur was heard at the age of 4, and at that time her heart was found enlarged on chest X-ray. These findings were attributed to rheumatic heart disease despite a negative history. She had occasional sore throats during the winter months and has been taking prophylactic penicillin. She had experienced dyspnea on exertion for the previous six months.

Physical Examination. BP 120/70 mm Hg. P. 90/min. regular. Respiration 16. No cyanosis or edema. The neck veins are not distended. The arterial pulsations are normal. The lungs are clear on percussion and auscultation. The abdomen is normal.

Heart. No enlargement on percussion. The second heart sound is split with respiratory variations. There is a third heart sound. A grade II, early- and midsystolic murmur is present along the left sternal border and at the apex.

ECG. Normal tracing.

X-ray. Vertical heart, prominent pulmonary artery.

Phonocardiogram. Split II heart sound with respiratory variations. Ejection sound. Low frequency, wide III sound. Diamond-shaped systolic murmur that starts with a click in some tracings, and is recorded best at the 2nd left i.c.s. In the midprecordium, there is occasionally a midsystolic murmur with irregular vibrations. Amyl nitrite causes a marked increase in the amplitude of the murmur.

The apex cardiogram is suggestive of left ventricular hypertrophy.

Cardiac Catheterization. (Right and Left) No gradients, no increased output, no shunts.

In spite of the suggestive history and the atypical phonocardiographic findings, the final conclusion was that of an "innocent flow" murmur. However, doubts can still be entertained, and subsequent examinations are planned.

CASE 5, A.R.

This 13-year-old boy has been known to have a cardiac murmur since a year ago. No history of frequent upper respiratory infections or joint pains. He has complained of easy fatigability and occasional precordial pain.

Physical Examination. BP 110/70 mm Hg. P. 80/min. regular. Neck, chest, lungs—normal. Abdomen, normal. Pulses, normal. The second heart sound is loud and split. No third or fourth sounds. A grade II-III systolic, crescendo-decrescendo, murmur is heard at the 3rd left i.c.s with little radiation.

ECG. Normal tracing.

Chest X-Ray. Normal.

Phonocardiogram. An ejection-type systolic murmur is best recorded over the 2nd and 3rd left i.c.s. A large third heart sound is recorded in the low frequency tracings at the apex. The second heart sound is widely split but there are large respiratory variations.

Right Heart Catheterization. No gradient across the pulmonic valve. A 2.2 mm diastolic gradient is observed at the tricuspid level. No evidence of shunts.

In the absence of valve lesions or shunts, the final diagnosis of "innocent murmur" was made.

CASE 6, L.J.

This 16-year-old male developed precordial chest pain, shortness of breath, and dizziness while playing basketball 2 months ago. The pain and the exertional dyspnea lasted two weeks off and on.

Physical Examination. BP 130/80 mm Hg. P. 70/min. regular.

No cardiac enlargement is detected but a left parasternal lift is noted. The second heart sound is increased in intensity and there is a fixed splitting. A grade II midsystolic murmur is present at the left sternal border, and a grade I systolic murmur is heard over the entire precordium.

The other systems are normal.

ECG. Sinus rhythm. Borderline atrioventricular conduction (PR = 0.20 sec.). Increased voltage in the horizontal plane suggests left ventricular hypertrophy (this voltage might be normal for the age and build of the subject).

Chest X-ray. Enlarged cardiac shadow in the PA projection is probably due to "straight back" rather than actual enlargement. The aortic arch is relatively small.

Phonocardiogram. Pulmonary flow murmur. Minimal early-diastolic murmur at the base, possibly functional.

Cardiac Catheterization. No evidence of shunts. Systolic gradient (6.4 mm Hg) across the aortic valve, probably favored by high cardiac output (CO: 11.97 L/min; Index 7.82 L/min/m^2).

The presence of a high output in the absence of shunts or valve lesions may explain the murmur. The existence of a small gradient across the aortic valve is probable evidence that the "innocent" murmur was of left-sided origin.

CASE 7, T.M.

This 61-year-old male has been complaining of shortness of breath upon exertion and at night for one year duration. He also had a cough, weakness, dizziness, and ankle edema, and has been taking medication. He has elevation of systolic blood pressure.

Physical Examination. Patient in no acute distress. BP 170/90 mm Hg. P. 72/min. The heart is not enlarged. The heart sounds are normal. Grade III-IV harsh, pansystolic, decrescendo murmur at the apex, and more of ejection type at the aortic area with propagation to the neck.

ECG. Left ventricular hypertrophy and strain. Evidence of diffuse ischemia.

Chest X-ray. No significant enlargement of the ascending aorta. The pulmonary artery is enlarged.

Phonocardiogram. The first heart sound is within normal limits. There is a normal splitting of the second heart sound. Large fourth heart sound. There is an early and midsystolic, diamond-shaped murmur and a minimal early diastolic murmur of high frequency.

Interpretation. Heart sounds and murmurs are consistent with hypertensive heart disease.

Cardiac Catheterization and Angiocardiography. No evidence of aortic or pulmonic stenosis. Occlusion of the left coronary artery. No evidence of mitral insufficiency. Insignificant aoritic insufficiency.

In spite of the history and the presence of murmurs, no evidence of mitral insufficiency, caused by papillary muscle dysfunction, or aortic stenosis was found.

The final conclusion was "aortic flow murmur" related to chronic hypertension and probable aortic dilatation caused by atherosclerosis.

The minimal aortic insufficiency and the large fourth sound are also typical of hypertensive heart disease (Chapter 15).

The few representative cases that have been described show that, in several instances, even the most refined methods of study cannot give the final answer in regard to the cause of a systolic murmur. The complexity of the problem is revealed by Caceres and Perry's book, in which several cardiologists present various, often conflicting viewpoints, and in which a long list of causes for systolic murmurs is presented.

Chapter 27

Functional Diastolic Murmurs

DIASTOLIC MURMURS, NOT CAUSED by valve lesions, and thus termed "functional," have been described for a long time. Probably the first to be described was the "Austin Flint,' low frequency, apical rumble of aortic insufficiency. This was followed by the "Graham Steell," high frequency, basal blowing murmur caused by pulmonary hypertension in mitral stenosis, and by the midprecordial or apical "Carey Coombs murmur" of acute rheumatic fever. Since then, "functional" diastolic rumbles have been described in pericardial and hypertensive heart diseases (White), acute rheumatic fever without mitral stenosis (Bland *et al.*, Bramwell, Gibson), and myocarditis (Weinstein and Lev). Subsequently, phonocardiographic studies documented such murmurs in rheumatic, coronary, hypertensive, and congenital heart diseases (Ravin and Darley; Fishleder and Medrano; Likoff *et al.*). In particular, certain diastolic phenomena of mitral insufficiency (occasional opening snap, frequent third sound, and nearly as frequent mid-diastolic rumble) have a special importance because of possible misdiagnosis of double mitral defect (Bleifer *et al.*; Nixon *et al.*; Perloff and Harvey; Bloemer *et al.*). It is of interest that occasional diastolic murmurs (of both low and high frequency) can be found in graphic tracings of "normal" individuals even though such murmurs are not common on auscultation of "normal" individuals (Libman and Sad, Argano and Luisada).

In spite of repeated statements and publications on this subject including several from our group (Luisada and Perez Montes; Luisada *et al.*, 1955 and 1959; Argano and Luisada, 1971), teaching in several Medical Schools still persists in stating that "all apical diastolic rumbles are caused by mitral stenosis" perpetuating an erroneous concept. For this reason, a more recent contribution by our group was recently published (Luisada and Dayem).

Numerous examples of functional diastolic murmurs were presented in Chapters 7, 8, 9, 12, 13, and 16.

The normal heart of young people has already minimal, inaudible, low-frequency diastolic vibrations at the apex that can be recorded by means of adequate phonocardiographs. These vibrations may occasionally become larger and reach the threshold of audibility if there is an overactive cardiovascular system (exercise, after inhalation of amyl nitrite, hyperkinetic syndrome). The cause of these vibrations can only be speculated upon. They are most likely arising in the left ventricular wall during its distension, as suggested by the fact that they frequently start with a large third sound and represent aftervibrations of this sound.

The murmur of mitral stenosis is caused by the difference in caliber between mitral valve and left ventricle. Thus, it is related to eddies which start at the valve and spread from the inflow tract toward the apex. The vibrations of the blood are transmitted to the left ventricular wall and, from this, to the chest wall.

Functional diastolic rumbles are usually encountered when one of the ventricles is severely dilated (most commonly the left, occasionally the right); when its filling is abnormally large (valve insufficiency, shunts); when its pressure is high (hypertensive heart disease, cor pulmonale); and especially if the ventricular wall is damaged (myocarditis or cardiomyopathy; coronary heart disease, especially after infarcts, and even more if there is a ventricular aneurysm) or is in a state of failure.

The mechanism of these murmurs is still in question. More rapid and greater filling (valve insufficiency, shunts) would increase the diastolic vibrations of the wall. High pressure would often alter the diastolic tension (moderate or severe increase of the end-diastolic pressure, often associated with strain or failure). Left ventricular damage from any cause would cause dyssynergies with uneven filling of various sections of the left ventricle, and this may be a cause of murmurs. Whether the old concept of "relative mitral stenosis" (normal mitral opening, large ventricle) is still valid, we do not know. Probably such concept should be abandoned because the vibrations are usually found over a much larger area than in organic stenosis.

DIFFERENTIAL DIAGNOSIS BETWEEN FUNCTIONAL DIASTOLIC MURMURS AND THE MURMUR OF MITRAL STENOSIS

In the differential diagnosis, clinical and roentgenological data may be of help. However, the most important are the electrocardiographic and phonocardiographic data.

Electrocardiographic Data

Left Ventricular Hypertrophy. If a patient has no evidence of aortic stenosis or insufficiency or mitral insufficiency, and no systemic hypertension, the electrocardiographic evidence of left ventricular hypertrophy should be considered *against* the diagnosis of mitral stenosis. In the presence of a diastolic rumble, one should consider either a cardiomyopathy or coronary heart disease.

In our experience, *bundle branch block (BBB)* is seldom found in cases of mitral stenosis (MS). *Left BBB* is definitely against the diagnosis of MS. *Right BBB* is also against, though rare elderly cases of MS may have this complication through the association with coronary heart disease.

Phonocardiographic Criteria

(1) **Wide area of recording of a diastolic murmur** or recording it in the 2nd or 3rd left i.c.s. is in favor of a functional murmur. This fact can be explained by severe ventricular enlargement.

(2) **Wide splitting of the second sound** is against MS. Cor pulmonale, congenital shunts, or severe right heart failure should be considered instead.

(3) **High frequency of the murmur** (soft, blowing murmur on auscultation) is unusual in MS. In a special study devoted to this problem (Mehta and Luisada), we found high frequency

vibrations only in 12% of the cases studied by phonocardiography.

(4) The murmur of mitral stenosis is frequently initiated by an *opening snap*. This may be absent, however, in cases with very severe narrowing of the mitral valve, often revealed by the finding of calcification on x-ray. If there is no evidence of severe MS (clinical, roentgenological, electrocardiographic data plus typical history), lack of opening snap is against MS. An opening snap can be found in exceptional cases of pure mitral insufficiency. It is usually followed by a very large third sound. Considering the various exceptions pro and con, a moderate importance should be given the lack of opening snap.

(5) **High Voltage of the Murmur.** This is usually not appreciated by auscultation on account of the low frequency of the vibrations, which are often close to, or even below, the threshold of the auditory system. In certain cases, a barely audible murmur will appear in the tracing (with adequate filtration) nearly as large or larger than the second sound. Cases of MS with high voltage of the mid-diastolic murmur are extremely rare and are probably cases in which severe left ventricular damage, due to concurrent processes, adds a functional element to the organic murmur. High voltage is in favor of a functional rumble and against mitral stenosis.

(6) **Wide Separation of the Murmur from the Previous Second Sound.** The murmur of mitral stenosis starts with the beginning of left ventricular filling, be this marked or not by an opening snap. A functional rumble usualy starts later, and its onset is very often marked by a large third sound. Thus, the former starts generally from 60 to 120 msec. after the aortic component of the second sound while the latter starts from 150 to 180 msec. after such component. Late start of the mid-diastolic rumble is in favor of a functional rumble and against mitral stenosis.

(7) **Mid-diastolic Rumble without Presystolic Murmur in the Presence of Sinus Rhythm.** While functional presystolic murmurs may occur, a functional diastolic rumble without presystolic accentuation is more common. In mitral stenosis with sinus rhythm, if there is a mid-diastolic rumble, there is also a presystolic murmur. Thus, an isolated mid-diastolic rumble is against mitral stenosis unless there is atrial fibrillation.

(8) **Presystolic Murmur that is Not in Crescendo.** In mitral

stenosis with sinus rhythm, the presystolic murmur is typically in crescendo. An exception is represented by cases with very long conduction time. Functional murmurs in subjects with normal P-R interval may have a sharp separation from the following first sound.

(9) **Large Third Sound or Fourth Sound.** A patient with mitral stenosis nearly never has a large third sound. If there is a fourth sound at low frequency and a presystolic murmur at high frequency, the stenosis is mild. A functional diastolic murmur starts very often with a large or even gigantic third sound and may be followed by a fourth sound.

(10) **Effect of Amyl Nitrite Inhalation.** This functional test very often increases the murmur of mitral stenosis by causing tachycardia. It has been stated that it would decrease the Austin Flint murmur but we have found such decrease only in some cases. Therefore, it is difficult to assign a definite value to the results of this test.

Chapter 28

Unusual Sound Phenomena (Honk, Whoop, Crunch)

THE SYSTOLIC WHOOP WAS so named by McKusick because it resembles the noise of whooping cough. It is a loud noise occurring in systole (often in late systole) every few cycles, and frequently consists of the accentuation of an already existing systolic murmur (Fig. 208). It is a musical murmur, and has also been called "goose honk" or "precordial honk." It is *not* connected with the phases of respiration and thus the cause of the accentuation of the murmur is still unknown. It was thought in the past that this sound phenomenon was not related to a cardiac lesion. However, often, either a pansystolic murmur is disclosed by the phonocardiograph in the cycles without the "whoop" or, in subsequent studies, the whoop has disappeared and a more conventional pansystolic murmur is present in all cycles. Therefore, this unusual murmur seems related to some kind of dysfunction of the mitral valve. This interpretation was confirmed by the fact that a "whoop" has been noted during cardiac catheterization when a catheter is passed across the mitral valve (Caceres and Perry). As ballooning of the mitral valve was demonstrated in 8 cases (Behar *et al.*), deficiency of the muscular components of the mitral valve has been suggested. These components have been documented in man by Montiel.

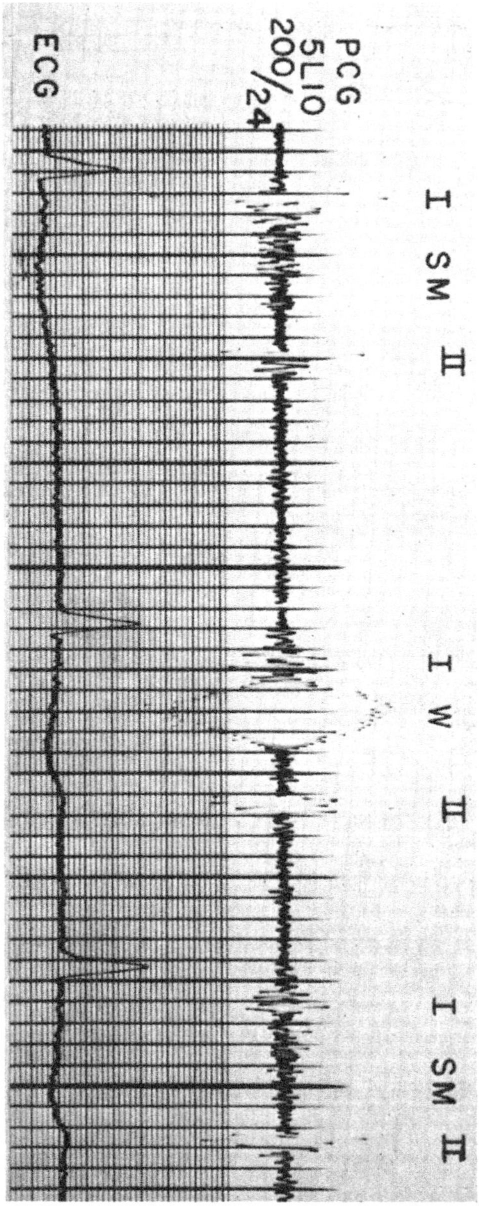

FIGURE 208. Systolic whoop (W) occurring occasionally in a patient with mitral insufficiency.

A systolic whoop has been described in left-sided *pneumothorax* and has been explained by cardiac action on air pockets trapped in the pleural cavity near the left ventricle (Boelandt *et al.*).

The *sternal crunch* or *xiphosternal crunch* is a sound, usually systolic but occasionally diastolic, having a "crunching or splitting" quality, heard best to the left and above the xiphoid process. In most cases, respiration does not affect the noise but occasionally expiration increases its loudness. Recumbency decreases the noise; leaning forward increases it. Pectum excavatum was found present in 25% of the cases (Schugoll). In one case of our observation, several clicking vibrations were recorded in some of the cycles but not in others (Fig. 209). The most satisfactory explanation is that of noises created by movement of the 7th left costal cartilage at its junction, produced by the heart action (Schwab *et al.*). Motions of the diaphragm can also be involved by changing the relationship between heart and chest wall.

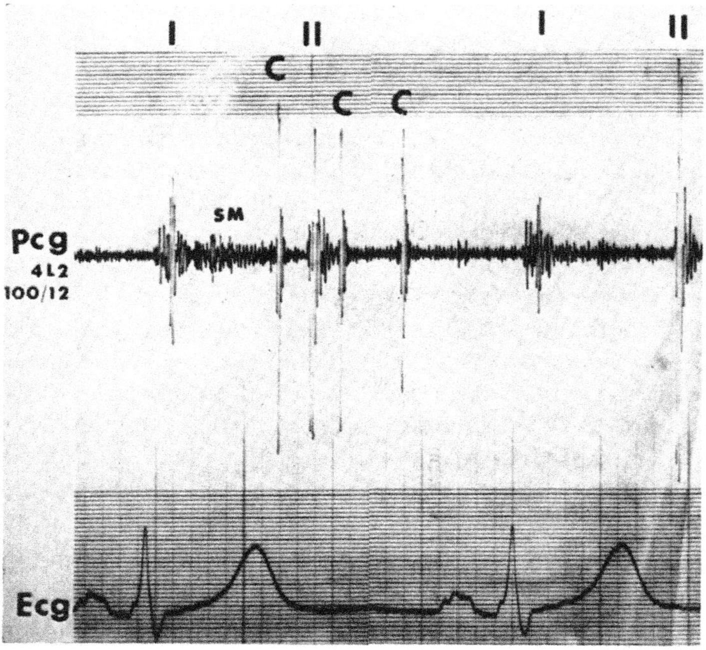

FIGURE 209. Sternal crunch in a 43-year-old man. The tracing, recorded over the lower part of the sternum, shows an early and mid-systolic murmur terminating in a click. Two other clicks are recorded in diastole. However, the next cycle shows no clicks and a smaller murmur.

Chapter 29

Sounds Caused by Artificial Valves and by Pacemakers

THE MODERN SURGICAL PROCEDURE of replacing heart valves with plastic ball valves in a cage of silicone-covered vitallium (Starr-Edwards valves) has resulted in the appearance of new sounds caused by the artificial valves themselves.

Physiologic deductions about time intervals cannot be safely made from the sound tracings of these patients because valvular opening and closure is different from that of normal valves. This results from the higher pressure necesary to cause valve movements in comparison with normal valves and by the different structure and relationship of the valve. Still, these tracings supply data of great interest for the individual patients.

In the last few years, several studies tried to detect *prosthetic valve malfunction* by means of phonocardiography. This adds to the importance of sound tracings in such cases.

NORMAL VALVES

Mitral Valve Replacement. The first component of the first sound is still present (Fig. 210). It may seem small because, due to the great magnitude of the mitral valvular sounds, small amplification is generally used. The mitral closing sound (MC) follows the first component after about 40 msec. and has great magnitude and high frequency. There are two components of

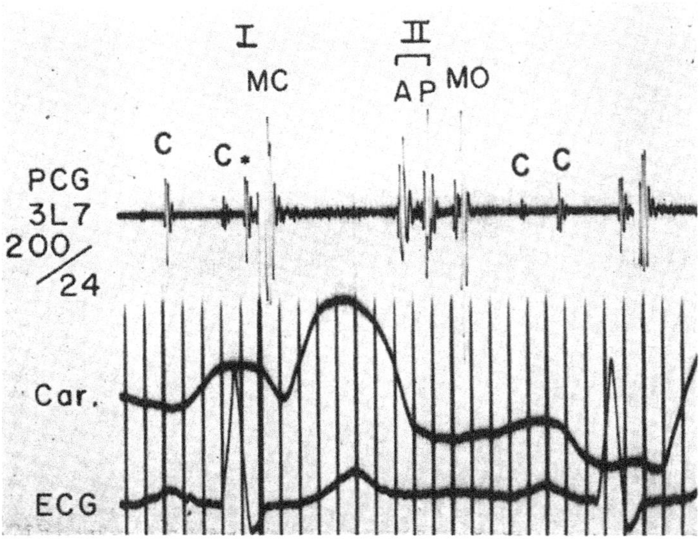

FIGURE 210. Clicks determined by a prosthetic mitral valve in a 32-year-old woman. Tracings over the third left i.c.s. Carotid tracing.
C,C = clicks in diastole
MC = "mitral" valve closure
MO = "mitral" valve opening
A,P = components of second sound
"Mitral" closure is preceded by the first component of the first heart sound (*).

the second sound and then a mitral opening sound (MO). This occurs from 80 to 110 msec. after the aortic component (A) of the second sound. As the A-MO interval is dictated by the gradient of pressure across the prosthetic valve, the shorter the A-MO interval, the higher will be the left atrial pressure. Additional clicking sounds are often heard and recorded in diastole; they are probably caused by brief reclosures of the valve. There is no explanation why they are not always present, and we do not know what they mean when they occur.

Aortic Valve Replacement. The first sound is small as a result of small amplification (Fig. 211). There is a large and high frequency aortic opening sound (AO). A few vibrations can be observed in systole because of either persisting aortic dilatation or a minor degree of stenosis, which is often unavoidable. There is also a large and high frequency aortic closing sound (AC). In the case represented in Figure 211, there was also a small

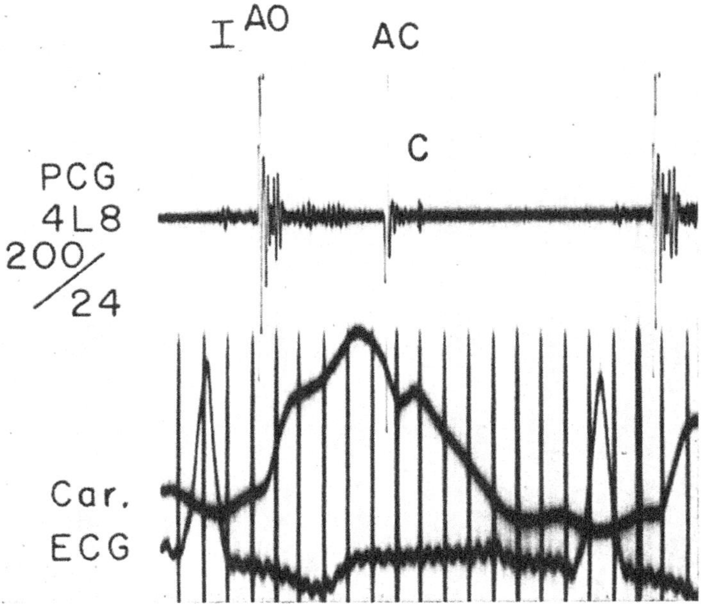

FIGURE 211. Tracings of a 52-year-old man with a Starr-Edwards valve replacing the aortic valve. The artificial valve causes an opening click (AO) and a closing click (AC) plus a small click (C) in early diastole. The opening click is 0.07 sec. after the first component of the first sound, which is small in this tracing.

click in early diastole, which could have been either the pulmonary component, a mitral opening sound, or a secondary click of the aortic valve.

Double Valve Mitral and Aortic Replacement. In the case illustrated in Figure 212, a combination of the two previously described patterns can be seen, each valve being responsible for two major clicking sounds. Additional small clicks are also present in systole.

Triple Valve Mitral, Aortic, and Tricuspid Replacement. This combination taxes the imagination of the phonocardiographer. The symbols given the various clicks are usually based on normal sequence of events and may not apply to an individual case due to higher right or left atrial pressure or still existing stenosis. Examples wil be given below (Figs. 216, 218).

MALFUNCTION OF PROSTHETIC VALVES

Several investigators have studied the phonocardiographic

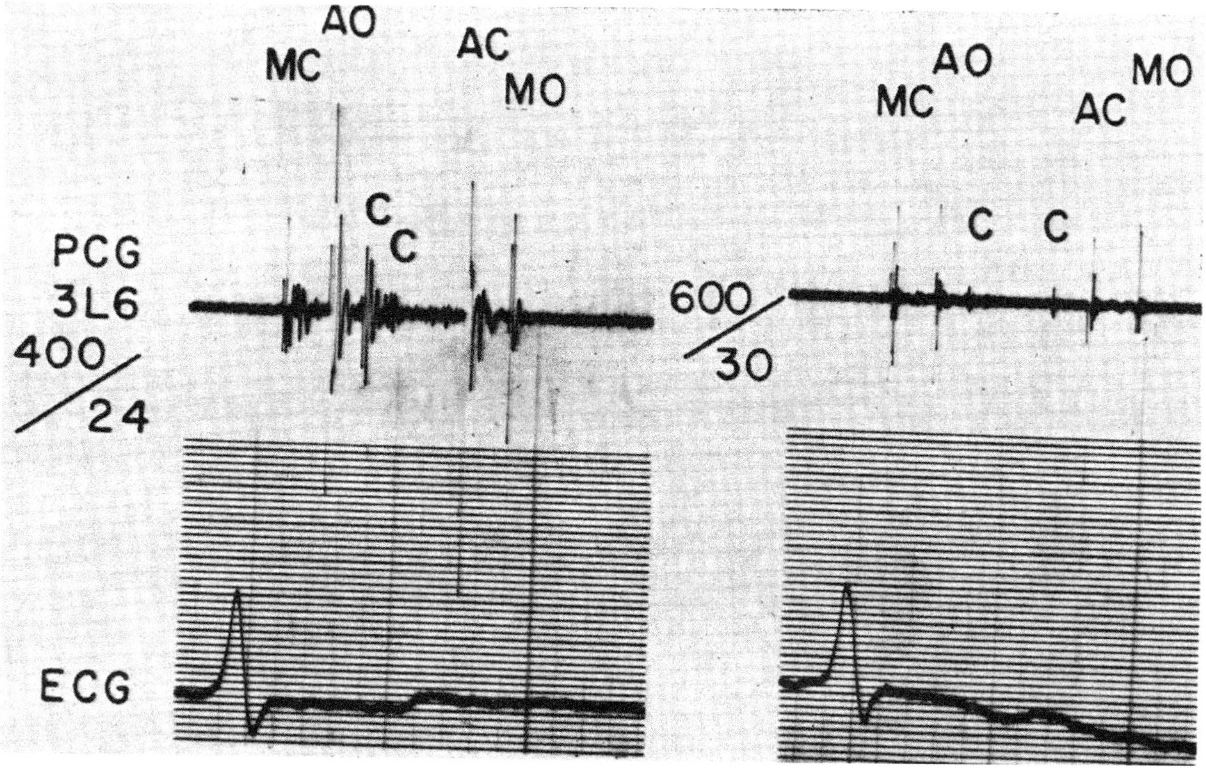

FIGURE 212. Tracings of a 28-year-old man with replacement of both the mitral and the aortic valves.
(A) With filtration 400, there are four clicks for each cycle; mitral closure (MC) followed by aortic opening (AO), then aortic closure (AC) followed by mitral opening (MO).
(B) With filtration 600, two small additional clicks (C and C) appear in systole.

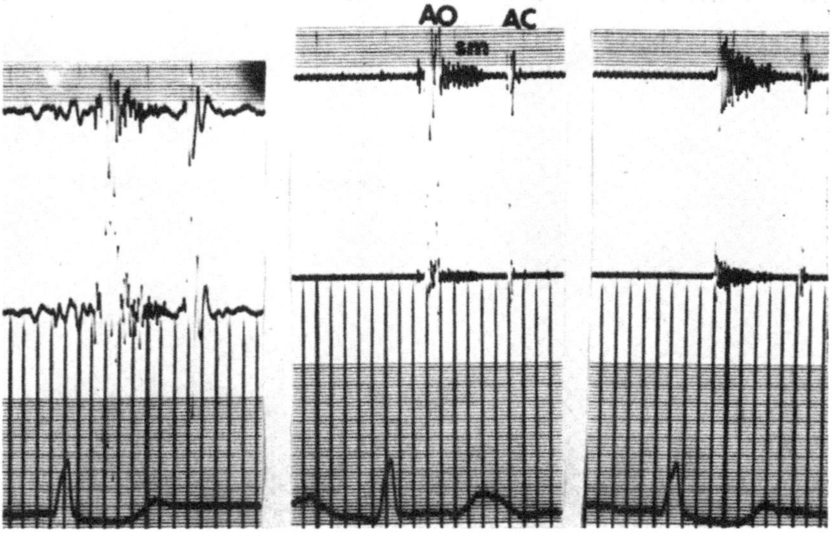

FIGURE 213. Tracings of a 52-year-old man with replacement of the aortic valve (same as in Figure 211 but 4 years later).
At left — velocity tracings at apex (no filtration).
Above = no filtration.
Below = 200 Hz high pass filters.
At center — velocity tracings at apex with 400 Hz (above) and 600 Hz (below) high pass filters.
At right — velocity tracings over 2nd left i.c.s. with same filters as at center.
Delayed opening click (AO); early-systolic murmur.

records of patients with artificial valves in order to detect possible alterations of their function. Willis Boicourt *et al.;* Hylen *et al.* (1968, 1970); Leachman and Corkinos deserve quotation.

Several problems need to be considered:

(a) *Residual stenosis*—this is more commonly observed in regard to an aortic valve, due to implantation of a small cage (Figs. 213-215).

(b) *Residual or newly-appeared insufficiency*—this may occur in either mitral or aortic valve replacement. The problem whether the insufficiency is caused by *dehiscence* or by *clots,* developed over the ball, is at times of difficult solution.

(c) *Mitral valve dehiscence* has been considered present when the opening click of a disk-valve prosthesis disappears.

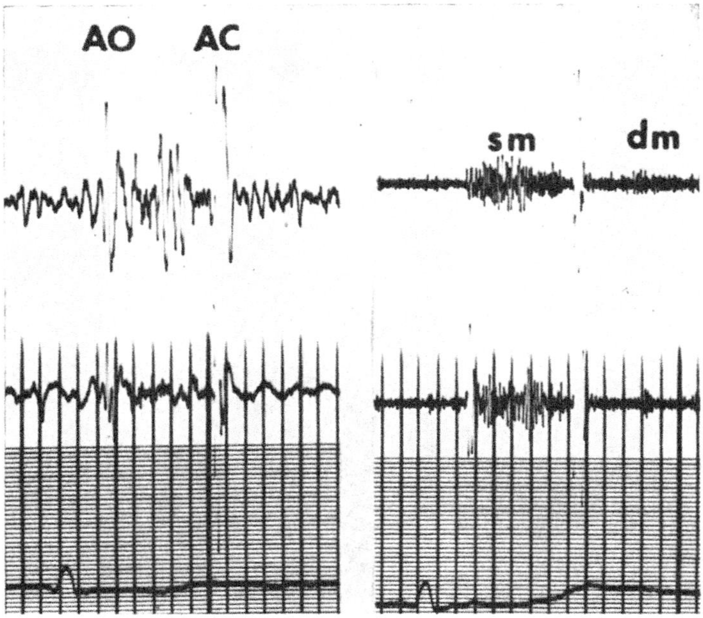

FIGURE 214. Tracings in a 52-year-old man, four years after replacement of the aortic valve.
At left — tracings with high pass filters at 50 Hz.
At right — tracings with high pass filters at 200 Hz.
Low frequency fourth and first sounds; large, but low frequency, aortic opening click; large, diamond-shaped systolic murmur; large aortic closing click; small, early-diastolic murmur.

Suture disruption was found on re-operation (Leachman and Corkinos).

Mitral insufficiency (whether from dehiscence or other cause) was demonstrated in one of our cases by an early and midsystolic murmur or a pansystolic murmur (Fig. 217).

Both *residual aortic stenosis* and *mitral insufficiency* existed in another case with triple valve replacement (Fig. 218).

(d) *Aortic ball variance* may be caused by either decreased or increased diameter of the poppet. Aortic ball variance has been considered present when the aortic opening click sound is less than one-half as large as the aortic closing sound over the 2nd right i.c.s. or when both clicks are

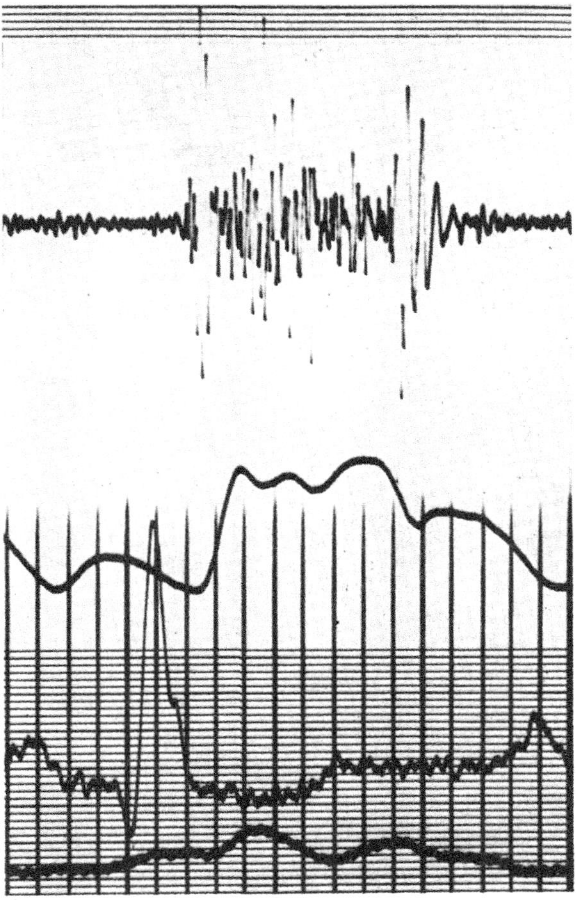

FIGURE 215. Tracings recorded over the 2nd left i.c.s. in a 50-year-old man with aortic valve replacement.
Velocity tracing with 50-300 band pass filter. Large valve clicks. Large, diamond-shaped systolic murmur.

decreased with loss of the high frequency components (Hylen *et al.*). In certain cases both left ventricular failure and ball variance may be suspected. In a case of our observation (Fig. 219), decreased ratio between opening and closing clicks, decreased rapidity of pulse rise, and decreased depth of the incisura were noted, together with a relative increase of the systolic murmur.

A *pacemaker sound* is occasionally found in both epicardial

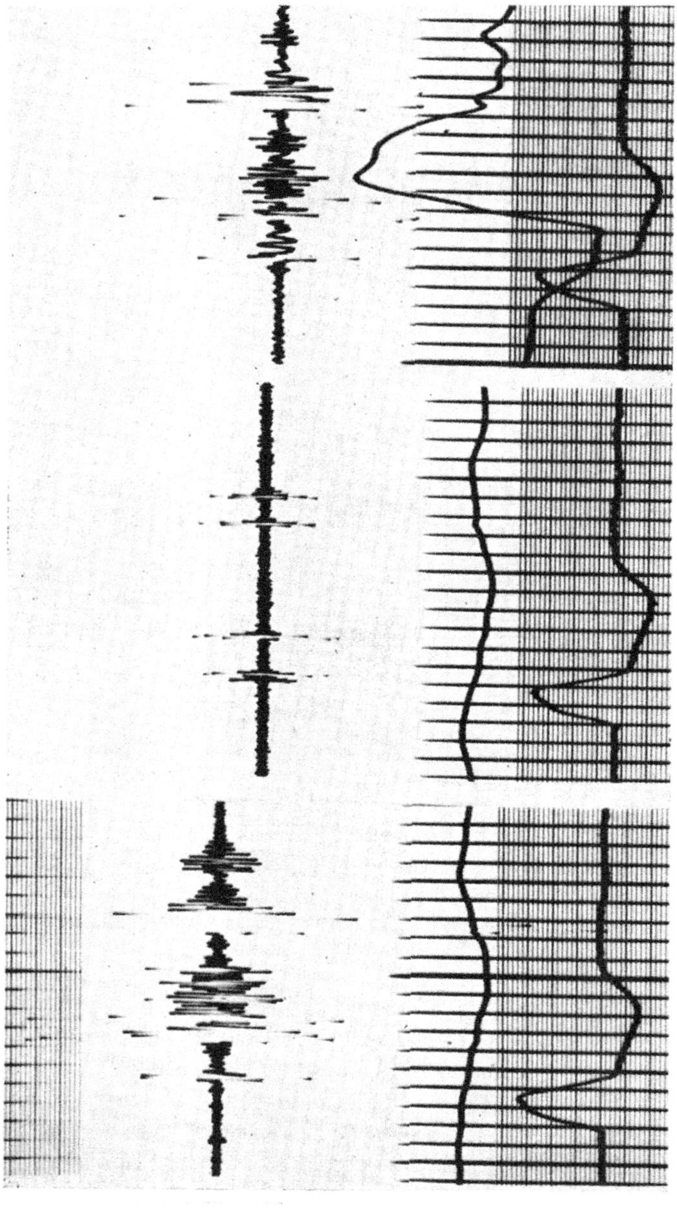

FIGURE 216. Tracings recorded in a 60-year-old man with triple valve replacement. At left—First and second left i.c.s.; velocity tracing 100-700 Hz. Center—Third and fourth left i.c.s.; acceleration tracing 300-600 Hz. At right—Third and fourth left i.c.s.; velocity tracing 40-600 Hz. Pulse tracing at suprasternal notch. Multiple valve clicks: aortic and mitral clicks are of high frequency (center); tricuspid clicks are of lower frequency and are probably present at left and right.

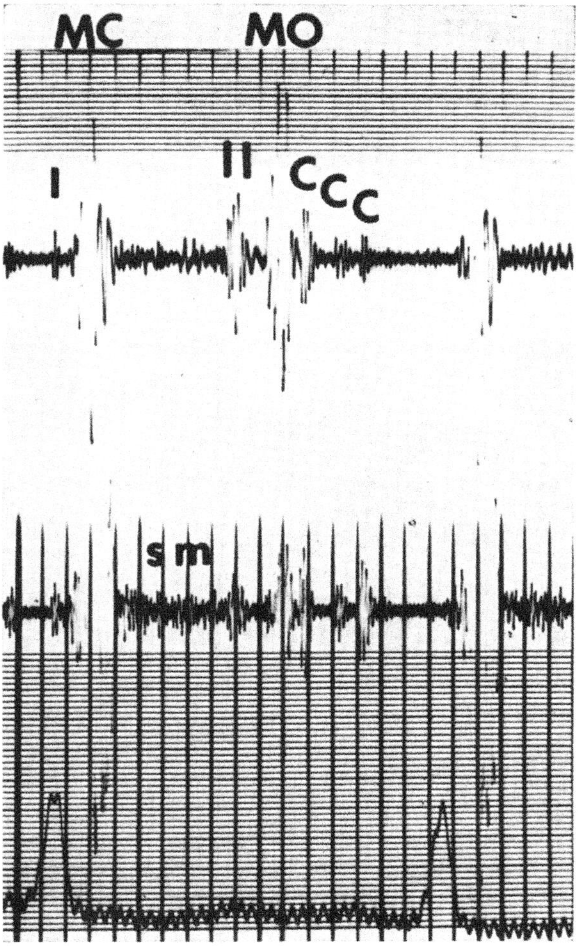

FIGURE 217. Tracings recorded in a 34-year-old man with mitral valve replacement.
Above — 2nd left i.c.s.
Center — apex
In addition to the usual clicks (MC, MO), there are additional clicks in diastole (C,C) and a pansystolic murmur at the apex.

and endocardial systems (Nager *et al.*), as well as in bipolar systems (Mower *et al.*) (Fig. 220). This sound precedes, not only the first heart sound, but also the onset of the QRS. It followed the pacemaker impulse by about 20 msec. in one of our cases (60 msec. in Harris' case). The most likely cause for this

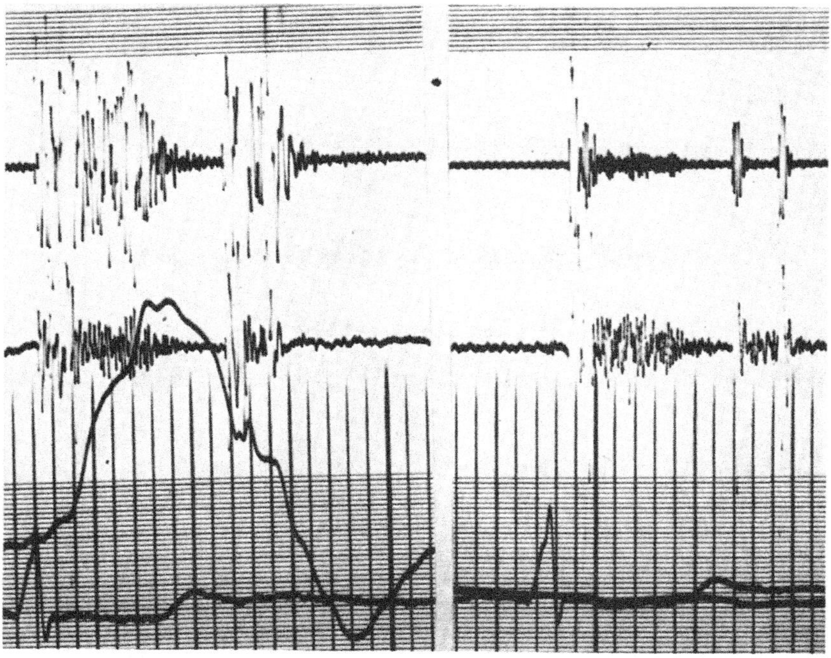

FIGURE 218. Tracings recorded in a 45-year-old woman 5 years after triple valve replacement.
At left — midprecordium; above, acceleration tracing at 50-400; center, velocity tracing, unfiltered; below, pulse tracing at suprasternal notch. At right — 2nd right i.c.s.; above, velocity tracing at 150-800; center, acceleration tracing, unfiltered.
There is a large early and mid-systolic murmur, probably originating at both the aortic and mitral valves. The aortic pulse has a marked anacrotic depression indicating persistence of aortic stenosis.

sound is a brief twitch of either the diaphragm or the intercostal muscles as no increase in ventricular pressure is noted during its occurrence. Partial penetration of the right ventricular myocardium has been suggested (Korn *et al.*) but not demonstrated. In the case of a bipolar system, current leakage has been postulated. If this is the case, excessive current drain might predispose to battery failure.

ARTEFACTS

Numerous artefacts can be found in the phonocardiograms as a result of external or extraneous noises or occasional noises

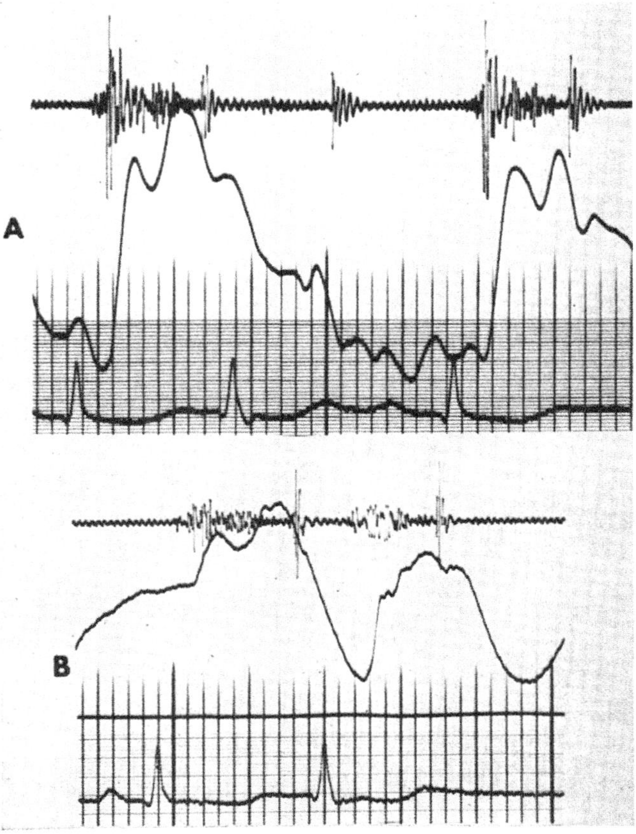

FIGURE 219. Tracings recorded in a young woman with replacement of the aortic valve. Velocity tracings at 40-400 Hz, carotid tracing.
(A) Patient asymptomatic. Moderate systolic murmur. Opening click larger than closing click. Rapid rise of the pulse, anacrotic notch, well marked incisura.
(B) A year later. Patient with clinical evidence of left ventricular embarrassment and recent history of hemiplegia. Small opening click, larger systolic murmur. Slower rise of the pulse, deeper and longer anacrotic notch, barely visible incisura.
In both tracings, there is a sinus beat followed by a supraventricular ectopic beat.

that occur in the patient's body but not in his heart. These are noises such as snoring, puffing, sibilant or musical rales, rhonchi, borborygmus, rubbing of a microphone against hair and many others.

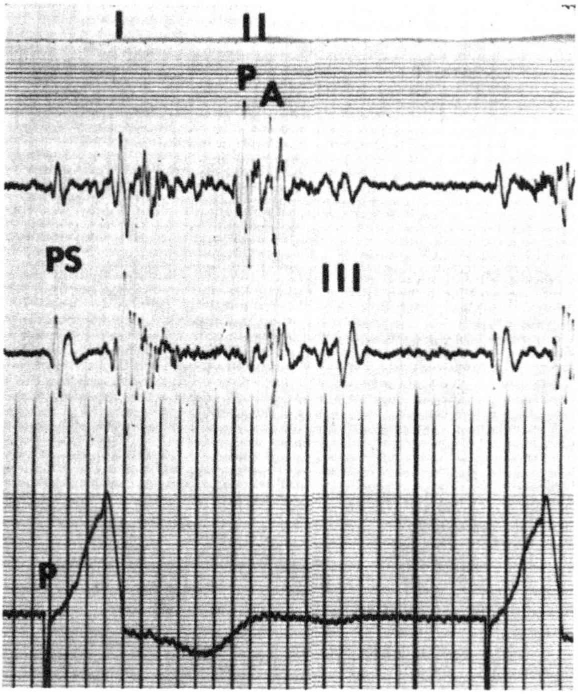

FIGURE 220. Pacemaker sound in a patient with transvenous pacemaker
P = pacemaker impulse.
PS = pacemaker sound.
There is a reverse splitting of the second sound and a double third sound.

A few examples are presented in Figure 221.

A trained observer has no difficulty in recognizing them because they occur sporadically. Occasionally, a special cardiac sound is not present in all cycles, like the "whoop." However, when it occurs, it has a definite relationship with the heart sounds.

As a rule, only repetitive phenomena, i.e., those occurring for each cardiac cycle, deserve to be evaluated and studied.

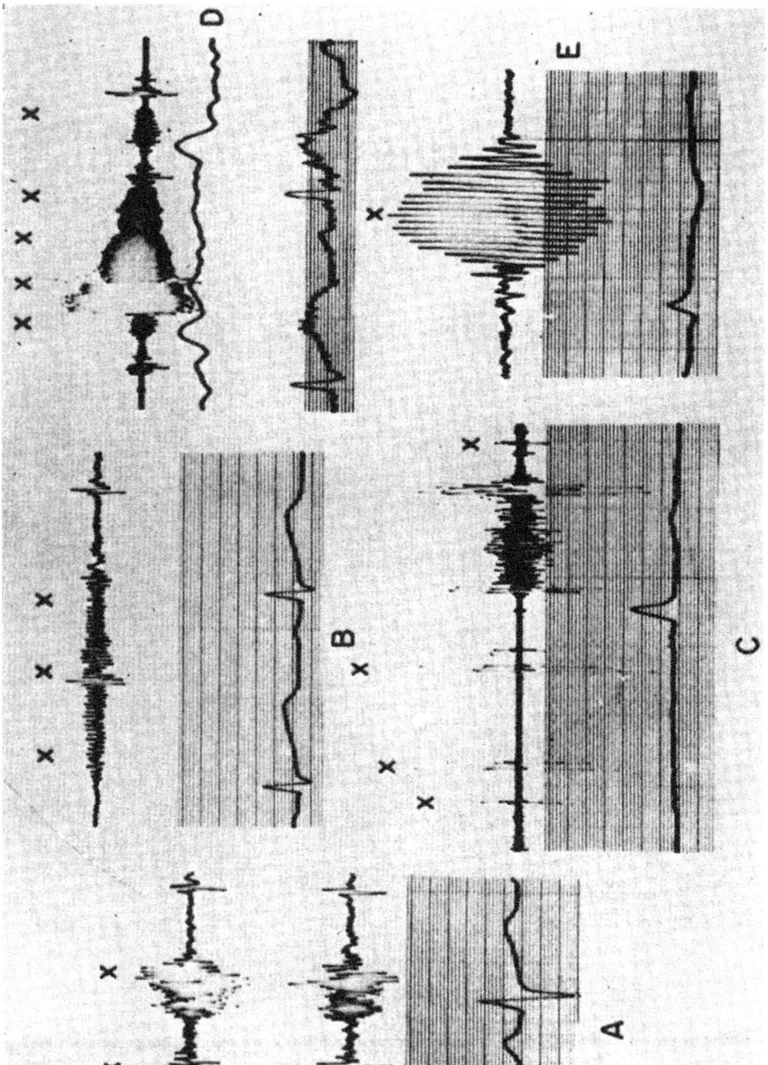

FIGURE 221. Examples of artifacts (X) that can be found in phonocardiograms.
(A) Noise made by a snoring child.
(B) Musical rale that occurred in a case of bronchitis.
(C) Clicks made by the chest piece rubbing against hair.
(D) Unexplained noise made by restless patient.
(E) Puffing noise made by child.

Bibliography

Abbott, J. A. and Whipple, G. H.: Paradoxic splitting of the second heart sound in systemic hypertension. The role of myocardial competence. Dis. Chest, *46*:304, 1964.

Abelman, W. H., Ellis, L. B. and Harken, D. E.: The diagnosis of mitral regurgitation. An evaluation of clinical criteria, fluoroscopy, phonocardiogram, auricular esophagogram and electrokymogram. Am. J. Med., *15*:5, 1953.

Aceves, S. and Carral, R.: The diagnosis of tricuspid valve disease. Am. Heart J., *34*:114, 1947.

Adams, C. W.: Atrial sound in presence of mitral stenosis. Dis. Chest, *44*:538, 1963.

Adams, C. W.: Varying Intensity of the first sound with atrial flutter. Dis. Chest, *51*:437, 1967.

Adams, J. C. L. and Hudson, R.: A case of Ebstein's anomaly surviving to the age of 79. Brit. Heart J., *18*:129, 1956.

Adolph, R. J. and Fowler, N. O.: The second heart sound: a screening test for heart disease. Mod. Conc. Cardiovasc. Dis. *39*:91, 1970.

Adolph, R. J., Fowler, N. O. and Tanaka, K.: Prolongation of isovolumic contraction time in left bundle-branch block. Am. Heart J., *78*:585, 1969.

Adolph, R. J., Stephens, J. F. and Tanaka, K.: The clinical value of frequency analysis of the first heart sound in myocardial infarction. Circul., *41*:1003, 1970.

Adolph, R. J., Tanaka, K. and Fowler, N. O.: Effect of sitting posture on audible expiratory splitting of second heart sound (Abstract). Circulation, *38*:Suppl. VI-31, 1968.

Agnew, T., Bucher, H., McDonald, L. and Seymour, J.: Delayed closure of the aortic valve in ischaemic heart disease. Brit. Heart J., *29*:775, 1967.

Agress, C. M., Wegner, S. and Nakakura, S.: Measurement of cardiac events by a precise technique. Comparison with vibrocardiogram. Jap. Heart J., *5*:414, 1964.

Agress, C. M., Wegner, S., Bleifer, D. J., Lindsey, A., Van Houston, J., Schroyer K. and Estrin, H.M.: The common origin of precordial vibrations, Am. J. Cardiol., *13*:226, 1964.

Ahuja, S. P. and Coles, J. C.: Further observations on the genesis of early systolic clicks. An intracardiac phonocardiographic study in a case of absence of the pulmonary valve. Am. J. Cardiol., *17*:291, 1966.

Ahuja, S. P., Gutierrrez, M. R. and Manning, G.W.: Value of apex cardiography in the assessment of cardiac aneurysms. Am. J. Cardiol., *19*:464, 1967.

Aixala, R.: The apex cardiogram and its utility in phonocardiography. Am. Heart J., *33*:703, 1947.

Akimov, Y. I. and Orlov, L. L.: Some aspects of phonocardiography in patients with mitral stenosis (Russian). Kardiologiya, *2*:71, 1966.

Aldinger, E. E.: Effects of several cardiovascular drugs on various phases of circulatory dynamics. Am. Heart J., *68*:55, 1964.

Algarra, V. J., Cebolla, R. R., Tormo, A. V. and Ferrer, R. J.: Patent ductus arteriosus with pulmonic valvular insufficiency. Two cases studied by intracardiac phonocardiography. Excerpta Medica (Cardiovascular Diseases), *13*:18, 1969.

Alimurung, M. M., Rappaport, M. B. and Sprague, H. B.: Variations in the first apical sound simulating the so-called "presystolic murmur of mitral stenosis." A phonocardiographic study. New Eng. J. Med., *241*:631, 1949.

Alimurung, M. M., Rappaport, M. B. and Sprague, H. B.: The auscultatory sign in rheumatic valvular disease. New England J. Med., *244*:1, 1951.

Allan, F. N.: Heart noises. New England J. Med., *218*:655, 1938.

Allen, N. and Mustian, V.: Origin and significance of vascular murmurs of the head and neck. Medic., *41*:227, 1962.

Alvarez, J. A. and Herrmann, G.: Unusual signs from an expansive Chiari network along with signs of a syphilitic aortic regurgitation. Am. J. Syph., *15*:532, 1931.

Anders, J. M.: Relative aortic incompetency of muscular origin. Bull. Johns Hopkins Hosp., *20*:205, 1909.

Angelino, P. F., Levi, V. and Brusca, A. M.: Pulmonic valvular insufficiency: a clinical and hemodynamic study. Am. Heart J., *54*:182, 1957.

Antonelli, G.: Sul significato del cosiddetto rumore presistolico nella stenosi mitralica, particolarmente nei casi con artimia completa. Cuore e Circ., *20*:162, 1936.

Aquaro, G., Grassi, T., Lavezzaro, G. and De Franceschi, A.: Click da eiezione endoventricolare destro. Malattie Cardiovasc., *8*:141, 1967.

Aquaro, G., Grassi, T., Lavezzaro, G., Politi, G. and Gobbi, G.: Le deuxième bruit dans la communication interauriculaire. Archives des Maladies du Coeur., *61*:1414, 1968.

Aquaro, G., Lavezzaro, G., and De Franceschi, A.: Il fonocardiogramma e la dinamica cardiaca dopo impianto di valvola di Starr-Edwards aortica.

Cardiologia Pratica., *19*:207, 1968.

Aravanis, C.: Different types of diamond-shaped basal systolic murmurs. In Symposium on Cardiovascular Sound. Circulation, *16*:431, 1957.

Aravanis, C.: Opening snap in relative tricuspid stenosis. Report of two cases of atrial septal defect. Am. J. Card., *12*:408, 1963.

Aravanis, C.: Silent mitral insufficiency. Am. Heart J., *70*:620, 1965.

Aravanis, C. and Harris, R.: The normal phonocardiogram of the aged. Dis. Chest, *33*:214, 1958.

Aravanis, C. and Luisada, A. A.: Obstructive and relative aortic stenosis. Differential diagnosis by phonocardiography. Am. Heart J., *54*:32, 1957.

Aravanis, C. and Michaelides, G.: Unusual cases of mitral regurgitation with loud and widespread murmur. Acta Cardol., *22*:587, 1967.

Aravanis, C., Feigen, L., and Luisada, A. A.: The duration and intervals of normal heart sounds in man. Amer. Heart J., *82*:187, 1971.

Arevalo, F. and Sakamoto, T.: On the duration of the isovolumetric relaxation period in dog and man. Am. Heart J., *67*:651, 1964.

Arevalo, F., Meyer, E. C., MacCanon, D. M. and Luisada, A. A.: Hemodynamic correlates of the third heart sound. Am. J. Physiol., *207*:319, 1964.

Aron, M., Gallavardin, L. and Froment, R.: Les donneés stethoscopiques et phonocardiographiques particulières à la maladie d'Ebstein. Arch. Mal. Coeur, *54*:181, 1961.

Asano, K., Washio, M. and Hoshino, K.: The significance of the continuous murmur in the Lutembacher syndrome. Jap. Heart J., *9*:134, 1968.

Ash, R.: The evolution of rheumatic heart disease in childhood. Penn. Med. J., *44*:484, 1941.

Auinger, W.: Ueber die veraenderliche Lautstarke des Vorhoftones bei Galopprhythmus. Wien. Zeit. inn. Med., *35*:163, 1954.

Auinger, W.: Phonokardiographische Veraenderungen nach Mitralstenosenoperation. Wien. Zeit. inn. Med., *36*:253, 1955.

Auinger, W.: Ueber die Lautstarke des 1. Herztones bei Arrhythmien. Zeit. Kreislauff., *44*:194, 1955.

Auinger, W.: Ueber den Einfluss der Druckanstiegszeit und des Blutdruckes auf die Lautstaerke des 1. Herztones. Ztschr. Kreislaufforsch., *45*:840, 1956.

Auinger, W.: The relation between the loudness of the first heart sound and the intracardiac pressure curve in atrial fibrillation. Wien. Z. Inn. Med., *48*:222, 1967.

Auinger, W.: Zur Differentialdiagnose ueberzaehliger Herztoene bei Mitralstenosen. Zeit. Kreislauff., *46*:865, 1957.

Aygn, M. M. and Braunwald, E.: The splitting of the second heart sound in normal subjects and in patients with congenital heart disease. Circul., *25*:328, 1962.

Aykent, Y., Thurmann, M. and Bussmann, D. W.: Continuous murmur in mitral stenosis. Am. J. Cardiol., *15*:715, 1965.

Baker, J. D., Del Banco, T. L., Price, J. B., Rogers, W. M., Salomon, P. F.

and McAllister, F. F.: Experimental study of murmurs in stenotic arteries. Surg., *59*:88, 1966.

Baker, L. A., Sprague, H. B. and White, P. D.: The clinical significance of loud aortic and apical systolic heart murmurs without diastolic murmurs. Am. J. Med. Sci., *206*:31, 1943.

Barber, J. M., Magidson, O. and Wood, P.: Atrial septal defect. Brit. Heart J., *12*:277, 1950.

Barbera, G. and Triolo, R.: Sulla genesi del rumore di trottola. Cuore e Circ., *33*:1, 1949.

Barlow, J. B. and Bosman, C. K.: Aneurysmal protrusion of the posterior leaflet of the mitral valve: an auscultatory-electrocardiographic syndrome. Am. Heart J., *71*:166, 1966.

Barlow, J. B. and Pocock, W. A.: The significance of aortic ejection systolic murmurs. Am. Heart J., *64*:149, 1962.

Barlow, J. and Shillingford, J.: Amyl nitrite in the differentiation of regurgitant and ejection systolic murmurs (Abstr.). Brit. Heart J., *19*:585, 1957.

Barlow, J. and Shillingford, J.: The use of amyl nitrite in differentiating mitral and aortic systolic murmurs. Brit. Heart J., *20*:162, 1958.

Barlow, J. B. and Pocock, W. A.: The isolated systolic murmur. S. Afr. Med. J., *39*:909, 1965.

Barlow, J. B., Pocock, W. A., Marchand, P. and Denny, M.: The significance of late systolic murmurs. Am. Heart J., *66*:443, 1963.

Barlow, J. B., Bosman, C. K., Pocock, W. A. and Marchand, P.: Late systolic murmurs and non-ejection ("mid-late") systolic clicks. An analysis of 90 patients. Brit. Heart J., *30*:203, 1968.

Barritt, D. W., Davies, D. H. and Jacob, G.: Heart sounds and pressures in atrial septal defect. Brit. Heart J., *27*:90, 1965.

Bass, M. H., Mond, H., Messeloff, C. R. and Oppenheimer, E. T.: Systolic murmurs in children. J.A.M.A., *101*:17, 1933.

Battro, A., Braun-Menendez, E. and Orias, O.: Asincronismo de la contracción ventricular en el bloqueo de rama: su demonstración mediante el registro optico de los fenomenos mecanicos de la actividad cardiaca. Rev. Arg. Cardiol., *3*:325, 1936.

Battro, A., Braun Menendez, E. and Orias, O.: L'asynchronisme ventriculaire dans les blocs de branche. Sa demonstration au moyen de l'enregistrement des phenomenes mecaniques de l'activité cardiaque. Arch. Mal. Coeur, *31*:250, 1938.

Battro, A., Quirno, N. and Gonzalez Segura, R.: Chasquido protodiastolico en un caso de calcificacion pericardica. Rev. Arg. Cardiol., *6*:173, 1939.

Bayer, O., Loogen, F. and Wolter, H. H.: The mitral opening snap in the quantitative diagnosis of mitral stenosis. Am. Heart J., *51*:234, 1956.

Bean, W. B.: Precordial noises heard at a distance from the chest. J.A.M.A., *152*:1293, 1953.

Beard, O. W. and Decherd, G. M., Jr.: Variations in the first heart sound in complete A-V block. Am. Heart J., *34*:809, 1947.

Beck, W. and Schrire, V.: An attempt at the assessment of myocardial contractility from phonocardiographically measured time components of ventricular contraction in patients with prosthetic aortic valves. (Asynchronous activation as a cause of diminished velocity of contraction). Circulation., *38*:489, 1968.

Beck, W., Schrire, V. and Vogelpoel, L.: Splitting of the second heart sound in constrictive pericarditis with observations on the mechanism of pulsus paradoxus. Am. Heart J., *64*:765, 1962.

Beck, W., Schrire, V. and Vogelpoel, L.: The hemodynamic effects of amyl nitrite and phenylephrine in patients with mitral stenosis and severe pulmonary hypertension. Am. Heart J., *64*:631, 1962.

Beck, W., Schrire, V. and Vogelpoel, L.: The value of phonocardiography in the assessment of the surgical closure of ventricular septal defect. Am. Heart J., *67*:742, 1964.

Beck, W., Schrire, V., Vogelpoel, L., Nellen, M. and Swanepoel, A.: Hemodynamic effects of amyl nitrite and phenylephrine on the normal human circulation and their relation to changes in cardiac murmurs. Am. J. Card., *8*:341, 1961.

Becker, M. C., Halpern, M. M. and Kent, D. S.: Systolic click. Variation of position with appearance in early diastole. Case report. J. Mt. Sinai Hosp., *15*:307, 1949.

Behar, V. S., Whalen, R. E. and McIntosh, H.D.: The ballooning mitral valve in patients with the "precordial honk" or "whoop". Am. J. Cardiol., *20*:789, 1967.

Bellet, S., Gouley, B., Nichols, C. F. and McMillan, T. M.: Loud, musical, diastolic murmurs of aortic insufficiency. Clinical and pathologic observations upon their cause and the mechanism of their production. Am. Heart J., *18*:483, 1939.

Bellettini, W.: Caratteristiche clinico-strumentali dello sdoppiamento del primo tono del cuore nell'iperteso senile. Giornale di Gerontologia, *8*:252 and 220, 1960.

Benchimol, A. and Dimond, E. G.: The apex cardiogram in ischaemic heart disease. Brit. Heart J., *24*:581, 1962.

Benchimol, A. and Dimond, E. G.: The normal and abnormal apexcardiogram; its physiologic variation and its relation to intracardiac events. Am. J. Card., *12*: 368, 1963.

Benchimol, A. and Dimond, E. G.: Apex cardiogram in the diagnosis of congenital heart disease. Am. J. Cardiol., *17*:63, 1966.

Benchimol, A. and Maroko, P.: The apex cardiogram. The value of the apex cardiogram in coronary artery disease. Dis. Chest, *54*:68, 1968.

Benchimol, A., Legler, J. F. and Dimond, E. G.: The carotid tracing and apex cardiogram in subaortic stenosis and idiopathic myocardial hypertrophy. Am. J. Card., *11*:427, 1963.

Benchimol, A., Wu, T. and Dimond, E. G.: Apex cardiogram in the diagnosis of congenital heart disease. Am. J. Cardiol., *17*:63, 1966.

Bennett, D. W. and Kerr, W. J.: A note on auricular sounds in a case of auricular flutter. Heart, *16*:109, 1932.

Benson, P. F., Bonham-Carter, R. E. and Smellie, J. M.: Transient and intermittent systolic murmurs in newborn infants. Lancet, *1*:627, 1961.

Bergland, J. M., Rucker, W. R., Reeves, J. T. and Surawicz, B.: Preexcitation as a cause of appearance and increased intensity of systolic murmurs. Circulation, *33*:131, 1966.

Besterman, E. M. M.: Phonocardiography in acute rheumatic carditis. Brit. Heart J., *17*:360, 1955.

Biehl, J. P. and Simon, D. L.: Auricular sounds in a case of auricular flutter. Am. J. Med., *15*:134, 1953.

Bishop, L. and Logue, R. B.: External rupture of heart causing systolic murmur and thrill. J.A.M.A., *144*:757, 1950.

Bjoerck, V. O. and Malmstroem, G.: The diagnosis of aortic stenosis. Am. Heart J., *50*:303, 1955.

Bland, E. F., White, P. D. and Jones T. D.: The development of mitral stenosis in young people with a discussion of the frequent misinterpretation of a middiastolic murmur at the cardiac apex. Am. Heart J., *10*:995, 1935.

Bland, E. F., Jones, T. D. and White, P. D.: Disappearance of the physical signs of rheumatic heart disease. J.A.M.A., *107*:569, 1936.

Blazek, W. V. and Bliss, H. A.: The second heart sound in uncomplicated ventricular septal defect. Med. Clin. No. Amer., *50*:111, 1966.

Bleich, A., Lewis, J. and Marcus, F. I.: Aortic regurgitation in the elderly. Am. Heart J., *71*:627, 1966.

Bleifer, S., Kahn, M., Grisham, A. and Donoso, E.: Systolic triple rhythms. Am. J. Cardiol., *4*:492, 1959.

Bleifer, S., Donoso, E. and Grisham, A.: The auscultatory and phonocardiographic signs of ventricular septal defects. Am. J. Card., *5*:191, 1960.

Bleifer, S., Dack, S., Grisham, A. and Donoso, E.: The auscultatory and phonocardiographic findings in mitral regurgitation. Am. J. Card., *5*:836, 1960.

Bloom, H. J. G.: Venous hums in hepatic cirrhosis. Brit. Heart J., *12*,343, 1950.

Blumenthal, J. S.: The systolic murmur. Ann. Int. Med., *17*:637, 1942.

Boicourt, O. W., Bristow, J. D., Starr, A. and Griswold, H. E.: A phonocardiographic study of patients with multiple Starr-Edwards prosthetic valves. Brit. Heart J., *28*:531, 1966.

Bondi, S.: Die Entstehung der Herzgeraeusche. Erg. Inn. Med. Kinderh. *50*:308, 1936.

Bonham Carter, R. E. and Walker, C. H. M.: Continuous murmurs without patent ductus arteriosus. Lancet, *1*:272, 1955.

Borchart, P. R. and Groedel, F. M.: The phonocardiogram before and after pneumothorax. Exp. Med. & Surg., *9*:148, 1951.

Bose, H. K. and Das, P. C.: Importance of auscultation of the second heart sound in chronic cor pulmonale. Indian Heart J., *18*:45, 1966.

Bourne, G.: Cardiac signs in young adults, with special reference to functional murmurs. Lancet, *2*:779, 1946.

Bourne, G.: Physical signs of functional heart disease. Minnesota Med., *39*:617, 1956.

Bousvaros, G. A.: Effect of norepinephrine on the phonocardiographic, auscultatory and hemodynamic features of congenital and acquired heart disease. Am. J. Card., *8*:328, 1961.

Bousvaros, G. A.: Diagnostic auscultatory complex in coarctation of the aorta. Brit. Heart J., *29*:443, 1967.

Bousvaros, G. and Palmer, W.: Phonocardiographic features of the systolic murmur in pulmonary artery stenosis. Brit. Heart J., *27*:374, 1965.

Bousvaros, G. A. and Stubington, D.: Some auscultatory and phonocardiographic features of tricuspid stenosis. Circulation, *29*:26, 1964.

Bouvrain, Y., Bourthomieux, A. and Nezry, R.: Souffles systolique à irradiations axillaires et retrecissement des branches de l'artère pulmonaire. Arch. Mal. Coeur, *54*:999, 1961.

Boyer, S. H. and Chisholm, A. W.: Physiologic splitting of the second heart sound. Circulation, *18*:1010, 1958.

Brachetti, D., Pascucci, E., Ambrosioni, E. and Mazzanti, G.: Il fonocardiogramma e la meccanica cardica nei blocchi di branca completi. G. Clin. Med., *46*:864, 1965.

Bramwell, C.: Graham Steell. Brit. Heart J., *4*:115, 1942.

Braun, H. A. and Comeau, W. J.: Importance of a high-pitched squeaking systolic murmur in the diagnosis of aortic stenosis and calcification of the aortic valve. New Eng. J. Med., *244*:507, 1951.

Braun, H. A., Reynolds, W. A., Diettert, G. A. and McCarthy, G. G.: Auscultatory characteristics of innocent and stenotic cervical bruits (Abstract). Circulation, *34*:III-64, 1966.

Braun-Menendez, E. and Solari, L. A.: Ventricular asynchronism in bundle-branch block. Arch. Intern. Med., *63*:830, 1939.

Braunwald, E., *et al.*: Timing of electrical and mechanical events of the left side of the human heart. J. Appl. Physiol., *8*:309, 1955.

Braunwald, E.: Physiological principles regulating the timing of closure of the heart valves. In *The Theory and Practice of Auscultation* (Segal, Editor). F. A. Davis, Phila., 1964.

Braunwald, E., Aygen, M. M. and Ross, J.: Auscultatory and phonocardiographic findings in atrial septal defect. In *The Theory and Practice of Auscultation* (Segal, Editor). F. A. Davis, Phila., 1964.

Braunwald, E., Donoso, E., Sapin, S. O. and Grishman, A.: Right bundle-branch block: Hemodynamic, vectorcardiographic and electrocardiographic observations. Circulation, *13*:866, 1956.

Braunwald, E., Goldblatt, A., Aygen, M. M., Rockoff, S. D. and Morrow, A. G.: Congenital aortic stenosis. I. Clinical and hemodynamic findings in 100 patients.

Braunwald, E., Fishman, A. P. and Cournand, A.: Time relationship of dynamic events in the cardiac chambers, pulmonary artery and aorta in man. Circulation Res., *4*:100, 1956.

Braunwald, E., Lambrew, C. T., Rockoff, S. D., Ross, J. and Morrow, A. G.: Idiopathic hypertrophic subaortic stenosis. Circulation, *30*, (Suppl. IV), 3, 1964.

Braunwald, E., Moscovitz, H. L., Amram, S. S., Lasser, R. P., Sapin, S. O., Himmelstein A., Ravitch, M. M. and Gordon, A. J.: Timing of electrical and mechanical events of the left side of the human heart. J. Appl. Physiol., *8*:309, 1955.

Braunwald, E., and Morrow, A. G.: Sequence of ventricular contraction in human bundle-branch block. Amer. J. Med., *23*:205, 1957.

Braunwald, E., Morrow, A. G., Cornell, W. P., Aygen, M. M. and Hilbish, T. F.: Idiopathic hypertrophic subaortic stenosis. Am. J. Med., *29*:924, 1960.

Brest, A. N., Udhoji, V. and Likoff, W.: A re-evaluation of the Graham-Steell murmur. New England J. Med., *263*:1229, 1960.

Bricaud, H., Dallocchio, M., Besse, P., Sagardiluz, J., Kerdelhue, H., Martin, P. L. and Broustet, P.: Les bruits protodiastoliques chez les mitraux. Arch. Mal. Coeur, *59*:625, 1966.

Brigden, W. and Leatham, A.: Mitral incompetence. Brit. Heart J., *15*:55, 1953.

Bruns, D. L.: A general theory of the causes of murmurs in the cardiovascular system. Am. J. Med., *27*:360, 1959.

Burch, G. E., De Pasquele, N. P. and Phillips, J. H.: Clinical manifestations of papillary muscle dysfunction. Arch. Int. Med., *59*:508, 1963.

Burch, G. E., De Pasquale, N. P. and Phillips, J. H.: The syndrome of papillary muscle dysfunction. Am. Heart J., *75*:399, 1968.

Cabot, R. C. and Locke, E. A.: On the occurrence of diastolic murmurs without lesions of the aortic or pulmonary valves. Bull. Johns Hopkins Hosp., *14*:115, 1903.

Caceres, G. A. and Perry, L. W.: *The Innocent Murmur.* Little, Brown and Co., Boston, 1967.

Calo, A.: Ritmi cardiaci a quattro tempi (Studio fonocardiografico). Cuore e Circ., *21*:422, 1937.

Calo, A.: Ascoltazione ed iscrizione dei toni atriali in un caso di flutter auricolare. Cardiol., *3*:203, 1939.

Calo, A.: Contribution à l'étude phonocardiographique des extrasystoles. Rev. Méd. Franç., p. 249, 1942.

Calo, A.: Sur une varieté peu connue et essentiellement benigne du rythme à trois temps; le claquement systolique pleuro-pericardique. Tunisie Med., *35*:31, 1947.

Calo, A.: Le souffle presystolique musical de l'insuffisance aortique. Arch. mal. coeur, *40*:137, 1947.

Calo, A.: Il quinto tono cardiaco. Cuore e Circ., *33*:208, 1949.

Calo, A.: Contribution à l'etude phonocardiographique des aortites chroniques. Cardiologia, *14*:279, 1949.

Calo, A.: Étude phonocardiographique d'un acces de tachycardie paroxystique infranodale. Arch. Mal. Coeur, *42*:934, 1949.

Calo, A.: *Les Bruits du Coeur et des Vaisseaux.* Masson, Paris, 1950.

Calo, A.: Phonocardiography in arrhythmias and blocks. In *Cardiology* (Luisada, Edit.), McGraw-Hill-Blakiston, New York, Vol. III, Part 11, Chap. 8, p. 55, 1959.

Caniggia, A., Bertelli, G. and Lodone, M.: High speed phonocardiography in aortic insufficiency. Cardiol., *38*:371, 1961.

Cassels, D. E.: Cardiovascular murmurs in infants and children. M. Clin. N. Am., *41*:75, 1957.

Castle, R. F.: Clinical recognition of innocent cardiac murmurs in children. J.A.M.A., *177*:1, 1961.

Castle, R. F.: Variables affecting the splitting of the second sound in atrial septal defect. Am. Heart J., *73*:468, 1967.

Castle, R. F. and Jones, K. L.: The mechanism of respiratory variation in splitting of the second heart sound. Circulation, *24*:180, 1961.

Caulfield, W. H. (Captain), Smith, R. H. (Major), Franklin, R. B. (Colonel): The second heart sound in coronary artery disease. A phonocardiographic assessment. Am. Heart J., *77*:187, 1969.

Cecil, R. C., Parker, C. P., Jr. and Porter, W. B.: Bacterial endocarditis. Report of a case in which a true musical diastolic murmur appeared and disappeared. Am. Heart J., *36*:934, 1948.

Cheng, T. O.: Early systolic ejection click associated with supravalvular pulmonary stenosis. Am. J. Cardiol., *18*:292, 1966.

Cheng, T. O.: Diastolic murmur caused by coronary artery stenosis. Ann. of Int. Med., *72*:543, 1970.

Chiche, P., Kalmanson, D. and Veyrat, C.: Apport de la phonocardiographie à selection de frequences pour l'étude des insuffisances aortiques. Semaine des Hôp., *42*:2214, 1966.

Claypool, J. G., Ruth, W. and Lin, T. K.: Ventricular septal defect with aortic incompetence simulating patent ductus ateriosus. Am. Heart J., *54*:788, 1957.

Cobbs, Jr., B. W., Logue, R. B. and Dorney, E. R.: The second heart sound in pulmonary embolism and pulmonary hypertension. Am. Heart J., *71*:843, 1966.

Coblentz, B., Gerbaux, A., Anjuere, J., Bourdarias, J. P. and Lenegre, J.: Phonocardiographie et pouls carotidien dans la cardiomyopathie obstructive du ventricule gauche. Arch. Mal. Coeur, *58*:766, 1965.

Cohen, L. S., Mason, D. T. and Braunwald, E.: Significance of an atrial gallop sound in mitral regurgitation. A clue to the diagnosis of ruptured chordae tendineae. Circulation, *35*:112, 1967.

Cohn, K. E. and Hultgren, H. N.: The Graham-Steell murmur re-evaluated. New England J. Med., *274*:486, 1966.

Contro, S. and Luisada, A. A.: Modifications of the heart sounds in bundle-branch block. J. Mount Sinai Hosp., N.Y., *19*:70, 1952.

Cossio, P. and Berconsky, I.: El primer ruido cardiaco y el soplo presistolico en la estrechez mitral con fibrilacion auricular. Rev. Arg. Cardiol., *10*:162, 1943.

Cossio, P. and Braun-Menendez, E.: Estudio fonocardiografico del bloqueo total auriculo ventricular. Rev. Arg. Cardiol., *2*:1, 1935.

Cossio, P. and Orias, O.: Chasquido de apertura de la mitral. Rev. Arg. Cardio., *1*:450, 1935.

Cossio, P., Berconsky, I. and Dambrosi, R. G.: Auricular and ventricular pericardial frictions. Am. Heart. J., *24*:223, 1942.

Cossio, P., Cossio, P. M. and Cossio, P. R.: Kinetocardiovision y kinetocardioestetoscopio. Pren. Med. Argent., *54*:1216, 1967.

Cossio, P., Dambrosi, R. G., Warnford-Thomson, H. F.: The first heart sound in auricular and ventricular extrasystoles. Brit. Heart J., *9*:275, 1947.

Cossio, P.: El ruido auricular en la taquicardia paroxistica nodal. Alternancia parcial del primer ruido cardiaco. Rev. Arg. Cardiol., *5*:242, 1938.

Cossio, P., Lascalea, M. and Fongi, E. G.: Alternation of the heart sounds. Arch. Int. Med., *58*:812, 1936.

Counihan, T., Messer, A. L. Rappaport, M. B. and Sprague, H. B.: The initial vibrations of the first heart sound. Circulation, *3*:730, 1951.

Craige, Ernest: *Gallop Rhythm. Physical Diagnosis in Cardiovascular Disease.* Edited by Charles K. Friedberg, 1969. page 140.

Craige, E.: Phonocardiographic studies in mitral stenosis. New England J. Med., *257*:650, 1957.

Craige, E.: Phonocardiogram in interventricular septal defects. Am. Heart J., *60*:51, 1960.

Crampton, R. S.: Reversed splitting of the second heart sound in ambulatory coronary heart disease. Clin. Res., *13*:525, 1965.

Criley, J. M., Humphries, J. O., Lewis, K. B. and Ross, R. S.: Mitral valve prolapse: clinical and cineangiographic findings. (Abstr.) Am. J. Card., *15*:128, 1965.

Dack, S., Bleifer, S., Grishman, A. and Donoso, E.: Mitral stenosis. Auscutatory and phonocardiographic findings. Am. J. Cardiol., *5*:815, 1960.

Dalsgaard-Nielsen, T.: Studies on intracranial vascular sounds. Acta psychiat. et neurol., *14*:69, 1939.

Davidsen, H. G.: Pulmonary hypertension and incompetence with holodiastolic murmur in atrial septal defect. Acta Med. Scandinav., *160*: 177, 1958.

Davison, E. T. and Friedman, S. A.: Significance of systolic murmurs in the aged. New Eng. J. Med., *279*:225, 1968.

Dawson, A. A. and Palmer, K. N. V.: The significance of cardiac murmurs in anemia. Am. J. Med. Sci., *252*:555, 1966.

Decherd, G. M., Jr. and Beard, O. W.: Functional mitral diastolic murmurs. Texas Rep. Biol. & Med., *4*:119, 1946.

Delman, A. J., Gordon, G. M., Stein, E. and Escher, D. J. W.: The second sound-mitral opening snap (A2-OS) interval during exercise in the evaluation of mitral stenosis. Circulation, *33*:399, 1966.

DePasquale, N. P., Burch, G. E. and Phillips, J. H.: The second heart sound. Am. Heart J., *76*:419, 1968.

De Sautelle, W. T. and Grey, E. G.: The relation of the papillary muscles to mitral regurgitant murmurs. Arch. Int. Med., *8*:734, 1911.

Detweiler, D. K. and Patterson, D. F.: Abnormal heart sounds and murmurs of the dog. J. Small Animal Practice., *8*:193, 1967.

Di Bartolo, G., Nunez-Dey, D. and Bendezu-Prieto, J.: Left heart studies in mitral stenosis with special reference to intracardiac phonocardiography. Am. J. Card., *10*:93, 1962.

Di Bartolo, G., Nunez-Dey, D., Muiesan, G., MacCanon, D. M. and Luisada,

A. A.: Hemodynamic correlates of the first heart sound. Am. J. Physiol., *201*:888, 1961.

Dini, S. and Mattoli, G.: Possibilità di differenziare, can l'iscrizone fonocardiografica, i rumori sistolici funzionali da quelli organici non congeniti. Riforma Med., *56*:103, 1940.

Di Perri, T., Bertelli, G. and Fabrizi, G.: Il fonocardiogramma nelle extrasistoli. Folia Cardio., *17*:281, 1958.

Dock, W.: Mode of production of the first heart sound. Arch. Int. Med., *51*:737, 1933.

Donnelly, G. L. and Vandenberg, E. A.: Early systolic sounds in aortic valve stenosis. Brit. Heart J., *29*:246, 1967.

Dressler, W.: Precordial heave on the right related to left-atrial enlargement. J.A.M.A., *205*:96, 1968.

Dressler, W. and Rubin, R.: Complex shape and variability of the diastolic murmur of aortic regurgitation. Am. J. Cardio., *18*:616, 1966.

Duchosal, P. W.: Le double canon du bloc a-v en cas de bloc de branche. Cardiologia, *15*:208, 1949.

Dumery, R.: Le souffle continu thyroidien superieur dans le basedowisme. Paris, Thesis, 1931.

Edwards, E. A. and Levine, H. D.: Auscultation in the diagnosis of compression of the subclavian artery. New England J. Med., *247*:79, 1952.

Edwards, E. A. and Levine, H. D.: The murmur of peripheral arteriovenous fistula. New England J. Med., *247*:502, 1952.

Edwards, E. A. and Levine, H. D.: Peripheral vascular murmurs. Mechanism of production and diagnostic significance. A. M. A. Arch. Int. Med., *90*:284, 1952.

Edwards, P. W. and Simpson, T.: Observations on 'pericardial knock'. Tubercle, *20*:426, 1939.

Eguchi, A., Imai, T., Yoshioka, N., Kitkuta, R., Hanawa, T., Sato, T., Ohyama, Y., Saito, T., Hosono, K., Mashima, T. and Oguro, C.: Studies on the ventricular asynchronism in right bundle branch block. Japanese Circul. J., *25*:369, 1961.

Ehlers, K. H., Engle, M. A., Farnsworth, P. B. and Levin, A. R.: Wide splitting of the second heart sound without demonstrable heart disease. Am. J. Cardiol., *23*:690, 1969.

Eisenberg, R. and Hultgren, H. N.: Phonocardiographic features of atrial septal defect. Circulation, *20*:490, 1959.

Eldridge, F. L. and Hultgren, H. N.: Production of heart sounds by the cardiac catheter. Circulation, *19*:557, 1959.

Eldridge, F., Selzer, A. and Hultgren, H.: Stenosis of a branch of the pulmonary artery. An additional cause of continuous murmurs over the chest. Circulation, *15*:865, 1957.

Ellis, L. B. and Faulkner, J. M.: The heart in anemia. New Eng. J. Med., *220*:943, 1939.

Engle, M. A.: The syndrome of apical systolic click, late systolic murmur, and abnormal T waves. Circulation, *39*:1, 1969.

Evans, W.: Mitral systolic murmurs. Brit. Med. J., *1*:8, 1943.

Facquet, J., Alhomme, P. and Raharison, S.: Sur la signification du souffle frequemment associé au claquement telesystolique. Acta Cardiol., *19*: 417, 1964.

Farinelli, A., Zanardi, F., Alvisi, V. and Sbrighi, V.: Rilievi policardiografici su di un fenomeno vibratorio mesodiastolico rilevabile in cuori giovanili. Minnesota Med., *57*:3369, 1966.

Ferrero, C. and Humair, L.: Diagnostic de l'origine gauche ou droite d'un bruit de galop par la manoeuvre de Valsalva. Cardiol., *46*:160, 1965.

Ferrero, C., Doret, J. P., Foex, P., Eichenberger, G. and Collart, F.: Le quatrieme bruit du coeur, signe de l'insuffisance ventriculaire gauche latente de l'ischemie du myocarde. Cardiol., *49*:173, 1966.

Ferrero, C., Maeder, J. and Scherrer, J. R.: Signification hemodynamique des variations d'intensité du premier bruit dans le retrecissement mitral. Arch. Mal. Coeur, *64*:231, 1971.

Feruglio, G. A.: Intracardiac phonocardiography: A valuable diagnostic technique in congenital and acquired heart disease. Am. Heart J., *58*:827, 1959.

Feruglio, G. A.: Aortic diastolic musical murmurs: A conventional and intracardiac phonocardiographic study. Malattie Cardiovasc., *3*:181, 1962.

Feruglio, G. A. and Gunton, R. W.: Intracardiac phonocardiography in ventricular septal defect. Circulation, *21*:49, 1960.

Feruglio, G. A., Gunton, R. W., Sreenivasan, A. and Heimbecker, R. O.: Diagnostic value of intracardiac phonocardiography in some complicated cardiac problems. Ann. Surg., *152*:29, 1960.

Fishleder, B. L.: *Exploracion Cardiovascular y Fonomecanocardiografia Clinica.* Prensa Medica Mexicana, Mexico, 1966.

Fishleder, B. L. and Friedland, C.: La fonocardiografia en las cardiopatias congenitas. Princ. Cardiol., *4*:285 and 355, 1957.

Fishleder, B. L. and Gomez, A. R.: Los soplos sistolicos no patologicos. Sus caracteristicas y criterio de clasificacion. Princ. Card., *6*:325, 1959.

Fishleder, B. L. and Medrano, G. A.: Retumbo apical en la persistencia del conducto arterial. Arch. Inst. Card. Mexico, *22*:292, 1952.

Fishleder, B. L., Friedland, C. and Medrano, G. A.: El primer ruido en la estenosis mitral. Estudio fonocardiografico. Arch. Inst. Cardiol. Mexico, *24*:567, 1954.

Fishleder, B. L., Bermudez, F. and Friedland, C.: Estenosis subaortica dinamica. Su diagnostico clinico y por metodos graficos externos. Arch. Inst. Card. Mexico, *32*:430, 1962.

Fishleder, B. L., Serra, C., Prati, P. L. and Friedland, C.: La persistencia del conducto arterial con hipertension pulmonar: variedad "diastolica ruda." Arch. Inst. Card. Mexico, *32*:610, 1962.

Fishleder, B. L., Bonetti, P. F., Bermudez, F., Alvarez, V. R., Hamdan, G. and Friedland, C.: El chasquido protodiastolico de la arteria pulmonar (estudio de un caso de dilatacion idiopatica de la arteria pulmonar). Arch. Inst. Cardiol. Mexico, *33*:175, 1963.

Fletcher, G. F. and Hurst, J. W.: An intermittent "cooing" diastolic murmur due to a torn aortic valve cusp. Am. Heart J., *75*:537, 1968.

Flint, A.: On the mitral presystolic and a mitral diastolic heart murmur. Lancet, *1*:418, 1884.

Fogel, D. H.: The innocent systolic murmur in children: a clinical study of its incidence and characteristics. Am. Heart J., *59*:844, 1960.

Fowler, N. O. and Gause, R.: The cervical venous hum. Am. Heart J., *67*:135, 1964.

Fowler, N. O. and Marshall, W. J.: The supraclavicular arterial bruit. Am. Heart J., *69*:410, 1965.

Freilich, J. K. and Szatkowski, J.: Pelvic vein thrombosis with pulmonary vascular lesions simulating primary pulmonary hypertension. Am. J. Card., *7*:297, 1961.

Friedland, C. and Fishleder, B. L.: El estudio fonomecanocardiografico de pacientes con infarto del miocardio no reciente. Arch. Inst. Card. Mexico, *30*:467, 1960.

Friedlander, R. D. and Brown, M. G.: The systolic murmur: further observations on its clinical significance. Ann. Int. Med., *8*:893, 1935.

Friedman, B. and Hathaway, B. M.: Fenestration of the semilunar cusps, and "functional" aortic and pulmonic valve insufficiency. Am. J. Med., *24*:549, 1958.

Friedman, S. and Harris, T. N.: Apical diastolic heart sounds in active rheumatic fever. J. Pediat., *3*:603, 1949.

Friedman, S., Robie, W. A. and Harris, T. N.: Occurrence of innocent adventitious cardiac sounds in childhood. Pediatrics, *4*:782, 1949.

Froment, R., Gallavardin, L. and Cahen, P.: Claquements et vibrances des pericardites constrictives. A propos d'un nouveau cas protosystolique. Arch. Mal. Coeur, *42*:925, 1949.

Froment, R., Richard, M., Gallavardin, L. and Brette, R.: Souffles diastoliques anemiques par insuffisance sigmoidienne fonctionnelle. J. Med. Lyon, *30*:883, 1949.

Froment, R., Saint-Pierre, A., Gallavardin, L. and Age, C.: Syndrome de Lutembacher avec souffle systolo-diastolique parasternal. Archives des Maladies du Coeur., *61*:1703, 1968.

Froment, R., Tourniaire, A., Gonin, A. and Gallavardin, L.: Un nouveau signe de pericardite calcaire: la vibrance proto ou meso-systolique. Arch. Mal. Coeur, *38*:188, 1945.

Furuse, A., Mizuno, A., Ito, K. and Saigusa, M.: Two cases of cardiac murmur caused by deformity of spinal column. Jap. Heart J., *8*:670, 1967.

Fyler, D. C., Gallaher, M. F. and Nadas, A. S.: Auscultation in the evaluation of children with heart disease. Progr. Cardiovasc. Dis., *10*:363, 1968.

Gamboa, R. and Willis, K.: The systolic murmur of combined pulmonary stenosis and infundibular stenosis. Am. J. Cardio., *19*:880, 1967.

Gamboa, R., Hugenholtz, P. G. and Nadas, A. S.: Accuracy of the phonocardiogram in assessing severity of aortic and pulmonic stenosis. Circulation, *30*:35, 1964.

Garb, S.: The relationship of blood viscosity to the intensity of heart murmurs. Am. Heart J., *28*:568, 1944.

Garvin, C. F.: Functional aortic insufficiency. Ann. Int. Med., *13*:1799, 1940.

Gibson, D. G.: Abnormal splitting of the second heart sound in renal failure. Brit. Med. J., *4*:141, 1967.

Gilston, R. J. and McPhaul, J. J., Jr.: Physiologic murmurs in pregnancy: a reemphasis. M. Ann. District of Columbia, *24*:461, 1955.

Giorgetti, C., Pistelli, G. and Chiesa, F.: Gallop rhythm in acute myocardial infarction. A clinical and phonocardiographic study of 45 cases. Il ritmo di galoppo nell'infarto del miocardio in fase acuta. Studio clinico e fonocardiografico di 45 casi. Cardiol. Prat., *17*:398, 1966.

Giraud, G., Barati, H. and Loubatieres, R.: Inscription phono-arteriographique du souffle de l'anevrisme arterio-veineux. C. R. Soc. Sci. Med. Biol. Montpellier, June 27, 1941.

Giuliana, E. R.: Mitral valve incompetence due to flail anterior leaflet. A new physical sign. Am. J. Cardiol., *20*:784, 1967.

Gleason, W. L. and Braunwald, E.: Studies on the first derivative of the ventricular pressure pulse in man. J. Clin. Invest., *41*:80, 1962.

Gola, E. and Vitolo, E.: Il tono atriale nell'infarto miocardico acuto: considerazioni cliniche. Cuore e Circol., *50*:128, 1966.

Goldberg, L. M. and Uhland, H.: Heart murmurs in pregnancy: a phonocardiographic study of their development, progression and regression. Dis. Chest, *52*:381, 1967.

Goldblatt, A., Aygen, M. M. and Braunwald, E.: Hemodynamic-phonocardiographic correlations of the fourth heart sound in aortic stenosis. Circulation, *26*:92, 1962.

Goodwin, J. F., Harrison, C. V. and Wilcken, D.E.L.: Obliterative pulmonary hypertension and tromboembolism. Brit. Med. J., *1*:777, 1963.

Gouley, B. A.: Constriction of the pulmonary artery by adhesive pericarditis. Am. Heart J., *13*:470, 1937.

Gould, L., Ettinger, S. J. and Lyon, A. F.: Intensity of the first heart sound and arterial pulse in mitral insufficiency. Dis. Chest, *53*:545, 1968.

Graf, W., Moeller, T. and Mannheimer, E.: The continuous murmur. Incidence and characteristics in different parts of the human body. Acta Med. Scandinav., *196*:167, 1947.

Grant, R. P.: A precordial systolic murmur of extracardiac origin during pregnancy. Am. Heart J., *52*:944, 1956.

Grassi, T., Aquaro, G. and Andreone, A.: La dinamica del cuore umano analizzata mediante le derivate I e II delle curve pressorie endoventricolari. Min. Cardioang., *13*:123, 1965.

Gray, I. R.: Paradoxical splitting of the second heart sound. Brit. Heart J., *18*:21, 1956.

Grayzel, J.: Gallop rhythm of the heart. I. Atrial gallop, ventricular gallop and systolic sounds. Am. J. Med., *28*:578, 1960.

Grayzel, J.: Gallop rhythm of the heart. II. Quadruple rhythm and its relation to summation and augmented gallops. Circulation, *20*:1053, 1959.

Greene, J. A.: Unusual sounds emanating from the chest. Cause and diagnostic significance of bubbling, clicking, crunching, knocking and tapping sounds; with a report of two cases of interstitial emphysema of lung and mediastinum. Arch. Int. Med., *71*:410, 1943.

Griffith, J. P. C.: Mid-systolic and late-systolic mitral murmurs. Am. J. Med. Sci., *104*:285, 1892.

Groedel, F. M. and Miller, M.: The nature and origin of the so-called systolic gallop rhythm. Exp. Med. Surg., *3*:107, 1945.

Groom, D. and Boone, J. A.: Venous hum in cardiac auscultation. J.A.M.A., *159*:639, 1955.

Groom, D., Chapman, W., Francis, W. W., Bass, A. and Sihvonen, Y. T.: The normal systolic murmur. Ann. Int. Med., *52*:134, 1960.

Gunewardene, H. O.: The cardiac complications of ankylostoma infection with special reference to a presystolic murmur occurring in these cases. J. Trop. Med., *36*:49, 1933.

Guenther, K. H. and Munster, W.: Intracardiac phonocardiography and hemodynamics in pure mitral insufficiency. Cardiologia, *53*:321, 1968.

Guenther, K. H.: *Vergleichende extrakardiale und intrakardiale Phonokardiographie auf hemodynamischer Grundlage.* Akademie Verlag, Berlin, 1969.

Gwyn, N. B.: On some venous murmurs found in hepatic cirrhosis and their confusion with murmurs of congenital heart disease. Am. J. M. Sc., *180*:525, 1930.

Groom, D. and Boone, J. A.: The "dove-coo" murmur and murmurs heard at a distance from the chest wall. Ann. Int. Med., *42*:1214, 1955

Haber, E. and Leatham, A.: Splitting of heart sounds from ventricular asynchrony in bundle-branch block, ventricular ectopic beats, and artificial pacing. Brit. Heart J., *27*:691, 1965.

Halloran, K. H., Talner, N. S. and Browne, M. J.: A study of ventricular septal defect associated with aortic insufficiency. Am. Heart J., *69*: 320, 1965.

Hamburger, L. P.: Head murmurs. Am. J. M. Sc., *181*:756, 1931.

Hamby, R. I. and Gulotta, S. J.: Pulmonic valvular insufficiency, etiology, recognition, and management. Am. Heart J., *74*:110, 1967.

Hammond, J. H. and Eisinger, R. P.: Carotid bruits in 1,000 normal subjects. Arch. Int. Med., *109*:563, 1962.

Hancock, E. W.: The ejection sound in aortic stenosis. Am. J. Med., *40*: 569, 1966.

Hancock, E. W. and Cohn, K.: The syndrome associated with midsystolic click and late systolic murmur. Am. J. Med., *41*:183, 1966.

Haring, O. M., Aravanis, C., Liu, C. K., Gamna, C., Trace, H. D. and Luisada, A. A.: The mitral patient before and after surgery. Diagnostic-graphic investigations. Am. Heart J. *52*:18, 1956.

Haring, O. M., Liu, C. K. and Trace, H.: The left atrial pressure pulses in experimental mitral valve lesions. Circulation, *4*:381, 1956.

Haring, O. M., Luisada, A. A. and Gasul, B.: Phonocardiography in patent ductus arteriosus. Circulation, *10*:501, 1954.

Harris, A.: Pacemaker 'heart sound.' Brit. Heart J., *29*:608, 1967.

Harris, T. N. and Friedman, S.: Phonocardiographic differentiation of vibratory (functional) murmurs from those of valvular insufficiency: further observations and application to the diagnosis of rheumatic heart disease. Am. Heart J., *43*:707, 1952.

Harris, T. N., Friedman, S., Tahsin Tuncali, M. and Hallidie-Smith, K. A.: Comparison of innocent cardiac murmurs of childhood with cardiac murmurs in high output states. Pediatrics, *33*:341, 1964.

Harvey, P. W. and Perloff, J. K.: Some Recent Advances in Clinical Auscultation of the Heart. In *Physical Diagnosis in Cardiovascular Disease* (Friedberg, Editor). Grune and Stratton, New York, 1969.

Harvey, W. P.: Abnormalities of the first and second heart sounds in the diagnosis of heart disease. In *The Theory and Practice of Auscultation* (Segal, Editor). Davis, Philadelphia, 1964, p. 117.

Harvey, W. P. and Corrado, M. A.: Multiple sounds in paroxysmal ventricular tachycardia. An aid in diagnosis by auscultation. New England J. Med., *257*:325, 1957.

Harvey, W. P. and Levine. S. A.: The changing intensity of the first sound in auricular flutter, an aid to the diagnosis by auscultation. Am. Heart J., *35*:924, 1948.

Harvey, W. P. and Perloff, J. K.: The auscultatory findings in primary myocardial disease. Am. Heart J., *61*:199, 1961.

Harvey, W. P., Corrado, M. A. and Perloff, J. K.: "Right-sided" murmurs of aortic insufficiency (diastolic murmurs better heard to the right of the sternum rather than to the left). Am. J. Med. Sci., *245*:533, 1963.

Hecht, H. H. and Myers, G. B.: Auricular heart sounds in auricular flutter. Am. Heart J., *29*:610. 1945.

Heikkila, J.: Mitral incompetence as a complication of acute myocardial infarction. Acta Med. Scand. Suppl., 475, 1967.

Heller, S. J. and Carleton, R. A.: Abnormal left ventricular contraction in patients with mitral stenosis. Circulation, *42*:1099, 1970.

Henry, J. N.: Report of a case of loud venous hum heard over the xiphoid cartilage in cirrhosis of the liver. Autopsy. Am. J. Med. Sci., *143*:72, 1912.

Holldack, K.: Beitrag zur Diagnostik der Aortenisthmusstenose Zeit. Kreislauff., *38*:466, 1949.

Holldack, K.: Die Phonokardiographie, ihre Bedeutung fuer die sinnesphysiologischen Grundlagen der Herzauskultation und ihre diagnostische Verwendung. Erg. Inn. Med., *3*:407, 1952.

Holldack, K. and Kuhn, E.: Gefaessgeraeusche und periphere Durchblutung; die Bedeutung der Gefaessgeraeusche in der Aorta abdominalis und den Iliacae fuer die Diagnose der peripheren Durchblutungsstoerungen. Deutsch. Med. Wchnschr., *78*:842, 1953.

Holldack, K. and Loehr, B.: Die diagnostische Bedeutung der Endform systolischer Geraeusche. Zeit. Kreislaufforsch., *44*:388, 1955.

Holldack, K. and Mielke, F.: Funktionelle diastolische Geraeusche. Klin. Wchnschr., *30*:719, 1952.

Holldack, K., Fuchs, G., Zapfe, H. and Maehler, L.: Die inverse Spaltung des II Herztones bei Myocardinfarkt. Zeitschr. Kreislauff., *56*:541, 1967.

Holldack, K., Heller, A. and Groth, W.: The pericardial friction rub in the phonocardiogram. Am. J. Card., *4*:351, 1959.

Holldack, K., Yoshimura, S., Okamura, T. and Zapfe, H.: Functional diastolic murmur in 2:1 A-V block. Jap. Heart J., *6*:313, 1965.

Hollman, A., Morgan, J. J., Goodwin, J. F. and Fields, H.: Auscultatory and phonocardiographic findings in ventricular septal defect. Circulation, *28*:94, 1963.

Holloway, Jr., D. H., Whalen, R. E. and McIntosh, H. D.: Systolic murmur developing after myocardial ischemia or infarction. Differential diagnosis. J.A.M.A., *191*:888, March 15, 1965.

Howell, H.: Cardiac murmurs in old age: a clinico-pathological study. J. Am. Geriat. Soc., *15*:509, 1967.

Hubbard, T. F., Dunn, F. L. and Neis, D. D.: A phonocardiographic study of the apical diastolic murmurs in pure mitral insufficiency. Am. Heart J., *57*:223, 1959.

Hultgren, H. N.: Intensity of the first heart sound in mitral stenosis with atrial fibrillation. Clin. Res., *6*:84, 1958.

Hultgren, H. N. and Hubis, H.: A phonocardiographic study of patients with the Starr-Edwards mitral valve prosthesis. Am. Heart J., *69*: 306, 1965.

Hultgren, H. N. and Leo, T. F.: Phonocardiographic features of combined mitral stenosis and insufficiency. Medicine, *38*:103, 1959.

Humphries, J. O. and McKusick, V. A.: The differentiation of organic and "innocent" systolic murmurs. Prog. Cardiov. Dis., *5*:152, 1962.

Hurst, J. W., Staton, J. and Hubbard, D.: Precordial murmurs during pregnancy and lactation. New England J. Med., *259*:515, 1958.

Hylen, J. C., Kloster, F. E., Herr, R. H., Hull, P. Q., Ames, A. W., Starr, A. and Griswold, H. E.: Phonocardiographic diagnosis of aortic ball variance. Circulation, *38*:90, 1968.

Hylen, John C., Kloster, Frank E., Herr, Rodney H., Starr, Albert and Griswold, Herbert E.: Sound spectrographic diagnosis of aortic ball variance. Circulation, *39*:849, 1969.

Ikram, H., Makey, A. R. and Bliss, B. P.: Genesis of diastolic sounds in mitral incompetence. Brit. Heart J., *31*:762, 1969.

Jacoby, Jr., W. J., Tucker, D. H. and Sumner, R. G.: The second heart sound in congenital pulmonary valvular insufficiency. Am. Heart J., *69*:603, 1965.

Jacono, A. and Friedland, C.: Frequency characteristics of extra sounds. Am. J. Cardiol., *4*:207, 1959.

Jona, E., Gamna, G., Messina, D., Dughera, L. and Actis-Dat, A.: Il fonocardiogramma epicardico dell'uomo. II. Nella comunicazione interatriale. Minerva Med., *49*:47, 1958.

Jonsson, B. J.: The split second heart sound and its correlation to the pulmonary vascular resistance. Cardiol., *32*:155, 1958.

Joseph, G.: Geschichte der Physiologie der Herztoene. Janus, *2*:1-39, 345-374, 505-527, 1847.

Julius, S. and Stewart, B. H.: Diagnostic significance of abdominal murmurs. New England J. Med., *276*:1175, 1967.

Kasparian, H., Segal, B. L. and Novack, P.: Experiences with left heart intracardiac phonocardiography. In *The Theory and Practice of Auscultation* (Segal, Editor). Davis, Philadelphia, 1964, p. 535.

Katz, L. N. and Feil, H. S.: Clinical observations on the dynamics of ventricular systole. III. Aortic stenosis and insufficiency. Heart, *12*: 171, 1925.

Kaufmann, G., Rutishauser, W. and Hegglin, R.: Heart sounds in atrial tumors. Am. J. Cardiol., *8*:350, 1961.

Kay, C. F., Joyner, C. R., Helwig, J. and Raymond, T. H.: The "late systolic heartbeat" of pericardial effusion. Am. Heart J., *72*:7, 1966.

Kerber, A. E., Isaeff, D. M. and Hancock, E. W.: Echocardiographic patterns in patients with the syndrome of systolic click and late systolic murmur. New England J. Med., *284*:691, 1971.

Kesteloot, H. and van Houte, O.: On the origin of the telesystolic murmur preceded by a click. Acta Cardiol., *20*:197, 1965.

Kiger, R. G.: Differentiation of Austin Flint and mitral stenosis murmurs by amyl nitrite (Abstr.). Clin. Res., *11*:24, 1963.

Kincaid-Smith, P., Barlow, J.: The atrial sound in hypertension and ischaemic heart disease. Brit. Heart J., *21*:479, 1959.

King, J. T.: Clinical recognition and physical signs of bundle branch block. Amer. Heart J., *3*:505, 1928.

King, J. T. and McEachern, D.: Nature of the physical signs of bundle-branch block. Amer. J. Med. Sci., *183*:445, 1932.

Klinefelter, H. F.: The heart in sickle cell anemia. Am. J. Med. Sci. *203*: 34, 1942.

Kontos, H. A., Shapiro, W., Kemp, V. E.: Observations on the atrial sound in hypertension. Circulation, *28*:877, 1963.

Kossmann, C. E.: The opening snap of the tricuspid valve: a physical sign of tricuspid stenosis. Circulation, *11*:378, 1955.

Kramer, D. H., Moss, A. J. and Shah, P. M.: Mechanisms of extracardiac pacemaker sound. (Abstract). Circulation, *38*: Suppl. VI-119, 1968.

Kumar, S. and Luisada, A. A.: On the mechanism of changes of the second heart sound in aortic stenosis. Am. J. Card., *28*:162, 1971.

Kumar, S. and Luisada, A. A.: The second heart sound in atrial septal defect. Am. J. Cardiol., *28*:168, 1971.

Kuo, P. T., Schnabel, T. G., Jr., Blakemore, W. S. and Whereat, A. F.: Diastolic gallop sounds, the mechanism of production. J. Clin. Invest., *36*:1035, 1957.

Kurz, H., Slodki, S. J. and Luisada, A. A.: Left ventricular hemodynamics in mitral stenosis. Am. J. Cardiol., *19*:183, 1967.

Kusukawa, R., Bruce, D. W., Sakamoto, T., MacCanon, D. M. and Luisada, A. A.: Hemodynamic determinants of the amplitude of the second heart sound. J. Appl. Physiol., *21*:938, 1966.

Kusukawa, R., Sakamoto, T., MacCanon, D. M. and Luisada, A. A.: Hemodynamic determinants of the amplitude of the second heart sound. J. Appl. Physiol., *21*:938, 1966.

Lancaster, J. F., Leon, D. F. and Leonard, J. J.: Carcinoid heart disease with tricuspid stenosis and pulmonic insufficiency. A hemodynamic and intracardiac phonocardiographic study. Am. J. Cardiol., *17*:265, 1966.

Laubry, C. and Bouvrain, Y.: Remarques sur l'aspect phonocardiographique des extrasystoles ventriculaires. Arch. Mal. Coeur, *34*:1, 1941.

Leachman, R. D. and Cokkinos, D. V. P.: Absence of opening click in dehiscence of mitral-valve prosthesis. New England J. Med., *281*:461, 1969.

Leatham, A.: Phonocardiography. Brit. Med. Bull., *8*:333, 1952.

Leatham, A.: Rheumatic aortic incompetence with delayed diastolic murmurs on auscultation. Proc. Roy. Soc. Med., *43*:309, 1950.

Leatham, A.: The phonocardiogram of aortic stenosis. Brit. Heart J., *13*: 153, 1951.

Leatham, A.: Splitting of the first and second heart sounds. Lancet, *2*:607, 1954.

Leatham, A.: A classification of systolic murmurs. Brit. Heart J., *17*:574, 1955.

Leatham, A.: The second heart sound. Key to auscultation of the heart. Acta Cardiol., *19*:395, 1964.

Leatham, A. and Gray, I.: Auscultatory and phonocardiographic signs of atrial septal defect. Brit. Heart J., *18*:193, 1956.

Leatham, A. and Segal, B.: Auscultatory and phonocardiographic signs of ventricular septal defect with left-to-right shunt. Circulation, *25*:318, 1962.

Leatham, A. and Towers, M.: Splitting of the second heart sound in health. Brit. Heart J., *13*:575, 1951.

Leatham, A. and Vogelpoel, L.: The early systolic sound in dilatation of the pulmonary artery. Brit. Heart J., *16*:21, 1954.

Leatham, A. and Weitzman, D.: Auscultatory and phonocardiographic signs of pulmonary stenosis. Brit. Heart J., *19*:303, 1957.

Leatham, A., Segal, B. and Shafter, H.: Auscultatory and phonocardiographic findings in healthy children with systolic murmurs. Brit. Heart J., *25*:451, 1963.

LeBauer, E. J., Perloff, J. K. and Keliher, T. F.: The isolated systolic click with bacterial endocarditis. Am. Heart J., *73*:534, 1967.

Lefcoe, N. M., Brien, F. S. and Manning, G. W.: An opening snap recorded in a case of tumor of the left atrium. New England J. Med., *257*:178, 1957.

Leo, T. and Hultgren, H.: Phonocardiographic characteristics of tight mitral stenosis. Medicine, *38*:85, 1959.

Leon, D. F., Leonard, J. J., Lancaster, J. F., Kroetz, F. W. and Shaver, J. A.: Effect of respration on pansystolic regurgitant murmurs as studied by biatrial intracardiac phonocardiography. Am. J. Med., *39*:429, 1965.

Leon, D. F., Leonard, J. J., Kroetz, F. W., Page, W. L., Shaver, J. A. and

Lancaster, J. F.: Late systolic murmurs, clicks, and whoops arising from the mitral valve. Am. Heart J., *72*:325, 1966.

Leonard, J. J., Weissler, A. M. and Warren, J. V.: Observations on the mechanism of atrial gallop rhythm. Circulation, *17*:1007, 1958.

Levine, S. A. and Harvey, W. P.: *Clinical Auscultation of the Heart.* W. B. Saunders Co., Philadelphia, 1959, 2nd Edition.

Levine, S. A. and Likoff, W. B.: Some notes on the transmission of heart murmurs. Ann. Int. Med., *21*:298, 1944.

Lewis, T.: *The Mechanism and Graphic Registration of the Heart Beat.* Shaw and Sons, London, 1925.

Lian, C.: Remarques phonocardiographiques sur les frottements pericardiques. Arch. Mal. Coeur, *36*:5, 1943.

Lian, C. and Deparis, M.: Le claquement mesosystolique pleuro-pericardique. Bull. Mem. Soc. Hop. Med. Paris, *49*:496, 1933.

Lian, C. and Welti, J. J.: Le claquement arteriel pulmonaire protosystolique. Arch. Mal. Coeur, *30*:946, 1937.

Lian, C., Minot, G. and Welti, J. J.: *Phonocardiographie. Auscultation Collective.* Paris, Masson. 1941.

Liebman, J. and Sood, S.: Diastolic murmurs in apparently normal children. Circulation, *38*:755, 1968.

Linhart, J. W. and Taylor, W. J.: The late systolic murmur, clinical, hemodynamic and angiocardiographic observations. Am. J. Cardiol., *18*: 164, 1966.

Little, R. C.: Effect of atrial systole on ventricular pressure and closure of the A-V valves. Am. J. Physiol., *166*:289, 1951.

Little, R. C., Hilton, J. G., Schaefer, R. D.: The first heart sound in normal and ectopic ventricular contractions. Circ. Res., *2*:48, 1954.

Littmann, D.: Cardiac Alternation. Alternation of heart sounds and murmurs. Am. J. Cardiol., *14*:420, 1964.

Liu, C. K. and Jacono, A.: Phonocardiography in atrial septal defect. External and intracardiac phonocardiograms. Am. J. Cardiol., *2*:714, 1958.

Liu, C. K., Aravanis, C., Testelli, M. R. and Luisada, A. A.: Intracardiac phonocardiography in mitral and aortic valve lesions. Am. J. Cardiol., *1*:379, 1958.

Logue, R. B., Cobbs, B. W. and Dorney, E. R.: Second heart sound in pulmonary embolism and pulmonary hypertension. Trans. Am. Clin. Climat. Asso., *78*:38, 1966.

Lopez, J. F., Linn, H. and Shaffer, A. B.: The apical first heart sound as an aid in the diagnosis of atrial septal defect. Circulation, *26*: 1296, 1962.

Lucardie, S. M.: Phonocardiographical aspects of obstructive cardiomyopathy. Acta Cardiol., *19*:478, 1964.

Lucardie, S. M. and Durrer, D.: The late systolic murmur. Arch. Kreislauff., *53*:174, 1967.

Luisada, A. A.: *From Auscultation to Phonocardiography.* St. Louis, C. V. Mosby. 1965.

Luisada, A. A.: Gallop rhythm; alternans. In *Cardiology* (A. A. Luisada,

Edit.), McGraw-Hill-Blakiston, New York, Vol. 3, Part 5, p. 36, 1962 (Suppl. 1).

Luisada, A. A.: Graphic data in pericarditis. In *Cardiology.* (Luisada, Editor). New York, McGraw-Hill, Blakiston (v. 3, part 8, pages 38 and 54), 1959.

Luisada, A. A.: On the apical sounds and murmurs in aortic regurgitation. Am. Heart J., *28*:156, 1944.

Luisada, A. A.: The cardiac effects of hypertension. Trans Am. Coll. Cardiol., *5*:50, 1955.

Luisada, A. A.: The diastolic sounds of the heart in normal and pathological conditions. Acta Med. Scandinav. (Suppl. 266), *142*:685, 1952.

Luisada, A. A.: *The Heart Beat.* P. B. Hoeber, New York, 1953.

Luisada, A. A.: The second sound in normal and abnormal conditions. Am. J. Cardiol., *28*:150, 1971.

Luisada, A. A.: *The Sounds of the Normal Heart.* Warren H. Green, St. Louis, 1972.

Luisada, A.: Variable interval between electric and acoustic phenomena in auricular fibrillation. Am. Heart J., *22*:245, 1941.

Luisada, A. A. and Alimurung, M. M.: The systolic gallop rhythm. Acta Cardiol., *4*:309, 1949.

Luisada, A. A. and Argano, B.: The phonocardiogram in systemic hypertension. Chest, *58*:598, 1970.

Luisada, A. A. and Argano, B.: Bizarre murmurs in a case of ventricular aneurysm. Dis. Chest., *55*:489, 1969.

Luisada, A. A. and Cortis, B.: The dynamic events of the normal heart in man. Acta Cardiol., *25*:203, 1970.

Luisada, A. A. and Liu, C. K.: Intracardiac phonocardiography in mitral and aortic valve lesions. Circulation, *18*:989, 1958.

Luisada, A. A. and Fleischner, F. G.: Tracings of the left ventricle in myocardial infarction. Acta Cardiol., *3*:308, 1948.

Luisada, A. A. and Liu, C. K.: *Intracardiac Phenomena in Right and Left Heart Catheterization.* Grune and Stratton, New York, 1958.

Luisada, A. A. and MacCanon, D. M.: The phases of the cardiac cycle. Am. Heart J., *83*,705, 1972.

Luisada, A. A. and Mautner, H.: Experimental studies on functional murmurs and extra-sounds of the heart. Exp. Med. Surg., *1*:282, 1943

Luisada, A. A. and Montes, L. P.: A phonocardiographic study of apical diastolic murmurs simulating those of mitral stenosis. Ann. Int. Med., *33*:56, 1950.

Luisada, A. A. and Roitman, M.: Caracteristicas graficas de los extratonos y galopes diastolicos, frecuencia, importancia clinica y audibilidad en sujetos normales y cardiacos. Arch. Inst. Cardiol. Mexico, *18*:345, 1948.

Luisada, A. A. and Shah, P. M.: Controversial and changing aspects of auscultation. I. Areas of auscultation. II. Normal and abnormal first and second sounds. Am. J. Cardiol., *11*:774, 1963.

Luisada, A. A. and Shah, P. M.: Controversial and changing aspects of auscultation. III. Diastolic sounds. IV. Intervals. V. Systolic sounds. Am. J. Cardiol., *13*:243, 1964.

Luisada, A. A. and Wolff, L.: The significance of the pulmonary diastolic murmur in cases of mitral stenosis. Am. J. Med. Sci., *209*:204, 1945.

Luisada, A. A. and Zilli, A. B.: The phonocardiogram in rheumatic fever and rheumatic heart disease. *Cardiology* (A. A. Luisada, Edit.). McGraw-Hill-Blakiston, New York, Vol. 3, Part 7, Chap. 9, p. 84, 1959.

Luisada, A. A., Gagnon, G. and Ikeda, H.: The first heart sound and the dynamic events in ventricular ectopic beats and in paced beats. Am. J. Cardiol., *25*:529, 1970.

Luisada, A. A., Haring, O. M. and Zilli, A. B.: Apical diastolic murmurs simulating mitral stenosis. II. Graphic differentiation. Ann. Int. Med., *42*:644, 1955.

Luisada, A. A., Haring, O. M., Aravanis, C., Cardi, L., Jona, E. and Zilli, A. B.: Murmurs in children: a clinical and graphic study in 500 children of school age. Ann. Int. Med., *48*:597, 1958.

Luisada, A. A., Kurz, H., Slodki, S. J., MacCanon, D. M. and Krol, B.: Normal first heart sounds with nonfunctional tricuspid valve or right ventricle: Clinical and experimental evidence. Circulation, *35*:119, 1967.

Luisada, A. A., Liu, C. K. and Slodki, S. J.: Intracardiac phonocardiography in 172 cases studies by left or right heart catheterization. Acta Cardiol. (Brussels), *18*:533. 1963.

Luisada, A. A., MacCanon, D. M., Bruce, D. W., Worthen, M., Argano, B., Siwadlowski, W. and Kurz, H.: Heart sounds of the right heart. J. Appl. Physiol., *25*:362, 1968.

Luisada, A. A., MacCanon, D. M., Feigen, L., Griffen, P. M. and Darrel, B.: Design and first results of a new phonocardiograph. Am. J. Cardiol., *28*:134, 1971.

Luisada, A. A., MacCanon, D. M. and Slodki, S. J.: Intracardiac phonocardiography. A new simplified system. Circulation, *32*:563, 1965.

Luisada, A. A., Mendoza, F. and Alimurung, M. M.: The duration of normal heart sounds. Brit. Heart J., *11*:41, 1949.

Luisada, A. A., Slodki, S. J. and Krol, B.: Double (mitral and tricuspid) opening snap in patients with valvular lesion. Am. J. Cardiol., *16*: 800, 1965.

Luisada, A. A., Slodki, S. J., Mehta, S. and Gupta, P. D.: Two mechanisms of split second heart sound in a patient with atrial septal defect. Am. J. Cardiol., *18*:788, 1966.

Luisada, A. A., Szatkowski, J., Testelli, M. R. and Bendezú-Prieto, J.: Apical diastolic and presystolic murmurs of proved functional nature. Am. J. Cardiol., *4*:501, 1959.

Luisada, A. A., Kumar, S. and Pouget, M. J.: On the causes of the changes of the second heart sound in left bundle branch block. Japan Heart J., *13*:281, 1972.

MacCanon, D. M., Arevalo, F. and Meyer, E. C.: Direct detection and timing of aortic valve closure. Circulation Res., *14*:387, 1964.

MacCanon, D. M., Bruce, D. W., Lynch, P. R. and Nickerson, J. L.: Mass excursion parameters of first heart sound energy. J. Appl. Physiol., *27*:649, 1969.

McFadzean, A. J. S. and Gray, J.: Hepatic venous hum in cirrhosis of liver, Lancet, *2*:1128, 1953.

McGaff, C. J., Heller, J. A., Leight, L. and Towery, B. T.: Subvalvular pulmonary stenosis due to constriction of the right ventricular outflow tract by a pericardial band. Am. J. Med., *34*:142, 1963.

McGinn, F. X., Gould, L. and Lyon, A. T.: The phonocardiogram and apexcardiogram in patients with ventricular aneurysm. Am. J. Cardiol., *21*:467, 1968.

McKusick, V. A.: The diagnosis of organic mitral stenosis in the presence of sickle-cell anemia. Am. Heart J., *46*:467, 1953.

McKusick, V. A.: The genesis of musical murmurs. IRE Trans. Med. Electr., *PGME-9*:11, 1957.

McKusick, V. A.: *Cardiovascular Sound in Health and Disease.* Williams and Wilkins, Baltimore, 1958.

McNair, J. D.: The pathologic third heart sound in angina pectoris. Cardio., *51*:79, 1967.

Macpherson, A. I. S. and Morton, E. V. B.: Continuous venous hum in a case of portal cirrhosis. Brit. Heart J., *17*:105, 1955.

Magri, G., Caruzzo, C., Gamna, G. and Oddone, I.: Oesophageal phonocardiography. Cardiol., *31*:407, 1957.

Magri, G., Jona, E., Messina, D. and Actis-Data, A.: Direct recording of heart sounds and murmurs from the epicardial surface of the exposed human heart. Am. Heart J., *57*:449, 1959.

March, H. W., Selzer, A. and Hultgren, H. N.: The mechanical consequences of anomalous atrioventricular excitation (WPW) syndrome. Circulation, *23*:582, 1961.

Marcus, F. I., Perloff, J. K. and DeLeon, A. C.: The use of amyl nitrite in the hemodynamic assessment of aortic valvular and muscular subaortic stenosis. Am. Heart J., *68*:468, 1964.

Marienfeld, C. J., Telles, N., Silvera, J. and Nordsieck, M.: A 20-year follow-up study of "innocent" murmurs. Pediatrics, *30*:42, 1962.

Massumi, R. A., Hernandez, T., Just, G. and Tawakkol, A. A.: The audible sounds of atrial tachyarrhythmia (flutter?). A note on their genesis. Circulation, *33*:607, 1966.

Master, A. M. and Friedman, R.: A phonocardiographic study of the heart sounds in acute coronary occlusion. Am. Heart J., *24*:196, 1942.

Mattioli, L. Takagi, M. and Moghadam, A. N.: Intravascular phonocardiography in clinical and experimental peripheral pulmonary artery stenosis. Dis. Chest, *51*:625, 1967.

Meadows, W. R., Van Praagh, S., Indreika, M. and Sharp, T. J.: Premature mitral valve closure. A hemodynamic explanation for the absence of the first sound in aortic insufficiency. Circulation, *28*:251, 1963.

Mehta, S. and Luisada, A. A.: High-pitched murmur in mitral stenosis. Indian Heart J., *18*:320, 1966.

Mercer, E. N., Frye, R. L. and Giuliana, E. R.: Late systolic click in non-obstructive cardiomyopathy. Brit. Heart J., *32*:691, 1970.

Merendino, K. A. and Hessel, E. A.: The "murmur on top of the head" in acquired mitral insufficiency. Pathological and clinical significance. J.A.M.A., *199*:142, 1967.

Messer, A. L., Counihan, T. B., Rappaport, M. B. and Sprague, H. B.: The effect of cycle length on the time of occurrence of the first heart sound and the opening snap in mitral stenosis. Circulation, *4*:576, 1951.

Meyer, E. C., Slodki, S. J. and Siegel, I.: Phonocardiography during pregnancy. J.A.M.A., *185*:242, 1963.

Meyer, W.: The aortic and pulmonary components of the second heart sound in mitral stenosis, atrial septal defect and tetralogy of Fallot. Cardiol., *50*:65, 1967.

Miles, B. E.: The clinical significance of gallop rhythm in hypertension. Brit. Heart J., *13*:327, 1951.

Montiel, M. M.: Muscular apparatus of the mitral valve in man and its involvement in left-sided hypertrophy. Am. J. Card., *26*:341, 1970.

Moreyra, E., Segal, B. L. and Shimada, H.: The murmurs of mitral regurgitation. Dis. Chest, *55*:49, 1969.

Mori, M., Shah, P. M., MacCanon, D. M. and Luisada, A. A.: Hemodynamic correlates of the various components of the second heart sound. Cardiol. (Basel), *44*:65, 1964.

Morrow, A. G., Goldblatt, A. and Braunwald, E.: II. Surgical treatment and the results of operation. Circulation, *27*:426, 1963.

Mounsey, P.: The opening snap of mitral stenosis. Brit. Heart J., *15*:135, 1953.

Mounsey, P.: The early diastolic sound of constrictive pericarditis. Brit. Heart J., *17*:143, 1955.

Mounsey, P. and Brigden, W.: The apical systolic murmur in mitral stenosis. Brit. Heart J., *16*:255, 1954.

Mouquin, M. and Py, J.: Phonocardiographie des cardiopathies congenitales. III. Communications interauriculaires. Arch. Mal. Coeur, *55*: 397, 1962.

Movitt, E. R. and Gerstl, B.: Pure mitral insufficiency of rheumatic origin in adults. Ann. Int. Med., *38*:981, 1953.

Moyer, J. H., Segal, B. L., Tabesh, E., *et al.*: Auscultatory abnormalities associated with systemic hypertension in *The Theory and Practice of Auscultation* (C. Segal, Editor). Philadelphia, F. A. Davis, 1964.

Muiesan, G., Innocenti, P. F., Toso, M., Duranti, G. F. and Todisco, T.: Il tono atriale nell'ipertensione arteriosa. Cuore e Circ., *49*:187, 1965.

Murdock, M. I., Meyers, B. A. and Bacos, J. M.: Auscultatory clicks produced by pacemaker catheters. Ann. Int. Med., *68*:1320, 1968.

Myers, J. D., Murdaugh, H. V., McIntosh, H. D. and Blaisdell, R. K.: Observations on continuous murmurs over partially obstructed arteries. An explanation for the continuous murmur found in the aortic arch syndrome. Arch. Int. Med., *97*:726, 1956.

Myler, R. K. and Sanders, C. A.: Normal splitting of the second heart sound in atrial septal defect. Am. J. Cardiol., *19*:874, 1967.

Nadas, A. S.: Phonocardiography in congenital heart disease. In *Cardiology* (Luisada, Editor), Vol. III, Part 6, Chap. 6, p. 89, 1959.

Nadas, A. S. and Alimurung, M. M.: Apical diastolic murmurs in congenital heart disease. The rarity of Lutembacher's syndrome. Am. Heart J., *43*:691, 1952.

Nadas, A. S. and Ellison, R. C.: Phonocardiographic analysis of diastolic flow murmurs in secundum atrial septal defect and ventricular septal defect. Brit. Heart J., *29*:684, 1967.

Nadas, A. S., Hauck, A. J. and Gross, R. E.: Ventricular septal defect with aortic regurgitation. Circulation, *26*:766, 1962.

Najmi, M. and Segal, B. L.: Auscultatory and phonocardiographic findings in patients with prosthetic ball-valves. Am. J. Cardiol., *16*:794, 1965.

Nasser, W., Tavel, M. E., Feigenbaum, H. and Fisch, C.: Austin-Flint murmur versus the murmur of organic mitral stenosis. New England J. Med., *275*:1007, 1966.

Neill, C. and Mounsey, P.: Auscultation in patent ductus arteriosus with a description of two fistulae simulating patent ductus. Brit. Heart J., *20*:61, 1958.

Nellen, M. and Gotsman, M. S.: The systolic murmur in hypertrophic obstructive cardiomyopathy and the nature of the obstruction. Post-grad. Med. J., *44*:89, 1968.

Nellen, M., Gotsman, M. S., Vogelpoel, L., Beck, W. and Schrire, V.: Effects of prompt squatting on the systolic murmur in idiopathic hypertrophic obstructive cardiomyopathy. Brit. Med. J., *3*:140, 1967.

Neporent, L. M. and Da Silva, J. A.: Heart sounds in atrial flutter-fibrillation. Am. J. Cardiol., *19*:301, 1967.

Nixon, P. G. F.: The third heart sound in mitral regurgitation. Brit. Heart J., *23*:677, 1961.

Nixon, P. G. F. and Wagner, G. R.: The duration of left ventricular systole in mitral incompetence. Brit. Heart J., *24*:464, 1962.

Nixon, P. G. F., Wooler, G. H. and Radigan, L. R.: Mitral incompetence caused by disease of the mural cusp. Circulation, *19*:839, 1959.

Nixon, P. G. F., Wooler, G. H. and Radigan, L. R.: The opening snap in mitral incompetence. Brit. Heart J., *22*:395, 1960.

Nuzum, F. R.: Auscultatory evidence of cardiac rupture. Am. Heart J., *39*:909, 1950.

Olesen, K. H. and Warburg, E.: Fourteen cases of loud apical systolic murmurs in patients under 30 years (a follow-up study). Acta Cardiol., *11*:164, 1956.

Ongley, P. A., Sprague, H. B. and Rappaport, M. B.: The diastolic murmur of mitral stenosis. New England J. Med., *253*:1049, 1955.

Ongley, P. A., Sprague, H. B., Rappaport, M. B. and Nadas, A. S.: *Heart Sounds and Murmurs*. Grune and Stratton, New York, 1960.

Ongley, P. A., Rahimtoola, S. H., Kincaid, O. W. and Kirklin, J. W.: Continuous murmurs in tetralogy of Fallot and pulmonary atresia with ventricular septal defect. Am. J. Cardiol., *18*:821, 1966.

Orias, O. and Braun-Menendez, E.: *The Heart Sounds in Normal and Pathological Conditions*. London, Oxford University Press, 1939.

Oravetz, J., Wissner, S., Argano, B. and Luisada, A. A.: Dynamic analysis of heart sounds in right and left bundle-branch blocks. Circulation, *36*:275, 1967.

Oriol, A., Palmer, W. H., Nakhjavan, F. and McGregor, M.: Prediction of left atrial pressure from the second sound-opening snap interval. Am. J. Cardiol., *16*:184, 1965.

Oriol, A., Papp, O. and Sanz, G.: Splitting of second heart sound in ventricular ectopic beats. Brit. Heart J., *30*:757, 1968.

Osmundson, P. J., Callahan, J. A. and Edwards, J. E.: Mitral insufficiency from ruptured chordae tendineae simulating aortic stenosis. Proc. Staff Meet. Mayo Clin., *33*:235, 1958.

Parkinson, J., Bedford, D. and Thompson, W.: Cardiac aneurysm. Quart. J. Med., *7*:455, 1938.

Pearson, H. E. S.: Apical diastolic murmurs in severe anaemia. Brit. Heart J., *11*:98, 1949.

Penny, J. L., Gregory, J. J. and Ayres, S. M.: Audible atrial sounds in a case of atrial flutter. Am. J. Cardiol., *19*:305, 1967.

Perloff, J. K.: Auscultatory and phonocardiographic manifestations of pulmonary hypertension. Prog. Cardiov. Dis., *9*:303, 1967.

Perloff, J. K.: Differential diagnosis of the innocent (functional) murmur. Heart Bull., *9*:93, 1960.

Perloff, J. K. and Lebauer, E. J.: Auscultatory and phonocardiographic manifestations of isolated stenosis of the pulmonary artery and its branches. Brit. Heart J., *31*:314, 1969.

Perloff, Joseph K.: Second heart sound in cardiac diagnosis. Heart Bull., *18*:28, 1969.

Perloff, J. K. and Harvey, W. P.: Mechanisms of fixed splitting of the second heart sound. Circulation, *18*:998, 1958.

Perloff, J. K. and Harvey, W. P.: Auscultatory and phonocardiographic manifestations of pure mitral regurgitation. Prog. Cardiovasc. Dis., *5*:172, 1962.

Perloff, J. K., Caulfield, W. H. and De Leon, A. C.: Peripheral pulmonary artery murmur of atrial septal defect. Brit. Heart J., *29*:411, 1967.

Perrin, A., Saint-Pierre, A. and Froment, R.: Diagnostic des souffles systolo-diastoliques thoraciques. Actual. Cardiol. Angeiol. Internat., *14*:261, 1965.

Plass, R., Schmidt, K. H. and Guenther, H.: Intracardiac heart sounds and murmurs in atrial septal defect. Am. J. Cardiol., *28*:173, 1971.

Poggi, L., Bory, M., Serradimigni, A. and Audier, M.: Étude du deuxième bruit du coeur. Apport de la phonocardiographie. Coeur et Med. Int., *6*:49, 1967.

Potain, P. C.: Note sur les dedoublements normaux des bruits du coeur. Bull. Mem. Soc. Med. Hop., Paris, *3*:138, 1867.

Price, W. H. and Brown, A. E.: Alterations in intensity of heart sounds after myocardial infarction. Brit. Heart J., *30*:835, 1968.

Puchner, T. C., Houston, J. H., Hellmuth, G. A.: Heart sounds and murmurs in arterial hypertension. Am. J. Cardiol., *6*:630, 1960.

Puchner, T. C., Huston, J. H. and Hellmuth, G. A.: Aortic valve insufficiency in arterial hypertension. Am. J. Cardiol., *5*:758, 1960.

Rackley, C. E., Whalen, R. E., Floyd, W. L., Orgain, E. S. and McIntosh, H. D.: The precordial honk. Am. J. Cardiol., *17*:509, 1966.

Raftery, E. B.: The atrial systolic murmur in mitral stenosis. Am. J. Cardiol., *15*:735, 1965.

Ravin, A.: *Auscultation of the Heart.* Year Book Med. Publ., Chicago, 1958.

Ravin, A. and Bershof, E.: The intensity of the first heart sound in auricular fibrillation with mitral stenosis. Am. Heart J., *41*:539, 1951.

Ravin, A. and Darley, W.: Apical diastolic murmurs in patent ductus arteriosus. Ann. Int. Med., *33*:903, 1950.

Rawlings, M. S.: The straight back syndrome. Am. J. Cardiol., *5*:333, 1960.

Reeves, T. J., Hefner, L. L., Jones, W. B., Goghlan, C., Prieto, G. and Carroll, J.: The hemodynamic determinants of the rate of change in pressure in the left ventricle during isometric contraction. Am. Heart J., *60*:745, 1960.

Rivero-Carvallo, J. M. and Garza de los Santos, A.: Real and apparent apical impulse in tricuspid lesions. Differentiation of mitral and tricuspid murmurs and sounds. Am. J. Cardiol., *4*:367, 1959.

Roberts, W. C.: Anomalous left ventricular band. Am. J. Cardiol., *23*:735, 1969.

Robinow, M. and Harper, H. T., Jr.: Functional mitral stenosis. Ann. Int. Med., *17*:823, 1942.

Rodin, P. and Tabatznik, B.: The effect of posture on added heart sounds. Brit. Heart J., *25*:69, 1963.

Roelandt, J., Willems, J., van der Hauwaert, L. G. and De Geest, H.: Clicks and sounds (whoops) in left-sided pneumothorax. Clinical and phonocardiographic study. Dis. Chest, *56*:31, 1969.

Rogers, W. R.: Reversed splitting of the second heart sounds without left ventricular obstruction or conduction delay. Circulation (Suppl. III), *39-40*:173, 1969.

Rosa, L. M., Constantino, J. P., Reich, R., Karsak, N. and Zezmer, B.: The precordial accelerogram in normal subjects and noncardiac patients. Exper. Med. Surg., *19*:207, 1961.

Rosa, L. M., Constantino, J. P. and Reich, R.: The precordial acceleration tracing in hypertensive patients. Am. J. Cardiol., *9*:26, 1962.

Rosa, L. M., Constantino, J. P., Karsak, N., Reich, R. and Zezmer, B.: The precordial accelerogram in ischemic heart disease. A study of middle-aged and old patients with angina pectoris; recent and old myocardial infarction. Am. J. Cardiol., *9*:534, 1962.

Ross, J., Jr., Braunwald, E., Mason, D. T., Braunwald, N. S. and Morrow, A. G.: Interatrial communication and left atrial hypertension. A cause of continuous murmur. Circulation, *28*:853, 1963.

Ross, R. S. and McKusick, V. A.: Aortic arch syndromes. Diminished or absent pulses in arteries arising from arch of aorta. Arch. Int. Med., *92*:701, 1953.

Rubenstein, H. J. and Johnson, R. B.: The effect of external compression on the murmur and thrill of the straight back syndrome. Am. Heart J., *74*:88, 1967.

Runco, V. and Booth, R. W.: Basal diastolic murmurs. Am. Heart J., *65*:697, 1963.

Runco, V., Levin, H. S., Vahabzadeh, H. and Booth, R. W.: Basal diastolic murmurs in rheumatic heart disease: intracardiac phonocardiography and cineangiography. Am. Heart J., *75*:153, 1968.

Rushmer, R. F.: *Cardiovascular Dynamics.* Philadelphia, Saunders, 1961 (2nd Ed.).

Rushmer, R. F. and Morgan, C.: Meaning of murmurs. Am. J. Cardiol., *21*:722, 1968.

Rytand, D. A.: An analysis of the variable amplitude of the first heart sound in complete heart block. Standord M. Bull., *6*:187, 1948.

Rytand, D. A.: The variable loudness of the first heart sound in auricular fibrillation. Am. Heart J., *37*:187, 1949.

Safonov, Y. D.: The biohydraulic mechanism producing heart sounds. Cor Vasa (Praha), *10*:295, 1968.

Sakai, A., Feigen, L. P. and Luisada, A. A.: Frequency distribution of the heart sounds in normal man. Cardiovas. Res. (London), *5*:358, 1971.

Sakamoto, T., Kaito, G. and Ueda, H.: Electrocardiographic and phonocardiographic studies in hypertension. Jap. Heart. J., *1*:213, 1960.

Sakamoto, T., Kawai, N., Uozumi, Z., Yamada, T., Inoue, K., Chang, S. Y. and Ueda, H.: The point of maximum intensity of aortic diastolic regurgitant murmur, with special reference to the "right-sided" aortic diastolic murmur. Jap. Heart J., *9*:117, 1968.

Sakamoto, T., Kusukawa, R., MacCanon, D. M. and Luisada, A. A.: Hemodynamic determinants of the amplitude of the first heart sound. Circulation Res., *16*:45, 1965.

Sakamoto, T., Kusukawa, R., MacCanon, D. M. and Luisada, A. A.: The amplitude of the first heart sound in experimentally induced atrial fibrillation. Dis. Chest, *48*:401, 1965A.

Sakamoto, T., Kusukawa, R., MacCanon, D. M. and Luisada, A. A.: First heart sound amplitude in experimentally induced alternans. Dis. Chest, *50*:470, 1966.

Sakamoto, T., Sato, C., Yamada, T., Uozumi, A. and Ueda, H.: Better resolution in clinical phonocardiography. II. The use of compressor phonocardiograph. Jap. Heart J., *7*:460, 1966.

Sakamoto, T., Uozumi, Z., Kawai, N., Yamada, T., Inoue, K., Horikoh, U. and Ueda, H.: QRS dependence of the split interval of the second heart sound in complete right and left bundle branch block. Jap. Heart J., *8*:459, 1967.

Sakamoto, T., Yamada, T., Uozumi, Z., Chang, S. Yu., Nakamura, H. and Ueda, H.: Reversed splitting of the second heart sound in arterial stenotic lesions. Jap. Heart J. *9*:539-551, 1968.

Sakamoto, T., Yamada, T., Uozumi, Z., Kawai, N., Inoue, K. and Ueda, H.: Methoxamine induced "non paradoxical" reversed splitting. Jap. Heart J., *8*:642, 1967.

Sakamoto, T., Yamada, T., Uozumi, Z., Chang, S. K., Nakamura, H. and

Ueda, H.: Reversed splitting of the second sound in arterial stenotic lesions. Jap. Heart J. *9*:539, 1968.

Sarewitz, A. B. and Muehsam, G. E.: Aortic insufficiency simulating combined aortic stenosis and insufficiency. Dis. Chest, *48*:291, 1965.

Sarnoff, S. J., Gilmore, J. P. and Mitchell, J. H.: Influence of atrial contraction and relaxation on closure of mitral valve. Observations on effects of autonomic nerve activity. Circulation Res., *11*:26, 1962.

Scadding, J. G. and Wood, P.: Systolic clicks due to left-sided pneumothorax. Lancet, *2*:1208, 1939.

Schaefer, R. D. and Little, R. C.: The first heart sound in ventricular contractions arising from the apex and base. Proc. Soc. Exper. Biol. & Med., *85*:639, 1954.

Scherf, D. and Brooks, A.: The murmurs of cardiac aneurysm. Am. J. Med. Sci., *218*:389, 1949.

Schrire, V.: The relation of the apical systolic murmur to mitral valve disease. Am. Heart J., *68*:305, 1964.

Schrire, V. and Vogelpoel, L.: The loud musical diastolic murmur of an abnormal rheumatic chorda. Am. Heart J., *62*:315, 1961.

Schrire, V. and Vogelpoel, L.: The role of the dilated pulmonary artery in abnormal splitting of the second heart sound. Am. Heart J., *63*: 501, 1962.

Schrire, V., Gotsman, M. S. and Beck, W.: Unusual diastolic murmurs in constrictive pericarditis and constrictive endocarditis. Am. Heart J., *76*:4, 1968.

Schrire, V., Vogelpoel, L., Nellen, M., Swanepoel, A. and Beck, W.: Silent mitral incompetence. Am. Heart J., *61*:723, 1961.

Schrire, V., Vogelpoel, L., Beck, W., Nellen, M. and Swanepoel, A.: The effects of amyl nitrite and phenylephrine on the intracardiac murmurs of small ventricular septal defects. Am. Heart J., *62*:225, 1961.

Schuetz, E.: *Physiologie des Herzens.* Berlin, Springer, 1958.

Schugoll, G. I.: The innocent systolic murmur. Med. Ann. D.C., *31*:71, 1962.

Schwab, L., Smiley, G. L. and Meyn, W. P.: Xiphosternal crunch. An analysis of 106 cases among 3224 army separatees. Ann. Int. Med., *31*:228, 1949.

Schwab, R. H. and Killough, J. H.: The phonocardiographic differentiation of pulmonic and aortic insufficiency. Circulation, *32*:352, 1965.

Sears, G. A., Movafagh, P. and Manning, G. W.: Intracardiac phonocardiography in ventricular septal defect. Am. J. Med. Sci., *243*:775, 1962.

Segal, B. L.: Innocent murmurs. In *The Theory and Practice of Auscultation* (Segal, Editor). Davis, Philadelphia, 1964, p. 168.

Segal, B. L. and Likoff, W.: Late systolic murmur of mitral regurgitation. Am. Heart J., *67*:757, 1964.

Segal, B. L. and Likoff, W.: *Auscultation of the Heart.* Grune and Stratton, New York, 1965.

Segal, B. L., Mason, D., Kasparian, H. and Likoff, W.: Late systolic murmurs (Abstr.). Circulation, *24*:1033, 1961.

Segal, B. L., Novack, P. and Kasparian, H.: Intracardiac phonocardiography. Am. J. Cardiol., *13*:188, 1964.

Seymour, J., Bucher, H. and McDonald, L.: Reversed splitting of the second heart sound in ischaemic heart disease. Brit. Heart J., *27*:952, 1965.

Shafter, H. A.: Splitting of the second heart sound. Amer. J. Cardiol., *6*:1013, 1960.

Shah, P. M., Gramiak, R., Kramer, D. H. and Yu, P. N.: Gallop sounds in primary myocardial disease. New England J. Med., *278*:753, 1968.

Shah, P., Slodki, S. J.: The Q-II interval. A study of the second heart sound in normal adults and in systemic hypertension. Circulation, *29*:551, 1964.

Shah, P. M., Slodki, S. J. and Luisada, A. A.: A revision of the "classic" areas of auscultation of the heart. A physiologic approach. Am. J. Med., *36*:293, 1964.

Shah, P. M., Slodki, S. J. and Luisada, A. A.: A physiologic concept of the areas of auscultation. Acta Cardiol., *19*:111, 1964.

Shah, P. M., Gramiak, R., Kramer, D. H. and Yu, P. N.: Determinants of atrial (S_4) and ventricular (S_3) gallop sounds in primary myocardial disease. New England J. Med., *278*:753, 1968.

Shah, P. M., Mori, M., MacCanon, D. M. and Luisada, A. A.: Hemodynamic correlates of the various components of the first heart sound. Circulation Res., *12*:386, 1963.

Shapiro, S., Clark, T. J. H. and Goodwin, J. F.: Delayed closure of the pulmonary valve in obliterative pulmonary hypertension. Lancet, *2*: 1207, 1965.

Skinner, Jr., N. S., Mitchell, J. H., Wallace, A. G. and Sarnoff, S. J.: Hemodynamic effects of altering the timing of atrial systole. Am. J. Physiol., *205*:499, 1963.

Slodki, S. J., Hussain, A. T. and Luisada, A. A.: The Q-II Interval. III. A study of the second heart sound in old age. J. Am. Geriat. Soc., *17*: 673, 1969.

Smith, H. L., Essex, H. E. and Baldes, E. J.: Study of the movements of the heart valves and of heart sounds. Ann. Int. Med., *33*:1357, 1950.

Smith, S. M.: Pericardial knock. Brit. Med. J., *1*:78, 1918.

Solis-Cohen, M.: The xiphosternal crunch and its incidence in healthy inductees. Am. J. M. Sc., *210*:333, 1945.

Soloff, L. A.: Pericardial sounds and ejection sounds. In *The Theory and Practice of Auscultation* (Segal, Ed.). Davis, Philadelphia, 1964, p. 144.

Somers, K. and Williams, A. W.: The phonocardiogram in endomyocardial fibrosis. Brit. Heart J., *22*:546, 1960.

Soulié, P., Baculard, P., Bouchard, F., Cornu, C., Laurens, P. and Wolff, **cardiaque."** Baillière, Paris, 1961.

F.: *Le Catheterisme du Coeur au Micromanomètre "Le Son Intra-*

Soulié, P., Joly, F., Carlotty, J., Acar, J. and Valty, J.: La varieté mitrale du souffle systolique en echarpe. Arch. Mal. Coeur, *60*:243, 1967.

Spencer, M. P. and Greiss, F. C.: Dynamics of ventricular ejection. Circulation Res., *10*:274, 1962.

Spodick, D. H.: Pericardial friction. New England J. Med., *278*:1204, 1968.

Spodick, D. H.: Acoustic phenomena in pericardial disease. Am. Heart J., *81*:114, 1971.

Still, G. F.: *Common Disorders and Diseases of Childhood.* Frowde, London, 1912 (2nd Ed.).

Sutton, G., Harris, A. and Leatham, A.: Second heart sound in pulmonary hypertension. Brit. Heart J., *30*:743, 1968.

Sutton, G. C. and Craige, E.: Clinical signs of severe acute mitral regurgitation. Am. J. Cardiol., *20*:141, 1967.

Tabatznik, B., Randall, T. W. and Hersch, C.: The mammary souffle of pregnancy and lactation. Circulation, *22*:1069, 1960.

Tavel, M. E.: *Clinical Phonocardiography and External Pulse Recording.* Year Book Medical Publishers, Chicago, 1967.

Tavel, M. E., Campbell, R. W., and Zimmer, J. F.: Late systolic murmurs and mitral regurgitation. Am. J. Cardiol., *15*:719, 1965.

Tavel, M. E., Feigenbaum, H. and Campbell, R. W.: A study of the Q-1 interval in atrial fibrillation with and without mitral stenosis. Circulation, *31*:429, 1965.

Tofler, O. B.: The pulmonary component of the second heart sound in Fallot's tetralogy. Brit. Heart J., *25*:509, 1963.

Toutouzas, P., Gupta, D., Samson, R. and Shillingford, J.: Second sound interval in acute myocardial infarction. Brit. Heart J., *31*:462, 1969.

Ueda, H., Kaito, G. and Sakamoto, T.: *Clinical Phonocardiography.* Nanzando Co., Tokyo, 1963.

Ueda, H., Sakamoto, T., Kawai, N., Watanabe, H., Uozumi, Z., Okada, R.. Kobayashi, T., Yamada, T., Inoue, K. and Kaito, G.: The Austin Flint murmur. Phonocardiographic and patho-anatomical study. Jap. Heart J., *6*:294, 1965.

Ueda, H., Sakamoto, T., Yamada, T., Uozumi, Z., Kobayashi, T.. Kawai, N., Inoue, K. and Kaito, G.: Quantitative assessment of obstruction of the aorta and its branches in "aortitis syndrome". The value of functional phonarteriography using vasoactive drugs. Jap. Heart J., *7*:3, 1966.

Ueda, H., Sakamoto, T., Takeda, T., Uozumi, Z., Yamada, T., Nakamura. H. and Tagawa, H.: Phonoarteriographic and arteriographic evaluation of abdominal murmurs in renovascular hypertension. Jap. Heart J., *9*:142, 1968.

Vakil, R. J.: Transitory precordial pulsation in coronary occlusion. Brit. Heart J., *18*:248, 1956.

Vakil, R. J.: Ventricular aneurysm of the heart. Preliminary report on some new clinical signs. Am. Heart J., *49*:934, 1955.

Van Bogaert, A., van Genabeek, A., Arnoldy, M., Wauters, J., van der Henst, H., Vandael, J. and Keirsebelik, M.: Influence d'une fuite systolique ventriculaire sur l'intensité du premier bruit du coeur. Application a l'insuffisance mitrale. Arch. Mal. Coeur, *55*:961, 1962.

Van Bogaert, A., van Genabeek, A., Arnoldy, M., Wauters, J., van der

Henst, H., Keirsebelik, M. and Vandael, J.: Comportement du premier bruit du coeur dans la stenose aortique clinique et expérimentale. Arch. Mal. Coeur, *56*:40, 1963.

Van Bogaert, A., van Genabeek, A., Arnoldy, M., Wauters, J., Van der Henst, H., Vandael, J. and Keirsebelik, M.: Modifications du premier bruit du coeur dans le bloc de branche. Arch. Mal. Coeur, *56*:1253, 1963.

Van Bogaert, A., van Genabeek, A., Arnoldy, M., Wauters, J., van der Henst, H., Keirsebelik, M. and Vandael, J.: Contribution a l'étude du premier bruit du coeur normal. Arch. Mal. Coeur, *55*:368, 1962.

Van Bogaert, A., van Genabeek, A., Arnoldy, M. and Wauters, J.: La composante presystolique et systolique du premier bruit dans la stenose mitrale. Arch. Mal. Coeur, *57*:1062, 1964.

Victorica, B. E., Gessner, I. H. and Schiebler, G. L.: Phonocardiographic findings in persistent truncus arteriosus. Brit. Heart J., *30*:812, 1968.

Victorica, B. E., Krovetz, L. J., Elliott, L. P., Lodewhy, H. S., van Mierop, M. D., Bartley, Th. D., Gessner, I. H. and Schiebler, G. L.: Persistent truncus arteriosus in infancy. Am. Heart J., *77*:13, 1969.

Vogelpoel, L. and Schrire, V.: Auscultatory and phonocardiographic assessment of Fallot's tetralogy. Circulation, *22*:73, 1960.

Vogelpoel, L. and Schrire, V.: Pulmonary stenosis with intact ventricular septum and Fallot's tetralogy: assessment of postoperative results by auscultation and phonocardiography. Am. Heart J., *59*:645, 1960.

Vogelpoel, L., Nellen, M., Swanepoel, A. and Schrire, V.: The use of amyl nitrite in the diagnosis of systolic murmurs. Lancet, *2*:810, 1959.

Vogelpoel, L., Schrire, V., Nellen, M. and Swanepoel, A.: The use of amyl nitrite in the differentiation of Fallot's tetralogy and pulmonary stenosis with intact ventricular septum. Am. Heart J., *57*:803, 1959.

Vogelpoel, L., Schrire, V., Beck, W., Nellen, M. and Swanepoel, A.: The atypical systolic murmur of minute ventricular septal defect and its recognition by amyl nitrite and phenylephrine. Am. Heart J., *62*:101, 1961.

Von Egidy, H. and Dudeck, J.: A new theory about the origin of the mitral opening snap. J. Appl. Physiol., *27*:566, 1969.

Wallace, A. G., Mitchell, J. H., Skinner, N. S. and Sarnoff, S. J.: Duration of the phases of left ventricular systole. Circulation Res., *12*:611, 1963.

Warren, J. V., Leonard, J. J. and Weissler, A. M.: Gallop rhythm. Ann. Int. Med., *48*:580, 1958.

Wang, Y.: Wide, fixed splitting of the second heart sound and right ventricular overload. Circulation, *36*: (Suppl. II), 262, 1967.

Watanabe, H. and Sakamoto, T.: Clinical and phonocardiographic study of aortic regurgitation. Jap. Heart J., *2*:7, 1961.

Weinstein, W. and Lev, M.: Apical diastolic murmurs without mitral stenosis. Am. Heart J., *23*:809, 1942.

Weissler, A. M., Harris, W. S. and Schoenfeld, C. D.: Systolic time intervals in heart failure in man. Circulation, *37*:149, 1968.

Wells, B.: The assessment of mitral stenosis by phonocardiography. Brit. Heart J., *16*:261, 1954.

Wells, B.: The graphic configuration of innocent systolic murmurs. Brit. Heart J., *19*:129, 1957.

Wells, B.: The sounds and murmurs in transposition of the great vessels. Brit. Heart J., *25*:748, 1963.

Wells, B. G., Rappaport, M. B. and Sprague, H. B.: The graphic registration of basal diastolic murmurs. Am Heart J., *37*:586, 1949.

Wells, B. G., Rappaport, M. B. and Sprague, H. B.: The sounds and murmurs in coarctation of the aorta. A study by auscultation and phonocardiography. Am. Heart J., *38*:69, 1949.

Wiggers, C. J.: *Modern Aspects of the Circulation in Health and Disease.* Philadelphia and New York, Lea and Febiger, 1915.

Wolferth, C. C. and Margolies, A.: Various types of "extra" heart sounds. Med. Clin. No. Am., *14*:897, 1931.

Wolferth, C. C. and Margolies, A.: Gallop rhythm and the physiological third heart sound. I. Characteristics of the sounds, classification, comparative incidence of the various types and differential diagnosis. Am. Heart J., *8*:441, 1933.

Wolferth, C. C. and Margolies, A.: Asynchronism in contraction of the ventricles in the so-called common type of bundle-branch block: its bearing on the determination of the side of the significant lesion and on the mechanism of split first and second heart sounds. Am. Heart J., *10*:425, 1935.

Wolferth, C. C. and Margolies, A.: The influence of varying As-Vs intervals on split first heart sounds: its bearing on the cause of split sounds and the mechanism of the first sound. J. Clin. Invest., *14*:605, 1935.

Wolferth, C. C. and Margolies, A.: Systolic gallop rhythm. Studies on its characteristics and mechanism. Am. Heart J., *19*:129, 1940.

Wolferth, C. C. and Margolies, A.: Extra heart sounds and frictions and their differential diagnosis. New Intern. Clin., *1*:186, 1940.

Wolferth, C. C. and Margolies, A.: Heart Sounds. In *Diagnosis and Treatment of Cardiovascular Disease*, Vol. I (Stroud, Ed.). Davis, Philadelphia, 1950, 4th Ed., p. 71.

Wood, P.: The Eisenmenger syndrome. Brit. Med. J., *2*:701, 1958.

Yoshida, T., Mori, M., Nimura, Y., Hikita, G., Takagishi, S., Nakanishi, K. and Satomura, S.: Analysis of heart motion with ultrasonic Doppler method and its clinical application. Am. Heart J., *61*:61, 1961.

Yurchak, P. M. and Gorlin, R.: Paradoxical splitting of the second heart sound in coronary heart disease. New England J. Med., *269*:741, 1963.

Zakrzewski, T. K., Slodki, S. J. and Luisada, A. A.: The first heart sound in atrial septal defect. Am. Heart J., *78*:476, 1969.

Zalter, R., Hardy, H. C. and Luisada, A. A.: Acoustic transmission characteristics of the thorax. J. Appl. Physiol., *88*:428, 1963.

Zuberbuhler, J. R. and Bauersfeld, S. R.: Paradoxical splitting of the second heart sound in the Wolff-Parkinson-White syndrome. Am. Heart J., *70*:595, 1965.

Zuberbuhler, J. R., Bauersfeld, S. R. and Pontius, R. G.: Paradoxic splitting of the second sound with transposition of the great vessels. Am. Heart J., *74*:816, 1967.

INDEX

Printed in Dunstable, United Kingdom

91343648R00226